CLIMATE AND CULTURE
Multidisciplinary Perspectives on a Warming World

How does culture interact with the way societies understand, live with and act in relation to climate change? While the importance of the exchanges between culture, society and climate in the context of global environmental change is increasingly recognised, the empirical evidence is fragmented and too often constrained by disciplinary boundaries. Written by an international team of experts, this book provides cutting-edge and critical perspectives on how culture both facilitates and inhibits our ability to address and make sense of climate change and the challenges it poses to societies globally. Through a set of case studies spanning the social sciences and humanities, it explores the role of culture in relation to climate and its changes at different temporal and spatial levels; illustrates how approaching climate change through the cultural dimension enriches the range and depth of societal engagements; and establishes connections between theory and practice, which can stimulate action-oriented initiatives.

GIUSEPPE FEOLA is Associate Professor of Social Change for Sustainability at Utrecht University (the Netherlands) and Visiting Fellow at the School of Archaeology, Geography and Environmental Science at the University of Reading (United Kingdom). His research examines how societies (can) change towards sustainability, where and why. Core empirical themes in Giuseppe's research are sustainability transitions in agricultural systems, social movements and post-capitalist transformations. Giuseppe is the recipient of a Starting Grant from the European Research Council and a Vidi Grant from the Netherlands Research Organization.

HILARY GEOGHEGAN is Professor in Human Geography at the University of Reading (United Kingdom). Hilary researches at the intersection of the social and natural sciences and explores the emotional and affective relations between people and the material world. Using the geographical concept of landscape, she has worked on the human geographies of climate change. Hilary is currently examining the social and more-than-human dimensions of forest management which result from climate change–induced movement of pests and diseases.

ALEX ARNALL is Associate Professor in Environment and Development at the University of Reading (United Kingdom). He specialises in the study of migration, movement, and displacement of people and things. His work is characterised by an environmental theme, including climate change, agricultural development and food systems. Much of Alex's empirical research has taken place in the Maldives and Mozambique. Alex's research has been funded by the Economic and Social Research Council, Department for International Development, British Academy and Norwegian Research Council.

CLIMATE AND CULTURE

Multidisciplinary Perspectives on a Warming World

Edited by

GIUSEPPE FEOLA
Utrecht University

HILARY GEOGHEGAN
University of Reading

ALEX ARNALL
University of Reading

CAMBRIDGE
UNIVERSITY PRESS

University Printing House, Cambridge CB2 8BS, United Kingdom

One Liberty Plaza, 20th Floor, New York, NY 10006, USA

477 Williamstown Road, Port Melbourne, VIC 3207, Australia

314–321, 3rd Floor, Plot 3, Splendor Forum, Jasola District Centre,
New Delhi – 110025, India

79 Anson Road, #06–04/06, Singapore 079906

Cambridge University Press is part of the University of Cambridge.

It furthers the University's mission by disseminating knowledge in the pursuit of education, learning, and research at the highest international levels of excellence.

www.cambridge.org
Information on this title: www.cambridge.org/9781108422505
DOI: 10.1017/9781108505284

© Cambridge University Press 2019

This publication is in copyright. Subject to statutory exception and to the provisions of relevant collective licensing agreements, no reproduction of any part may take place without the written permission of Cambridge University Press.

First published 2019

Printed in the United Kingdom by TJ International Ltd. Padstow Cornwall

A catalogue record for this publication is available from the British Library.

Library of Congress Cataloging-in-Publication Data
Names: Feola, Giuseppe, 1979– editor. | Geoghegan, Hilary, editor. | Arnall, Alexander H. (Alexander Huw), 1978– editor.
Title: Climate and culture : multidisciplinary perspectives on a warming world / edited by Giuseppe Feola (Utrecht University, The Netherlands), Hilary Geoghegan (University of Reading), Alex Arnall (University of Reading).
Description: Cambridge ; New York, NY : Cambridge University Press, 2019. | Includes bibliographical references and index.
Identifiers: LCCN 2019004232 | ISBN 9781108422505 (alk. paper)
Subjects: LCSH: Climatic changes – Social aspects. | Global warming – Social aspects.
Classification: LCC QC903 .C438275 2019 | DDC 304.2/5–dc23
LC record available at https://lccn.loc.gov/2019004232

ISBN 978-1-108-42250-5 Hardback

Cambridge University Press has no responsibility for the persistence or accuracy of URLs for external or third-party internet websites referred to in this publication and does not guarantee that any content on such websites is, or will remain, accurate or appropriate.

Contents

List of Contributors	*page* vii
Foreword	ix
LESLEY HEAD	

1 Climate and Culture: Taking Stock and Moving Forward   1
HILARY GEOGHEGAN, ALEX ARNALL AND GIUSEPPE FEOLA

Part I   Knowing Climate Change   19

2 Cultures of Prediction in Climate Science   21
MARTIN MAHONY, MATTHIAS HEYMANN AND GABRIELE GRAMELSBERGER

3 Visualising Climate and Climate Change: *A Longue Durée* Perspective   46
SEBASTIAN VINCENT GREVSMÜHL

4 Indigenous Knowledge Regarding Climate in Colombia: Articulations and Complementarities among Different Knowledges   68
ASTRID ULLOA

5 Thin Place: New Modes of Environmental Knowing through Contemporary Curatorial Practice   93
CIARA HEALY-MUSSON

Part II   Being in a Climate Change World   115

6 Multi-temporal Adaptations to Change in the Central Andes   117
JULIO C. POSTIGO

Contents

7 Not for the Faint of Heart: Tasks of Climate Change
Communication in the Context of Societal Transformation 141
SUSANNE C. MOSER

8 At the Frontline or Very Close: Living with Climate Change on
St. Lawrence Island, Alaska, 1999–2017 168
IGOR KRUPNIK

9 Localising and Historicising Climate Change: Extreme
Weather Histories in the United Kingdom 190
GEORGINA ENDFIELD AND LUCY VEALE

Part III Doing in a Climate Change World 217

10 From Denial to Resistance: How Emotions and Culture Shape
Our Responses to Climate Change 219
ALLISON FORD AND KARI MARIE NORGAARD

11 Effective Responses to Climate Change: Some Wisdom from
the Buddhist Worldview 243
PETER DANIELS

12 Creating a Culture for Transformation 266
KAREN O'BRIEN, GAIL HOCHACHKA AND IRMELIN GRAM-HANSSEN

13 Back to the Future? *Satoyama* and Cultures of Transition and
Sustainability 291
JOHN CLAMMER

14 Culture and Climate Change: Experiments and Improvisations – An
Afterword 309
RENATA TYSZCZUK AND JOE SMITH

Index 327

Contributors

Alex Arnall is Associate Professor in Environment and Development at the University of Reading, United Kingdom.

John Clammer is Professor of Sociology at O.P. Jindal Global University, Delhi, India, and he previously taught for over twenty years in Japan, at Sophia University and the United Nations University.

Peter Daniels is Senior Lecturer at the School of Environment, Griffith University in Brisbane, Australia.

Georgina Endfield is Professor of Environmental History and Associate Pro Vice Chancellor for Research and Impact in the Faculty of Humanities and Social Sciences at the University of Liverpool, United Kingdom.

Giuseppe Feola is Associate Professor of Social Change for Sustainability at Utrecht University, the Netherlands, and Visiting Fellow at the School of Archaeology, Geography and Environmental Science at the University of Reading, United Kingdom.

Allison Ford is doctoral candidate at the University of Oregon, United States.

Hilary Geoghegan is Professor in Human Geography at the University of Reading, United Kingdom.

Gabriele Gramelsberger is Professor for Philosophy of Science and Technology at RWTH Aachen University, Germany.

Irmelin Gram-Hanssen is PhD candidate at the University of Oslo, Norway.

Sebastian Vincent Grevsmühl is CNRS researcher at the Center for Historical Studies (Centre de recherches historiques, CRH-EHESS) in Paris, France.

Lesley Head is Professor and Head of the School of Geography in the Faculty of Science at the University of Melbourne, Australia.

Ciara Healy-Musson is Lecturer at Carlow Institute of Technology, Ireland.

Matthias Heymann is Associate Professor at the Centre for Science Studies at Aarhus University, Denmark.

Gail Hochachka is PhD candidate at the University of Oslo, Norway.

Igor Krupnik is Chair of Anthropology and Curator of Circumpolar Ethnology at the National Museum of Natural History, Smithsonian Institution in Washington, DC, United States.

Martin Mahony is Lecturer in Human Geography at the University of East Anglia, United Kingdom.

Susanne C. Moser is Director and Principal Researcher of Susanne Moser Research & Consulting, and Research Faculty at Antioch University New England, United States.

Kari Marie Norgaard is Associate Professor of Sociology and Environmental Studies at the University of Oregon, United States.

Karen O'Brien is Professor in the Department of Sociology and Human Geography at the University of Oslo, Norway, and co-founder of cCHANGE.

Julio C. Postigo is Professor at the Department of Geography, Indiana University, United States.

Joe Smith is Director of the Royal Geographical Society (with the Institute of British Geographers). He was formerly Professor of Environment and Society, and Head of Geography, at the Open University, United Kingdom.

Renata Tyszczuk is Professor of Architectural Humanities at the University of Sheffield, United Kingdom.

Astrid Ulloa is an anthropologist and Professor in the Department of Geography at the Universidad Nacional de Colombia.

Lucy Veale is REF Manager at the Faculty of Science at the University of Nottingham, United Kingdom.

Foreword

LESLEY HEAD

It is widely recognised that we need to shift some very big cultural frames – the importance of economic growth, the dominance of fossil fuel capitalism, the hope of modernity as unending progress – to deal adequately with the climate change challenge. Cultural research, with its focus on in-depth qualitative methods, deals with some apparently very small things, at the scale of the everyday. The importance of cultural framings in understanding climate change impacts and response is now widely recognised (McCright and Dunlap 2011; Adger *et al.* 2013; Crow and Boykoff 2014). Showing how such scales of analysis are connected, for example by illuminating common sense or taken-for-granted understandings and practices, is an ongoing challenge for such research. Bringing cultural research into political analyses provides important insights into why our high-carbon world is so entrenched, and identifies windows of possibility (Bulkeley *et al.* 2016: 3).

This book makes diverse contributions to these debates from diverse disciplines. I summarise those contributions here under five themes. Further, the book comes at a time when there is a sufficient body of cultural research on climate change to enable a level of meta-analysis (Head *et al.* 2016). Comparisons, connections and generalisations between a multitude of in-depth studies are now possible. The new insights from this collection are no longer just good ideas in the wind but are underpinned by sufficient critical mass to give them weight and gravitas. I conclude with three such insights.

Showing How Concepts Matter

Whether it is the concept of nature itself, or component parts such as wilderness or invasive species, concepts always matter in human relations with the wider world (Castree 2015). Concepts both reflect and reinforce particular understandings and associated behaviours and governance practices. Sebastian Vincent Grevsmühl (this volume) starts with the concept of climate/*klima* itself, using the famous Humboldt map as a turning point and showing how the concept of climate became global because it was visualised in this way. As he argues, even old maps look familiar because the zonal division of regional climates is deeply enshrined in the Western geographical imagination. But the concept of climate that we now understand, as a statistical representation of long-term weather patterns, has shifted from an earlier understanding of *klima* as the 'habitable world'. If those climates and zonal divisions are now 'becoming migrant', as Grevsmühl puts it, what does that do to our wider geographical imaginations and notions of habitable worlds?

Documenting Variability

Cultural research has always been good at documenting variability, with obvious relevance to how different cultural groups approach climate change or any other environmental issue. Documenting diverse environmental knowledges, and understanding how the environmental cultures of different community sectors might intersect, helps identify and articulate potential environmental conflicts. Much more work can be done to understand the role of cultural diversity in relation to climate change response, and migrants' perspectives have been virtually ignored in climate change research (de Guttry *et al.* 2016: 11). Cultural research also throws light on cultural capacities and resources that might be otherwise understood as vulnerabilities or deficits (Gibson *et al.* 2013; Correa-Velez *et al.* 2014).

Julio Postigo (this volume) shows how these issues can be understood temporally by examining adaptive responses to changing climates in a particular region over time, while also emphasising that climate is not a stand-alone phenomenon experienced separately from other socio-economic processes within society.

Allison Ford and Kari Marie Norgaard (this volume) show this variability through four different contemporary social groups: explicit climate denialists, implicit climate denialists (which is most of us), urban homesteaders and Karuk indigenous people. But they go further in showing how such variability matters and what it means for the bigger picture. This chapter builds on Norgaard's earlier work on climate change scepticism and denial as significant phenomena requiring cultural analysis, and on climate change emotions as socially structured, rather than experienced only by individuals. Ford and Norgaard provide what they call 'meso-level' connections between broader structures of society and the everyday. The example of the 'dropping out' urban homesteaders, who maintain a strong cultural valuation of the individual, shows how the individualisation of emotions such as anxiety, despair and guilt can impede more collective approaches to climate change alternatives.

Indigenous peoples are prominent in the cultural literatures on climate change, partly because a number of groups, particularly Arctic ones, are at the 'front line', as Igor Krupnik (this volume) puts it. It is important that, in learning from indigenous examples, we do not generalise too quickly, but instead maintain the focus on the specific contexts in each case. In discussing the Karuk example in northern California, Ford and Norgaard remind us that indigenous groups are already in the catastrophe that began with colonial dispossession. In this sense climate change is not a new interruption but a continuation of 'colonial violence that already interferes with preferred, traditional relationships between human communities and the environment'. In a parallel with Postigo's argument, climate change is not understood or experienced as something separate and anomalous from other parts of life. Astrid Ulloa's chapter (this volume) provides a useful review of studies in the field, also reminding us how many of them are outside the English-speaking world.

Krupnik's decades of research in Alaskan rural communities provide important insights into the 'visible, massive and indisputable' changes in the northern Bering Sea over that time frame. In particular, the increasingly poor quality of ice gives hunters no place to go for good winter hunting but creates new opportunities for whaling. As noted in other research, the interruption and rearrangement of relationships between different phenomena give rise to 'strange things' being noticed and discussed, for example birds in the middle of winter. Krupnik notes that people are actively discussing these changes, seeking explanations and solutions. Climate change 'is pushed into the realm of village talk', as he puts it – normalised and woven into the fabric of everyday life. Outcomes are diverse, with both shrinking and growing communities in the region.

In an important parallel with a number of Pacific countries (Farbotko and Lazrus 2012), Krupnik shows how people in Alaska resist the academic casting of them as victims, have strong adaptive capacities and want assistance with very specific things from government. This study among others shows the importance of longitudinal research in the ethnographic tradition, providing fine-grained analysis of spatial and temporal variability. This documentation of what is happening in the recent long moment has great potential for connection to archaeological or historical comparators, a point I return to below.

Understanding Environmental Norms and Practices

As the examples above show so well, understanding cultural differences facilitates understanding of what stands for common sense and normal behaviours and how these came to be.

Chapters 2 and 3 provide important further examples via their historical analyses. In showing 'how the prediction of future climate has always involved a struggle with complexity and uncertainty', Martin Mahony, Gabriele Gramelsberger and Matthias Heymann (this volume) provide important historical perspective on scientific uncertainty and what they call its domestication. As climatology became global and increasingly dependent on modes of computation, ideas of what constitutes certainty and uncertainty in good science also changed. This study has obvious relevance to the present, when much of the cultural politics of climate change denial is framed around these same concepts. The chapter also gives us pause to reflect on the claim on the future that we stake with different predictive capacities and ideas of prediction.

Reframing Human–Environment Relations: Exposing Alternatives

If Western modernity has framed human–environment relations in particularly destructive ways, cultural framings from different places and times offer a range of cultural resources in the necessary reframing. Peter Daniels (this volume) makes a similar argument in relation to Buddhism – that we can draw inspiration from it to do less, consume less and enhance our connectedness with the natural world.

Georgina Endfield and Lucy Veale's (this volume) discussion of the TEMPEST database provides a fascinating example here. The database builds an archive of extreme weather histories in the United Kingdom through detailed weather information contained in personal diaries, letters and papers, all of them understood carefully in their context. It is clear from the rich examples provided that people were very aware and expressive of their more-than-human lives and the ways they were embedded in a system of strong more-than-human agency (something becoming lost in modernist understandings of climate). There is an interesting juxtaposition between this chapter and Krupnik's chapter with regard to the way in which extremes can be both absorbed and normalised in an everyday sense, while also being seared into individual and collective memory. Endfield and Veale argue that 'these narratives and stories have great potential to engage people in weather and climate issues, the history of place, and to improve "weather memory"'.

John Clammer's chapter title 'Back to the Future' explicitly raises the question of how past practices might help us in the future. This is not just a question of maintaining tradition or continuity from the past, but potentially something more transgressive. *Satoyama* – a Japanese term for a mosaic of different ecosystem types and socio-agricultural organisation in which humans are symbiotically embedded – provides an example of a mixed agro-ecological integrated system that can also be imagined for the future. Clammer argues that this is not just an empirical example of climate responsible lifestyles or systems, but that it also throws out theoretical challenges.

He explicitly challenges the notion that such small-scale systems are too small and fragile to provide viable models of the future, reminding us that fossil fuel economies are also fragile – we just do not acknowledge them as such. As I and others have discussed in different contexts, such apparently small-scale experiments provide important candidates for survival. For Clammer, *satoyama* and similar systems simultaneously (i) are among the best candidates to 'survive a collapse of the fossil fuel–based economy', (ii) represent a culture of transition as well as exemplar practices and (iii) are not only resilient but 'actually generative, bringing fertility back to both nature and community'. This is a very important argument, echoed elsewhere, that prefigures the next theme.

Identifying Cultural Resources and Thresholds for Change

Resources for imagining alternative ways of doing things can emerge from unlikely places. Increasingly they emerge not from existing governments or institutions but from more diverse scales of governance and from small experiments. Encounters between different ideas and practices occur during the migration process, particularly from the global south to global north (Carter *et al.* 2013; Head *et al.* 2018). Practices undertaken for non-environmental reasons can have environmental benefits by reducing consumption (Maller and Strengers 2013; Waitt and Welland 2019). Migration provides a unique opportunity to disrupt the status quo by throwing light on practices taken for granted; see, for example, Klocker *et al.*'s (2015; Waitt *et al.* 2016) work on car ownership.

Arts practice has long been important in helping communities imagine other worlds. Ciara Healy-Musson's (this volume) 'thin place' is one such emergent experiment. Using the concept of a thin place as a 'marginal, liminal realm, beyond everyday human experience and perception, where mortals can pass into Otherworlds', Healy-Musson describes a gallery experiment in which ancient and contemporary world views are interwoven. As she describes it, Thin Curating is very much a pragmatic intervention in the world rather than an attempt to reveal the world. Her example is also very place-based: in Carmarthen, an economically deprived area with a rich animist history. The theme of considering relationships to place as encompassing both continuity and change is enriched here by a revival of the archaic and the use of visual culture.

The question of social transformation is also addressed by Susanne Moser (this volume), who takes climate change communication beyond its traditional discussion point of communicating science to considering what it tells us about how to communicate the transformative imperative. Moser's ten characteristics echo themes that have been brought out in the preceding chapters, including 'naming and framing the necessary change', 'orienting towards the difficult' and 'fostering generative engagement'.

Karen O'Brien, Gail Hochacka and Irmelin Gram-Hanssen (this volume) also address the issue of how to create a culture 'for' transformation. Their useful review of different theories of cultural transformation arises out of the insight that there is a need for greater understanding of relationships between individual and collective change (or stability). However it is understood, culture can be both a barrier and a catalyst for transformative change, and cultural change of different types may be beneficial or detrimental for climate change action. The teaching experiment cCHALLENGE documented in this chapter is particularly interesting because of how it gets students to reflect on the process of change involved in taking environmentally friendly steps for thirty days.

Conclusion

The value of a good edited book is in the potential for the whole to generate more than the sum of the parts, for the juxtapositions and conversations between chapters to provide new insights and frame action differently. Perhaps the most exciting dimension of this book – and the way the whole may generate more than the sum of the parts – is in the juxtaposition of examples from different times and places. This goes well beyond a way of documenting cultural variability but starts to take us towards examining alternatives and how we might get there. To my reading, the whole of this book generates three such insights.

First, to summarise the findings of the transformation chapters, we have come a long way (apparently) from a simple notion that climate change transitions are simply about communicating the science in a more effective way, as if received by a rational body politic who will act accordingly. There is now a much fuller understanding of the cultural framings within which different individuals and groups receive and process new and challenging information. This is well understood by the vested interests who have sought to distract attention from themselves by fostering a public sphere in which explicit climate change denial seeks to continually engage the science on the extent to which debates are settled. It is surely past time for most of us to waste no further time trying to convince this minority, but instead to focus on enacting alternative ways forward. As many of the chapters here explicate so well, and as O'Brien and colleagues are teaching their students, these ways forward involve sophisticated understandings of how culture and power are entwined at scales from the individual to the societal. That some useful points of intervention and agency are unexpected or surprising provides grounds for hope.

Second, the different temporalities in the book show that historical perspectives on climate change are now far more sophisticated than the simplistic framing of 'learning lessons' from the past, a framing embedded in linear time with primitivist views of the past and progressivist views of the future. Instead, we can look for kinship across time and space, as a number of chapters here do. Leaving behind (a linear linguistic framing itself!) historical narratives of progress and collapse, we can imagine much more interesting juxtapositions in which, for example, medieval standards of living and disease risk combine with smart phone availability. Or we might imagine how future everyday talk of weather and climate involves a continuous undercurrent of how to cope with extremes and the resources to do so. In this way futures connect to conversations that have occurred for thousands of years. Healy-Musson's 'thin places' concept, as a means to crack between different temporalities, is immensely powerful.

Third, the (also powerful) concept of cultural resources has risks of being extractivist, and we need to be attuned to these risks. Healy-Musson's attempt to return 'the multi-layered nature of their sacred past' to communities raises the question of how we might do this when the past is not shared or is shared in a way that is premised on dispossession, such as in the settler colonial context. To consider the past, or any other cultural dimension, as a resource should immediately alert us to questions of power. To go beyond, and to seek a process that is dynamic and dialogical, will take careful attention. It must be done with deliberation.

I commend this collection to the reader and congratulate the editors for bringing it all together.

Lesley Head
University of Melbourne

References

Adger W. N., Barnett J., Brown K., Marshall N. and O'Brien K. 2013. Cultural dimensions of climate change impacts and adaptation. *Nature Climate Change*, 3(2), 112–117.

Bulkeley H., Paterson M. and Stripple J. (eds.) 2016. *Towards a Cultural Politics of Climate Change*. Cambridge: Cambridge University Press.

Carter E. D., Silva B. and Guzman G. 2013. Migration, acculturation, and environmental values: The case of Mexican immigrants in central Iowa. *Annals of the Association of American Geographers*, 103(1), 129–147.

Castree N. 2015. *Making Sense of Nature*. London: Routledge.

Correa-Velez I., McMichael C., Gifford S. and Conteh A. 2014. 'I was used to worse in a refugee camp': Men from refugee backgrounds and their experiences during the 2011 Queensland floods. In: J. Palutikof, S. Boulter, J. Barnett & D. Rissik (eds.), *Applied Studies in Climate Adaptation*. Wiley Blackwell, pp. 250–257.

Crow D. A. and Boykoff M. T. 2014. *Culture, Politics and Climate Change: How Information Shapes Our Common Future*. New York: Routledge.

de Guttry C, Döring M. and Ratter B. 2016. Challenging the current climate change–migration nexus: Exploring migrants' perceptions of climate change in the hosting country. *DIE ERDE – Journal of the Geographical Society of Berlin*, 147(2), 109–118.

Farbotko C. and Lazrus H. 2012. The first climate refugees? Contesting global narratives of climate change in Tuvalu. *Global Environmental Change*, 22(2), 382–390.

Gibson C., Farbotko C., Gill N., Head L. and Waitt G. 2013. *Household Sustainability: Challenges and Dilemmas in Everyday Life*. Cheltenham: Edward Elgar.

Gibson C., Head L. and Carr C. 2015. From incremental change to radical disjuncture: Rethinking everyday household sustainability practices as survival skills. *Annals of the Association of American Geographers*, 105, 416–424.

Head L., Gibson C., Gill N., Carr C. and Waitt G. 2016. A meta-ethnography to synthesise household cultural research for climate change policy. *Local Environment*, 21, 1467–1481.

Head L., Klocker N. and Aguirre-Bielschowsky I. 2018. Environmental values, knowledge and behaviour: Contributions of an emergent literature on the role of ethnicity and migration. *Progress in Human Geography*, DOI:https://doi.org/10.1177/0309132518768407

Klocker N., Toole S., Tindale A. and Kerr S. 2015. Ethnically diverse transport behaviours: An Australian perspective. *Geographical Research*, 53(4), 393–405.

Maller C. and Strengers Y. 2013. The global migration of everyday life: Investigating the practice memories of Australian migrants. *Geoforum*, 44, 243–252.

McCright, A. M. and Dunlap, R. E. 2011. Cool dudes: The denial of climate change among conservative white males in the United States. *Global Environmental Change*, 21(4), 1163–1172.

Toole S., Klocker N. and Head L. 2016. Re-thinking climate change adaptation and capacities at the household scale. *Climatic Change*, 135, 203–209.

Waitt G., Kerr S. M. and Klocker N. 2016. Gender, ethnicity and sustainable mobility: A governmentality analysis of migrant Chinese women's daily trips in Sydney. *Applied Mobilities*, 1(1), 68–84.

Waitt G. and Welland L. 2019. Water, skin and touch: Migrant bathing assemblages. *Social and Cultural Geography*, 20(1), 24–42.

1

Climate and Culture
Taking Stock and Moving Forward

HILARY GEOGHEGAN, ALEX ARNALL AND GIUSEPPE FEOLA

Climate change is acknowledged by many as one of the greatest challenges of the twenty-first century, a challenge that potentially poses an existential threat to the lives and livelihoods of millions of people in different societies around the world (Tanner and Allouche 2011). At the same time, the ways in which climate change is understood has also broadened, with scholars viewing it as both a physical phenomenon and an idea. As the former, there is unequivocal evidence that the global climate system is warming at rates unprecedented in human history (Stocker 2014). As an idea, climate change is travelling or actively unfolding in different parts of the world, meeting different religions, politics and societies along the way (Hulme 2007). The pathways that these ideas of climate change follow, the new meanings that they take on and the new purposes that they serve in the process will be shaped and mediated by human culture to a considerable extent (Livingstone 2012; Arnall 2014; Hulme 2015).

Whilst there is growing interest in the discursive dimensions of the climate change phenomenon, as well as people's multiple and varying perceptions of it, there has been to date a paucity of scholarly work examining the significance and role of culture in relation to climate and its changes in a systematic, multidisciplinary manner. The aim of this edited volume is to help address this knowledge gap by exploring how culture mediates the relationship between the phenomenon of climate change and its spread as an idea. In other words, we are interested in how culture makes climate and climate change meaningful, but also determines what climate change means for us as it unfolds (Hulme 2007). This chapter, then, represents the first step towards this somewhat daunting task. In the sections that follow, we first introduce the structure of, and approach taken to, this book. Section 1.2 helps to set the scene for this volume by reviewing the distinct but interlinked topics of climate, climate change and culture. Finally, Section 1.3 suggests connections and common themes across chapters that can help to guide the reader through the collection. The themes concern: (a) the relationship between climate change and capitalism; (b) modes of knowing and alternative ontologies; (c) how cultures persist in the face of climate change and other stressors; and (d) methodological and data diversity.

1.1 This Book

Much climate–culture research to date has assigned culture the following roles in relation to climate and its changes:

- a *cause*. Scholars have argued that climate change is not just a side effect but also an inherent condition of modern capitalist societies and the limitless economic growth and cultures of consumption that have come to define such societies (Urry 2010; Ghosh 2016).
- as a *victim*, and therefore something to be protected. Many cultures are now identified as under threat in different parts of the world due to climate change, from remote marginal regions of the world, such as the Arctic, to late capitalist cultures located in the West (Strauss 2012).
- as a *means* to adapt. Culture has been recognised not only as something for making sense of and responding to climate change (Adger *et al.* 2012) but also as a potential barrier to adaptive action (Nielsen and Reenberg 2010; Jones and Boyd 2011).

Whilst these approaches are helpful in identifying important climate–culture interactions, they also leave us with a conundrum. On the one hand, climate change fundamentally challenges our culture as a cause of climate change, and yet, on the other hand, culture is adaptable and makes life in a climate change world liveable (or at least possible). In recognition of these difficulties, this book takes a different approach that cuts across the cause-victim-means classification by exploring and critiquing the place and role of culture in climate and climate change in terms of three ways of living in a warmer world: 'knowing', 'being' and 'doing'. In the text that follows, these are briefly described in turn.

First, in terms of 'knowing', this book interrogates the power and politics of climate knowledge. This is undertaken by examining the ways in which dominant scientific knowledge systems, such as those embodied, objectified and institutionalised by the Intergovernmental Panel on Climate Change (IPCC), potentially exclude alternative possibilities for understanding climate and climate change. We unpack how climate is performed through cultural practices in past and contemporary scientific cultures and epistemic communities. These have brought climate and climate change to life through computations, predictions, visualisations and other representations, and the figure of the expert and associated norms. Such understandings enable us to reflect on present and past efforts to understand and visualise climate scientifically. Examples of alternatives covered in this book are indigenous epistemologies and ontologies of climate and the environment in Latin America and new modes of environmental knowing through curatorial practice in Wales.

Second, in terms of 'being', the book exposes how people make sense of climate and its changes through embodied experiences, affective and emotional encounters, and everyday practices. It does this by discussing scientific, traditional and creative practices and by examining whether and how they enable the coexistence of communities in a world in which the physical manifestations of climate are moving ever more to an unknown realm. This section of the book also exposes the role of, for example, communication and collective memory in enabling the continuation of human life in a meaningful manner in the face of a potentially destabilising climate change. These dimensions are rarely presented together in the manner done so in this book.

Third, in terms of 'doing', this book examines the role of culturally determined ideas and emotions in how individuals and societies respond, or fail to respond, to climate change. Where culture may become a barrier for action, a number of later chapters in this book discuss concrete experiences of 'cultural work' that create possibilities for social and economic change. Whilst the magnitude of change needed to respond to climate change is considerable, this section of the book also shows that examples and principles that can inform such dramatic but necessary social and cultural change may exist in traditional and religious cultures that have persisted and are being rediscovered as sources of inspiration and evolving models of alternative social-ecological interaction.

In exploring the 'knowing', 'being' and 'doing' of climate and culture, this book's content is firmly multidisciplinary. Climate facts arise from impersonal observation whereas meanings emerge from embedded experience, and the environmental social sciences, arts and humanities are well positioned to foster a more complex understanding of humanity's climate predicament (Jasanoff 2010; Offen 2014). Researchers from these disciplines are already attending to societal responses to actual, predicted and imagined climate change, asking important questions about climate and its changes across time and space (Brace and Geoghegan 2010; Trexler and Johns-Putra 2011; Arnall and Kothari 2015). However, whilst the importance of the exchanges between culture, society and climate in the context of global environmental change is being increasingly recognised by different fields of enquiry, the empirical evidence is fragmented and too often constrained by disciplinary boundaries. Climate knowledge is no longer solely based on scientific data but also shaped by ideologies, worldviews and values (Hoffman 2015). Bringing the social sciences, humanities and culture into discussions of climate is important, but if not dealt and engaged with on a level playing field with the natural sciences, then it may mean divides between knowledges are reproduced rather than transformed (Castree *et al.* 2014; Lövbrand *et al.* 2015). These cultural discussions should no longer be seen merely as an appendage to the natural sciences but rather as an opportunity to explore the ambiguous meaning of climate today (Hoffman 2015).

In recognition of these issues, this book aims to reach out across subjects as varied as history, literature, sociology, anthropology, human geography, philosophy, environmental history, visual studies, history of science and technology, religious ethics, and urban design and theory. Accordingly, it explores examples as diverse as: how Andean populations have adapted culturally to difficult environmental and climatic conditions over hundreds of years (Postigo, this volume); the documentation of the weather as part of the national memory in the United Kingdom (Endfield and Veale, this volume); and the United States' cultural traits, such as those linked to the colonial past, that inform emotional responses to climate change (Ford and Norgaard, this volume). The ideas, approaches and examples that feature here, when taken together, help to fill important knowledge gaps that continue to hamper the effective contribution of the humanities and social sciences to studies of climate and climatic change. This is important if we are to overcome the tendencies towards essentialism and determinism sometimes witnessed in the climate change literatures (Hulme 2011), including in the environmental social sciences and humanities, and fully open up the range of 'imaginable climate change-influenced futures' that are possible for us (Jackson 2015:479).

Finally, this book also reveals and highlights existing and future research and curatorial and communications praxis that is required to think about climate and climate change. We are interested in opening up and recovering ways for people to engage with place and environment, not only in a non-rational manner, but also relationally, symbiotically and deeply, to effect widespread societal transformation and to reveal the limitations of the modern, rational, scientific and utilitarian paradigm. Ultimately, this book expands existing natural science-led approaches to climate–culture relations by drawing on other areas of scholarship to make sense of the climate–culture interface. Climate change represents a potentially monumental shift in how we live on planet Earth, and culture is key to how this shift plays out and what it means for diverse societies around the world.

1.2 Climate, Climate Change and Culture

1.2.1 Climate

As we have already established, climate has long been the domain of the natural sciences, being understood as the thirty-year average of weather for a particular region over time (Hulme *et al.* 2009). For scientists, climate incorporates observed averages of precipitation, temperature, sunshine, wind as well as other variables. Climate change in this sense might be described as unusual observed averages repeated over several years. The resulting data are then used in climate models to predict future climate. Hulme *et al.* (2009) relate this approach to the epistemological practices formed in the enlightenment period that reinforce scientifically constructed knowledge of climate, but they also point to how climate is something felt and experienced in everyday life. Thus, for Hulme *et al.*, 'normal' climate is socially and scientifically constructed and can impact the world in material and imaginative ways. Climate has been an object of study and source of inspiration for the natural sciences, social sciences, humanities and the arts for centuries (Hulme 2016). Whilst meteorological evidence informs a physical connotation of climate which is external to the human imagination, cultural connotations of climate are inside the human imagination and are formed through experiences and memories of past weather events.

Building on this idea, Brace and Geoghegan (2010) have highlighted some of the difficulties of thinking with 'climate' when what people experience is weather, and weather forecasts are regularly consulted in everyday decision-making. They write, 'There is a metaphysical and semiotic problem here with discussing in terms of a future date something that is made of the stuff of everyday life (for example, weather) but which is not, in and of itself, *that stuff*, but aggregated, averaged, modified, smoothed, stripped of its outliers, rendered in statistical ways that remain mysterious to the majority' (ibid.:291). Thus, at the intersection of climate and culture, weather looms large in the construction and shaping of both self (social and cultural identities) and place in everyday life and memory. As a result, work on climate and culture needs to encompass normal weather, as well as extremes and large-scale spectacular weather events (Gergis *et al.* 2010; Vannini *et al.* 2011; Veale and Endfield 2014). For example, Sturken (2001) offers a discussion of the climatic phenomenon of El Niño, which has a deep-rooted cultural history, and Endfield and Morris (2012) uncover amateur practices of recording, observing and being in the weather.

1.2.2 Climate Change

The IPCC defines climate change as 'any change in climate over time, whether due to natural variability or because of human activity' (IPCC 2018, unpaginated). This usage differs from that in the United Nations Framework Convention on Climate Change (UNFCCC), which defines climate change as 'a change of climate which is attributed directly or indirectly to human activity that alters the composition of the global atmosphere and which is in addition to natural climate variability observed over comparable time periods' (UNFCCC 2018, unpaginated). However, scientists have never been in sole possession of the term 'climate change'. This is because, as Watson and Huntington (2014) have asserted, all knowledge is a product of culture. Moreover, over the last 20 years, the climate change debate has become politically, socially and culturally charged, being related by Rudiak-Gould (2013) to 'a proxy war for a larger debate on scientific versus lay knowledge and the role of expertise in democratic society' (p.120).

As a result, some scholars have moved towards less inflammatory notions of climate change in an attempt to shift the debate forwards and towards action. Hulme highlights the numerous expressions used in this area, from 'climate change' to 'changes in climate' to 'climatic change' and suggests using 'the construction 'climate-change' to refer to the contemporary idea of human-caused global climatic change' (2016:xii, n1). Morton (2013) prefers the language of 'global warming' over climate change in his work on hyperobjects, whereby the scale of the temporalities and spatialities of the climate change issue defeats traditional attempts to understand it. For Morton, '*climate change* as a *substitute* for *global warming* is like "cultural change" as a substitute for *Renaissance*, or "change in living conditions" as a substitute for *Holocaust*' (2013:8, emphasis in original). In addition, associated with ideas of climate change are recent debates around the Anthropocene, defined as a time when human activity has been the main influence on climate and environment (Lövbrand *et al.* 2015; Bai *et al.* 2016). Whilst there is disagreement in the natural sciences regarding the ratification of this new geological epoch, social science and humanities scholars have embraced the term, although a full discussion of it is beyond the scope of this book.

1.2.3 Culture

Culture has long been recognised as crucial in mediating the relationships between humans and the natural environment (Sanderson and Curtis 2016). However, much like the term 'climate', pinning down what culture means remains a challenge across disciplines. Indeed, culture is a broad and contested term even within those fields that traditionally study the concept, such as anthropology (Strauss 2012). Yet, a range of disciplines have found culture to be a productive lens for enquiry. For instance, in social theory, a useful and relatively broad understanding is Bourdieu's notion of 'cultural capital' in which he identifies three distinct but interrelated states of culture (1987). These are: first, an embodied state, as a form of knowledge that resides within us (mind and body), and which concerns beliefs, values, morals, emotions or the way that we talk or express ourselves. For Bourdieu, this was partly experienced through high culture, in spaces such as museums. Second is an objectified form, which includes cultural goods, namely items such as artistic works. This goes beyond mere ownership in an economic sense, involving a person's ability to use and enjoy their cultural goods. Third is an institutionalised form, which is related to the way that institutions confer recognition onto people who are especially knowledgeable or authoritative, for example holding a PhD. These three states of culture can, in turn, be applied to climate and climate change. Thus, embodied culture relates to how climate change is thought and talked about, felt and imagined; objectified culture concerns entities such as climate data, software programmes that can model or mobilise climate change, models, maps and projections; and institutionalised culture refers to people such as climate change science and scientists, politicians and journalists, and (more recently) grassroots and indigenous groups recounting direct experiences of climate change.

Culture and its states are not static, but rather they are fluid, dynamic and multiple, operating independently and simultaneously, much like the weather and climate themselves, and across time and space. Culture's temporal and spatial qualities are evoked through collective memory, and local knowledges associated with key events located in particular places (Ulloa 2011). Historical occurrences and landmarks become part of local and national cultures and are an important factor in how people interpret, understand and experience weather and climate (Harley 2003). By understanding these dynamics, it is also possible to understand cultural persistence and change and to interpret current dynamic interactions of culture and the environment. Cultures can evolve over millennia in particular places and can also be accessed digitally at the click of a button anywhere in the world. We need a view of time and space that can accommodate and explore these differing intensities, relationships and interactions around climate–culture.

Moreover, these embodied, objectified, institutionalised aspects of culture (located in time and space) merge with human practices, sensory experiences and imaginations, and with the non-human world, to form hybrid cultures, recognised in religion, farming practices or scientific knowledge, among others. Take, for example, the contrasting religious views with respect to climate change; indigenous African religions believe in meta-physical causality, whereas Christian and Islamic groups have taken a stance that refers more to humankind's neglect of the earth (Golo and Awetori Yaro 2013; Haluza-DeLay 2014). As a result, climate knowledge no longer resides elsewhere but is being made and remade personally, locally and in the everyday; culture is continually created through human activity and practice, rather than something that is 'out there', pre-existing and thus potentially 'lost'. Hulme (2016:6) draws upon the work of Ingold to define culture in these terms:

As Tim Ingold says, 'We can never expect to encounter culture "on the ground" (Ingold 1994:330), just as no-one has ever 'seen' climate. Instead, what we find are '... people whose lives take them on a journey through space and time in environments which seem to them to be full of significance, who use both words and material artefacts to get things done and to communicate with others, and who, in their talk, endlessly spin metaphors so as to weave labyrinthine and ever-expanding networks of symbolic equivalence' (Ingold 1994:330). It is therefore more accurate to say that people 'live culturally' rather than that they 'live in cultures'.

In this way, a more expansive notion of culture that engages with 'living culturally' is embraced in this book than is traditionally the case in climate change studies. For example, attention is given to liminal places to revive/recover ways of engaging with place/environment that are non-rational. Here, culture becomes relational and symbiotic. People's connections with the places where they live influence current and future responses to climate change at the local scale, and climate change narratives can circle around seemingly insignificant structures in the landscape and yet have decisive weight in different actors' understandings of climate change (Geoghegan and Leyshon 2012; Matless 2018). This also reinforces the need to consider culture through the small-scale, mundane interactions that formulate social and political worlds (Farbotko *et al.* 2015).

1.3 Cross-cutting Themes from Multidisciplinary Perspectives

1.3.1 Cultural Change: Moving Away from Modern Capitalism

A first theme emerging from this collection is the embeddedness of climate change in a particular culture – modern capitalism. While that culture can be seen as a *cause* of climate change, *cultural* change is thus crucial to address climate change.

Climate change, as many of the chapters in this book remind us (Ulloa; Postigo; O'Brien *et al.*; Daniels; Clammer; Moser; Ford; Norgaard, this volume), is the outcome of a particular, although non-monolithic, culture: modern capitalism. This culture, imposed on much of the world through imperialism and colonialism, has not only expanded consumer societies and the environmental impacts of mass consumption (including climate change), but it has also spread the structures of exploitation, including resource extraction, which made the project of modernity and capitalism possible in the West (Ghosh 2016). In non-Western societies, climate change is often perceived as another form of dispossession of traditional communities and original peoples (Ulloa; Ford and Norgaard; this volume). This reinforces our observation that there is a fundamental connection between ways of knowing, being and doing in the world – driven by capitalism – and climate change as a side effect that has come to define that culture. While capitalism is a given for many, it is also a target of critique and struggles that have been renewed and have found further momentum in connection with climate change (e.g. Klein 2014; Feola and Jaworska 2018). Thus, as Head (this volume) argues, 'we need to shift some very big cultural frames – the importance of economic growth, the dominance of fossil fuel capitalism, the hope of modernity as unending progress – to deal adequately with the climate change challenge'.

If we accept that we cannot avoid exposing and challenging modern capitalism and the structures and cultural processes that reproduce it, then cultural work is needed to conceive, envision and make sense of possible alternative ways of living, and deliberately pursue such cultural change (Braun 2015). As argued by various authors, climate change is so difficult to come to terms with, and respond to, because it poses such profound challenges to modern capitalist identities: it forces us to realise that we need to be 'other' or 'differently human' and probably renounce a lot of what defines 'us' (Hulme 2010; Morton 2013; Beck 2016; Ghosh 2016). Many chapters in this book discuss experiences of envisioning, imagining and deliberately transforming towards a future beyond capitalism. They show that examples of what post-capitalist alternatives may look like already exist, or are re-emerging, within modern capitalist societies and at their margins in indigenous or traditional communities (see Ulloa; Postigo; Ford and Norgaard; Clammer, this volume), in religious worldviews (Daniels, this volume), and in experimental communication and visual, artistic and social change practices (Moser; O'Brien *et al.*, this volume). These contributions illustrate the ongoing efforts of cultural transformation by practitioners and scholars alike and highlight that 'doing' in a climate change world often entails challenging culturally established ways of acting in everyday life. Deliberate cultural transformation (O'Brien 2018) can inform responses to climate change and support alternative, low-carbon forms of societal development.

1.3.2 Modes of Knowing and Alternative Ontologies

A second theme that cuts across this collection is the importance of overcoming the nature-culture duality that informs modern understandings of the climate, as well as possible ways to achieve this. Current climate predictions in the form of models and graphs produce visions of compressed time and space, i.e. they fold past and future on the present (Mahony *et al.*; Grevsmühl, this volume). Not only are these predictions and visualisations socially constructed through practices, institutions, expertise and impacts of claims, but they are also culturally received, which has broader effects on the way society at large interacts with a changing climate or fails to do so (Ulloa; Moser, this volume). Therefore, the risk is that climate becomes yet another abstraction, disembodied from everyday life, and that visualisations stand in for real-life experiences of climate change, making it something that happens 'somewhere else'.

There is much research that critiques the prevailing culture of modern Western knowledge, still largely inflected with enlightenment traditions of rationality, objectivity and fact (Brace and Geoghegan 2010). Furthermore, scientific knowledge as the hegemonic way of knowing climate change has served to prioritise the claims of certain actors whilst silencing vulnerable communities (Rice *et al.* 2015), such as indigenous ones (Ulloa, this volume). Indigenous knowledge has often been regarded only to the extent that it could inform particular Western climate change agendas by providing meaningful 'data' (Watson and Huntington 2014).

Many of the chapters in this book, in contrast, attempt to identify culture and nature as intimately connected, entangled and mutually co-produced in multiple ways (Whatmore 2002; Haraway 2008). As such, climate change emerges as a composition of the natural and the cultural, including not only language, cognition and conscious thought but also a range of multisensory engagements, relational sensibilities and emotional responses with the physical environment (whether organic or inorganic matter). Novel communication and curatorial practices (Moser; Healy-Musson, this volume), social experiments (O'Brien *et al.*, this volume) and artistic improvisational engagement (Tyszczuk and Smith, this volume), among others, help us to appreciate different climate knowledges and to comprehend the ambiguous meaning of climate, functioning simultaneously as metaphor, environment, explanatory force and a condition of human experience (Brierley 2010; Brace and Geoghegan 2010).

Thus, climate change is not something that is simply 'out there', a phenomenon to be objectively isolated, observed and managed, but rather is entangled in people's day-to-day lives, routines, plans, expectations, hopes and fears (Head 2016). As Brace and Geoghegan (2010:296) have argued, 'Climate and its changes might not only be observed ... but also felt, sensed, apprehended emotionally, passing noticed and unnoticed as part of the fabric of everyday life in which acceptance, denial, resignation and action co-exist as personal and social responses to the local manifestations of a global problem.' As a result, climate change can be lived with and become normal, even when people are aware of it and express concern about it (Norgaard 2011).

1.3.3 Cultural Persistence

The third theme that connects the chapters in this volume is that of cultural persistence. A number of authors in this book consider the relationships between Western capitalist cultures and other cultures located in the global South to point out that the worldwide penetration of capitalism, and the speed of the changes that it is bringing about, might be limiting culture's adaptive function. For example, Postigo (this volume) argues that, in the Andes, economic liberalisation as a mechanism of modernisation and integration of peasant communities in global economic structures negatively affects rural communities' cultural adaptive practices and magnifies vulnerability to climate change. Thus, in different parts of the world, cultures that have developed over millennia now face multiple changes of unprecedented magnitude and pace, and therefore it is unclear whether such cultures can inform successful adaptation in the future (e.g. for the Andean region, see Perez *et al.* 2010; Sietz and Feola 2016). There are also examples, however, of certain cultures and cultural practices being reinvented or re-emerging as responses to the failures of modern capitalism to deliver well-being for all. To illustrate from this volume, Healy-Musson experimented with a curatorial practice for the recovery of almost forgotten aspects of culture to appreciate the changing environment, while Clammer demonstrates the persistence of the *satoyama* traditional system over time and its emergence in the face of future contests over transitions towards post-oil and post-affluent societies.

In addition to the broad struggles outlined above, other authors in this book are concerned with the more day-to-day mechanisms and processes through which cultures 'carry on', potentially adapting to climate change in the process or at least normalising its effects. To illustrate, for Krupnik (this volume), culture, whilst often being mobilised around certain 'core' cultural practices, is also adaptable. Culture allows communities to live with climate change, normalising it in everyday life through, for example, jokes, gossip and new expectations about the weather or animals' migrations. Endfield and Veale (this volume), in contrast, focus on culture in the form of everyday collective memory. They suggest that building a memory of survival practices and symbolic responses to extreme events, as well as a more general awareness of the 'normality' of extreme events in a country's history, can build confidence in society to overcome future extreme weather events. They provide illustrations of these adaptive practices, for example soup kitchens and other social enterprise activities, that have proved effective in the past and that could be recovered in the present or future. Similarly, the exercise of visualising climate can have a 'normalising' effect in society by objectifying climate and reducing the sense of surprise by building a shared representation that can inform collective action (Grevsmühl, this volume). Such collective action, for example in the form of traditional collective fishing practices, can help maintain cooperation within a community, which is a strength when action is urgently needed to respond to climate change (Krupnik, this volume).

Certainly, it is worth remembering in any discussion of climate and culture that climate change is by no means the only, nor always the most pressing, issue that people around the world are having to make sense of nor confront in their lives (O'Brien and Leichenko 2000; Feola 2017). Greater recognition, therefore, must be given to the multiple socioeconomic and environmental stressors impacting upon people, as well as their influence on people's understandings of the world around them. This is something that a number of chapters in this edited volume remind us of. For example, in Krupnik's contribution, climate change has become something to be accepted and lived with in light of other challenges, such as geographical isolation or the threat of changing fishing quotas. Indeed, as Postigo (this volume) contends, research is needed to understand adaptation beyond climate change, thereby avoiding its fetishisation as the only problem. As Moser (this volume) asks: 'How do these coalescing changes and crises shape the understanding of those involved? How is the totality of changes constellated and experienced in people's lives? Where are the opportunities and openings when everything seems to close in on people?' To some extent, this realisation, of the need to consider and deal with the multiple socioeconomic and environmental 'stressors' that people face in diverse contexts, explains the move in recent years away from adaptation science to the concept of resilience in global climate change research and policy (Cannon and Muller-Mahn 2010). This is because adaptation attempts to identify and isolate the climate change signal from the background 'noise' whereas resilience takes a more holistic view, looking at how actors exercise agency in relation to wider social structures around them (Arnall 2015).

1.3.4 Methodological and Data Diversity

In this volume, the use of a diverse range of methodologies and forms of data facilitates exploration of how cultural perspectives on climate and culture move between memories of the past, experiences of the present and expectations of the future. Of course, past, present and future are not separate categories but are instead fluid and interlocking (Ingold 2000; DeSilvey 2012). Thus, an understanding of our collective, institutional and personal memories of weather and climate – what we choose as societies to remember and forget – can help us imagine, make sense of and deal with our common futures. In this volume, for example, Endfield and Veale (this volume) consider how personal memories and narratives of past weather inform and influence how we respond to weather in the present. They argue that 'Histories, memories and vivid, vicarious or actual experiences of weather events could serve a powerful role in informing memory but may also influence judgement and popular understanding of both local and global climatic change'.

Such archival, textual and visual approaches (Grevsmühl; Mahony et al.; Endfield and Veale; Tyszczuk and Smith, this volume) can also be used to trace the emergence of key ideas and influential figures that underpin current discussions of climate and climate change. Of these approaches, the field of historical climatology is especially well versed in combining past sources to make sense of former weather and climates, although challenges and frustrations of working with such documents that may be missing or partial can exist (Endfield and Veale, this volume). Studying the visual also provides insights that might be obscured from view when analysing the use of language (Grevsmühl, this volume), although attention can also be paid to the text beneath or accompanying images in this form of analysis (Nerlich and Jaspal 2013). In this way, memories – as found in myths, religious and traditional practices, and archives ranging from scientific archives and official histories, to personal letters, religious relics, and traditional rituals – are recovered and made use of in the present and can also inform our approaches to the future.

Of course, research on climate and culture can be conducted in a range of ways, using a variety of methodologies. In this volume, some approaches are relatively conventional, such as the use of semi-structured interviews, for example, and contribute to cultural debates by rejecting wholly positivistic, quantifiable notions of climate knowledges and supplementing climate data derived from scientific instruments or models with more ethnographic and interpretive understandings (Geoghegan and Leyshon 2012). The ethnographic work featured here (Ford; Krupnik; Clammer; Ulloa; Postigo, this volume) enables more emotionally sensitive research, which is crucial when exploring people's sense of self and place (Vannini *et al.* 2011), including work on understanding responses to climate change (Ford and Norgaard, this volume). Other approaches in this volume are more experimental, such as O'Brien et al.'s cCHALLENGE programme and Healy-Musson's innovative curatorial concept. Practice-based art and curation, especially that which encourages experimentation and improvisation (Tyszczuk and Smith, this volume), require a rejection of the modern idea of a purely rational subjectivity in favour of engagement with a range of subjectivities in a more holistic fashion. It also demands that scholars accommodate other ways of presenting their research. For example, Healy-Musson in this volume introduces a novel concept of curatorial practice, called 'thin curating'. This supports new situated modes of knowing the environment and soon gives way to notions of structural complexity, radical uncertainty, negotiation and navigation of multiple worldviews (Haraway 2016). Moreover, as O'Brien *et al.* (this volume) demonstrate through their cCHALLENGE project conducted at the University of Oslo with eighty-two undergraduate students, it is also possible to make one's own everyday life a site of experimentation.

Due to this variety in approach and method, the contributions in this volume also illustrate a diverse range of what constitutes data in climate–culture research. Examples include: personal testimony and emotional encounters through interview transcripts; sensory experiences of weather through archival records; peer-reviewed papers; climate models and images; photographs of landscapes; natural resources (e.g. seeds); artworks; and ethnographic observations. Moreover, the subjects of the data range from a strict focus on climate and culture to other cultural experiences, such as with the weather, which provides a powerful means to access relations between culture and climate (Endfield and Veale, this volume; Brace and Geoghegan 2010). Engaging with and combining a rich mixture of data sources and types is valuable as it allows the researcher access to different perspectives, thus offering triangulation of physical weather patterns with social interpretations and responses (Byg and Salick 2009). It also enables engagement with different temporalities, from everyday (personal diaries), to weekly (estate records and letters), to yearly (parish records) accounts (see Endfield and Veale, this volume), as well as access to a range of data 'producers' or 'recorders', including expert climatologists or laypeople, professionals in a specific sector or general observers. Yet, as some of the chapters here have shown, researchers at the climate–culture interface must pay careful attention to the potential of new methods and data sources. This is particularly important in an age in which the future cultural objects of climate and its changes are often unknown and in which the value of sources of information, such as personal diaries, land records and fictional accounts, is constantly being newly evaluated (Adamson et al. 2018).

1.4 Conclusion

This book represents the first multidisciplinary collection to explore how human culture contributes to, shapes and mediates climate, its changes and the spread of climate change as an idea. Organised around ways of 'knowing', 'being' and 'doing' in the world, it shines a light on some of the cultural frames that support, maintain and challenge human ideas of, and responses to, climate and climate change, including sense-making, envisioning, recovering imaginaries and projecting possibilities. Ultimately, this book aims to help provide some of the theoretical and empirical building blocks required to transform, rather than simply re-produce, existing cultural narratives and move debates, research and ways of life forward in this area. The various contributions contained in this book suggest that no less than a radical rethink is needed on this front, one that aims to, inter alia, overcome disciplinary divisions, maintain and rediscover adaptive cultural practices, and challenge culturally established ways of acting in everyday life.

Acknowledgements

We would like to thank the Climate, Culture and Society Research Cluster at the University of Reading for inspiring this volume; Marco Grasso and Mike Goodman for their comments on an earlier version of this chapter; Eleanor Fletcher and Rinchen Lama for their contributions to this chapter; the University of Reading and the Walker Institute (at the University of Reading) for their funding contributions; and the editors at Cambridge University Press for their advice.

References

Adamson, G. C., Hannaford, M. J., and Rohland, E. J., 2018. Re-thinking the present: the role of a historical focus in climate change adaptation research. *Global Environmental Change*, 48, 195–205.

Adger, W. N., Barnett, J., Brown, C., Marshall, S., and O'Brien, G., 2012. Cultural dimensions of climate change impacts and adaptation. *Nature Climate Change*, 3, 112–117.

Arnall, A. H., 2014. A climate of control: Flooding, displacement and planned resettlement in the Lower Zambezi River Valley, Mozambique. *Geographical Journal*, 180, 141–150.

Arnall, A., 2015. Resilience as transformative capacity: Exploring the quadripartite cycle of structuration in a Mozambican resettlement programme. *Geoforum*, 66, 26–36.

Arnall, A. and Kothari, U., 2015. Challenging climate change and migration discourse: Different understandings of timescale and temporality in the Maldives. *Global Environmental Change*, 31, 199–206.

Ulloa, A. (ed.), 2011. *Perspectivas culturales del clima*. Bogotá: Universidad Nacional de Colombia.

Bai, X., Van Der Leeuw, S., O'Brien, K., Berkhout, F., Biermann, F., Brondizio, E. S., Cudennec, C., Dearing, J., Duraiappah, A., Glaser, M., and Revkin, A., 2016. Plausible and desirable futures in the Anthropocene: A new research agenda. *Global Environmental Change*, 39, 351–362.

Beck, U., 2016. *The Metamorphosis of the World*. London: Polity Press.

Bourdieu, P., 1987. *Distinction: A Social Critique of the Judgement of Taste*. Cambridge: Harvard University Press.

Brace, C. and Geoghegan, H., 2010. Human geographies of climate change: Landscape, temporality, and lay knowledges. *Progress in Human Geography*, 35, 284–302.

Braun, B., 2015. Futures: Imagining socioecological transformation – An introduction. *Annals of the Association of American Geographers*, 105, 239–243.

Brierley, G., 2010. Landscape memory: The imprint of the past on contemporary landscape forms and processes. *Area*, 42, 76–85.

Byg, A. and Salick, J., 2009. Local perspectives on a global phenomenon – climate change in Eastern Tibetan villages. *Global Environmental Change*, 19, 156–166.

Cannon, T. and Muller-Mahn, D., 2010. Vulnerability, resilience and development discourse in the context of climate change. *Natural Hazards*, 55, 621–635.

Castree, N., Adams, W. M., Barry, J., Brockington, D., Büscher, B., Corbera, E., Demeritt, D., Duffy, R., Felt, U., Neves, K., and Newell, P., 2014. Changing the intellectual climate. *Nature Climate Change*, 4, 763.

DeSilvey, C., 2012. Making sense of transience: An anticipatory history. *Cultural Geographies*, 19, 31–54.

Endfield, G. H. and Morris, C., 2012. Exploring the role of the amateur in the production and circulation of meteorological knowledge. *Climatic Change*, 113, 69–89.

Farbotko, C., Stratford, E., and Lazrus, H., 2015. Climate migrants and new identities? The geopolitics of embracing or rejecting mobility. *Social & Cultural Geography*, 17, 533–552.

Feola, G., 2017. Adaptive institutions? Peasant institutions and natural models facing climatic and economic changes in the Colombian Andes. *Journal of Rural Studies*, 49, 117–127.

Feola, G. and Jaworska, S., 2018. One transition, many transitions? A corpus-based study of societal sustainability transition discourses in four civil society's proposals. *Sustainability Science*, DOI:10.1007/s11625-018-0631-9.

Geoghegan, H. and Leyson, C., 2012. On climate change and cultural geography: Farming on the Lizard Peninsula, Cornwall, UK. *Climatic Change*, 113, 55–66.

Gergis, J., Garden, D., and Fenby, C., 2010. The influence of climate on the first European settlement of Australia: A comparison of weather journals, documentary data and Palaeoclimate records, 1788–1793. *Environmental History*, *15*, 485–507.

Ghosh, A., 2016. *The Great Derangement: Climate Change and the Unthinkable*. London: Penguin Books.

Golo, B. and Awetori Yaro, J., 2013. Reclaiming stewardship in Ghana: Religion and climate change. *Nature and Culture*, *8*(3), 282–300.

Haluza-DeLay, R., 2014. Religion and climate change: Varieties in viewpoints and practices. *Wiley Interdisciplinary Reviews: Climate Change*, *5*, 261–279.

Haraway, D. J., 2008. *When Species Meet* (Vol. 224). Minneapolis: University of Minnesota Press.

Haraway, D. J., 2016. *Staying with the Trouble: Making KIN in the Chthulucene*. Durham: Duke University Press.

Harley, T. A., 2003. Nice weather for the time of year: The British obsession with the weather. In: Strauss, S., and Orlove, B. (eds.). *Weather, Climate, Culture*. Oxford: Berg Publishers, pp. 103–120.

Head, L., 2016. *Hope and Grief in the Anthropocene: Re-conceptualising Human–nature Relations*. London: Routledge.

Hoffman, A., 2015. *How Culture Shapes the Climate Change Debate*. Palo Alto: Stanford University Press.

Hulme, M., Dessai, S., Lorenzoni, I., and Nelson, D. R., 2009. Unstable climates: Exploring the statistical and social constructions of 'normal' climate. *Geoforum*, *40*, 197–206.

Hulme, M., 2007. Geographical work at the boundaries of climate change. *Transactions of the Institute of British Geographers*, *33*, 5–11.

Hulme, M., 2010. Cosmopolitan climates. *Theory, Culture & Society*, *27*, 267–276.

Hulme, M., 2011. Reducing the future to climate change: A story of climate determinism and reductionism. *Osiris*, *26*, 245–266.

Hulme, M., 2015. Climate and its changes: A cultural appraisal. *Geo: Geography and Environment*, *2*, 1–11.

Hulme, M., 2016. *Weathered: Cultures of Climate*. London: Sage.

Ingold, T. (ed.), 1994. *Introduction to Culture In Companion Encyclopedia of Anthropology: Humanity, Culture and Social Life*. London: Routledge, pp. 329–349.

Ingold, T., 2000. *The Perception of the Environment*. London: Routledge.

IPCC, 2018. Climate Change 2021: Impacts, Adaptation and Vulnerability www.ipcc.ch/ipccreports/tar/wg2/index.php?idp=689

Jackson, M., 2015. Glaciers and climate change: Narratives of ruined futures. *WIREs Climate Change*, *6*, 479–492.

Jasanoff, S., 2010. A new climate for society. *Theory, Culture & Society*, *27*, 233–253.

Jones, L. and Boyd, E., 2011. Exploring social barriers to adaptation: Insights from western Nepal. *Global Environmental Change*, *21*, 1262–1274.

Klein, N., 2014. *This Changes Everything*. New York: Simon & Schuster.

Livingstone, D. N., 2012. Reflections on the cultural spaces of climate. *Climatic Change*, *113*, 91–93.

Lövbrand, E., Beck, S., Chilvers, J., Forsyth, T., Hedren, J., Hulme, M., Lidskog, R., and Vasileidou, E., 2015. Who speaks for the future of the Earth? How critical social science can extend the conversation on the Anthropocene. *Global Environmental Change*, *32*, 211–218.

Matless, D., 2018. Next the Sea: Eccles and the Anthroposcenic. *Journal of Historical Geography*, *62*, 71–84.

Morton, T., 2013. *Hyperobjects: Philosophy and Ecology after the End of the World*. Minneapolis: University of Minnesota Press.

Nerlich, B. and Jaspal, R., 2013. UK media representations of carbon capture and storage: Actors, frames and metaphors. *Metaphor and the Social World*, *3*, 35–53.

Nielsen, J. O. and Reenberg, A., 2010. Cultural barriers to climate change adaptation: A case study from northern Burkino Faso. *Global Environmental Change*, *20*, 142–152.

Norgaard, K. M., 2011. *Living in Denial: Climate Change, Emotions, and Everyday Life*. Cambridge, MA: MIT Press.

O'Brien, K., 2018. Is the 1.5°C target possible? Exploring the three spheres of transformation. *Current Opinion in Environmental Sustainability*, *31*, 153–160.

O'Brien, K. L. and Leichenko, R. M., 2000. Double exposure: Assessing the impacts of climate change within the context of economic globalization. *Global Environmental Change*, *10*, 221–232.

Offen, K., 2014. Historical geography III: Climate matters. *Progress in Human Geography*, *38*, 476–489.

Perez, C., Nicklin, C., Dangles, O., Vanek, S., Sherwood, S. G., Halloy, S., Garrett, K. A., and Forbes, G. A., 2010. Climate change in the high Andes: Implications and adaptation strategies for small-scale farmers. *The International Journal of Environmental, Cultural, Economic and Social Sustainability*, *6*, 71–88.

Rice, J., Burke, B., and Heynen, N., 2015. Knowing climate change, embodying climate praxis: Experiential knowledge in Southern Appalachia. *Annals of the Association of American Geographers*, *105*, 253–262.

Rudiak-Gould, P., 2013. 'We have seen it with our own eyes': Why we disagree about climate change visibility. *Weather, Climate and Society*, *5*, 120–132.

Sanderson, M. and Curtis, A., 2016. Culture, climate change and farm-level groundwater management: An Australian case study. *Journal of Hydrology*, *536*, 284–292.

Sietz, D. and Feola, G., 2016. Resilience in the rural Andes. *Regional Environmental Change*, *16*, 2163–2169.

Stocker, T. (ed.), (2014). *Climate Change 2013: The Physical Science Basis: Working Group I contribution to the Fifth Assessment Report of the Intergovernmental Panel on Climate Change*. Cambridge: Cambridge University Press.

Strauss, S., 2012. Are cultures endangered by climate change? Yes, but . . . . *WIREs Climate Change*, *3*, 371–377.

Sturken, M., 2001. Desiring the weather: El Nino, the Media, and California identity. *Public Culture*, *13*, 161–190.

Tanner, T. and Allouche, J., 2011. Towards a new political economy of climate change and development. *IDS Bulletin*, *42*, 1–15.

Trexler, A. and Johns-Putra, A., 2011. Climate change in literature and literary criticism. *Wiley Interdisciplinary Reviews: Climate Change*, *2*, 185–200.

Urry, J., 2010. Consuming the planet to excess. *Theory, Culture & Society*, *27*, 191–212.

UNFCCC, 2018 http://unfccc.int/resource/ccsites/zimbab/conven/text/art01.htm

Veale, L. and Endfield, G., 2014. The helm wind of cross Fell. *Weather*, *69*, 3–7.

Vannini, P., Waskul, D., Gottschalk, S., and Ellis-Newstead, T., 2011. Making sense of the weather. *Space and Culture*, *15*, 361–380.

Watson, A. and Huntington, O., 2014. Transgressions of the man on the moon: Climate change, indigenous expertise, and the posthumanist ethics of place and space. *GeoJournal*, *79*, 721–736.

Whatmore, S., 2002. *Hybrid Geographies: Natures Cultures Spaces*. London: Sage.

Part I

Knowing Climate Change

2

Cultures of Prediction in Climate Science

MARTIN MAHONY, MATTHIAS HEYMANN AND GABRIELE GRAMELSBERGER

2.1 Introduction

Prediction is a social practice which pervades all corners of our life-worlds. From the anticipation of the future of our personal relationships to the planning of political campaigns and the design of long-term nuclear waste repositories, reckoning with each other, and with the non-human world, always involves some form of claim upon the future, whether expressed in the confidence of quantified expectations or in the hesitancy of conditional arguments. Of course, this reckoning with the future has a particular valence in collective and individual efforts to come to terms with weather and climate, whether that be planning a barbecue or designing strategies to adapt key infrastructures to the expected climates of the next century. Anticipations of collective futures under climate change are informed and shaped by authoritative scientific projections of future environmental states, which oscillate between a disarming uncertainty about the near and far future, and a seductive offer of control over the global 'earth system'. But where do these influential predictive knowledges come from, and how are they shaped by social and cultural forces themselves? In this chapter, we develop the notion of 'cultures of prediction' to describe and chart the rise of predictive modelling and simulation in the environmental sciences and to situate these practices within local epistemic, institutional and political cultures. We propose four aspects of cultures of prediction which can be used as vectors of comparison in the history and sociology of climate change science, namely modes of computation, strategies of domesticating uncertainty, the institutional forms of predictive expertise and the social role and cultural impact of predictive claims. In each case, we examine the *technologies, practices* and *norms* which constitute distinct cultures of prediction. The chapter draws on a range of case studies from an interdisciplinary engagement with historical and contemporary cultures of prediction and aims to crystallise emerging conversations about the nature of predictive knowledge and about the growing import of authoritative visions of the future in the cultural politics of environmental change (see also Heymann *et al.* 2017a).

We propose the use of the term 'cultures of prediction' in order to capture both the broader cultural import of predictive practices and the distinctive scientific cultures from which they arise. The sociologist Gary Alan Fine introduced the term 'culture of prediction' in his ethnographic study of the US National Weather Service, using it to describe a microculture of scientific practices and identities through which the authority, uncertainty and utility of predictive knowledge are negotiated (Fine 2009). He shows how meteorologists occupy a tricky intermediary position – between data and modelling, between present and future, and between science and the public – meaning that they are constantly being pulled in different directions and placed under multiple, often conflicting demands. The local accommodations made to these demands in different forecast offices constitute specific occupational cultures, 'idiocultures', which are in many ways analogous to the positionality and cultures of climate change scientists (Fine 2009:69). We seek to develop Fine's concept in the case of climate change, pluralising it to take into account a range of dimensions which, we argue, constitute distinctive and comparable cultures of prediction in climate change science.

We begin by discussing contemporary climate prediction within the context of its historical antecedents and developments. In so doing, we show how the prediction of future climate has always involved a struggle with complexity and uncertainty. Nonetheless, the post-war rise of computation was a significant boon to meteorology and climatology, and they were among the first scientific disciplines to embrace the possibilities of numerical simulation. Understanding this lineage, and the detail of contemporary computational practices, is the base upon which we can build a deeper appreciation of the multiple cultures of prediction present in modern-day climate change science. In the second section, we focus on the different strategies which have been employed to 'domesticate' uncertainty – to make it knowable, calculable, legible and understandable, within scientific communities and, crucially, beyond. Different ways of tackling and communicating uncertainty often map onto different institutional cultures of climate science, and in the following section we examine the significance of different institutional settings for understanding the polyvalent character of climate prediction practices. Finally, we discuss the broader social role and cultural impact of the predictive claims of climate science, placing them in the context of a longer history of authoritative prophesy and environmental anxiety. This process of zooming out, from the code underlying climate simulations to the cultural politics of the future, enables us to offer a new account of cultures of prediction at various societal scales, which potentially enriches our understanding of the cultural politics of climate change, and of the relationships between science, politics and society more broadly.

2.2 Modes of Computation in Climate Science

Today's climate science is organised around computational practices. Among the most prominent products of this science are the measurement of the global annual surface temperature and the anomalies of anthropogenic climate change. Both are pure mathematical objects, which cannot be directly experienced by sensing human bodies in the same way as weather. Today's climate science has thus become a highly abstract, and abstracting, scientific discipline (Hulme 2010). Besides simply understanding the climate system, predictions of future climate have become a core topic of research, shaped by the sociopolitical expectation of reliable climate change prognoses. However, this was not always the case. In this section, we briefly outline the historical development that has resulted in today's understanding of climate and the computational practices, technologies and norms which sit at the heart of contemporary cultures of prediction.

2.2.1 Towards a Modern Understanding of Climate

Before the twentieth century, the sciences of climate were primarily concerned with the climates of Earth's various regions. In the ancient world, different climatic zones from torrid and temperate to frigid were classified and described, reflecting the term's origin in the Greek κλίμα (klima), referring to the angle (or incline) of a piece of land in relation to the sun. From the first century AD onwards, climate and latitude on maps were essentially synonymous, until degrees of latitude were introduced in the sixteenth century. From the eighteenth century, measurable variables such as temperature and precipitation were employed to indicate climate zones and began to pave the way to a modern understanding of climate and meteorology (Heymann 2010). The main meteorological measurement devices were, and still are, the barometer for measuring air pressure (invented in the 1640s by Evangelista Torricelli), the mercury thermometer (invented in the 1710s by Daniel Gabriel Fahrenheit) and the hygrometer for measuring humidity (invented in the 1780s by Horace-Bénédict de Saussure; see Middleton 1969). Equipped with these instruments, local series of measurements were begun, but only in 1780 was the first truly international network introduced by the Societas Meteorologica Palatina, centred in Mannheim, coordinating thirty-nine stations across fourteen countries (Daston 2008). In order to gain empirical knowledge about climates, such distributed measurement campaigns were urgently required. But local observations, even when organised into distributed networks, were, strictly speaking, measuring weather and not climate. 'Climate' had to be produced through the statistics of the weather over a longer period – over seasons, years or even decades – and these averages needed a new form of representation. Thus, in 1817, Alexander von Humboldt introduced isolines of equal temperature into the symbolic repertoire of cartography so that climatological patterns could begin to be depicted in all their spatial complexity (Schneider and Nocke 2014). This was a key moment in the emergence of a new sense not only of global patterning but also of global interconnection in natural phenomena (Grevsmühl 2016).

However, until the end of the nineteenth century climatology was still largely a science of regional climates. This began to change when the major paradigm, that climates were inherently stable, started to be questioned. In the middle of the nineteenth century debate began on the origins of geological anomalies, and researchers became increasingly convinced that several catastrophic climatic shifts must have happened in the past – such as the waxing and waning of ice ages – to produce the geological oddities to be found on the Earth's surface. Climate increasingly came to be seen as a variable phenomenon, and research turned towards the causal mechanism of climate in general and of climatic changes in particular (Oldroyd and Grapes 2008).

An eventual corollary of this new concept of climate as a variable system was the development of a physical and mechanical understanding of climate, as is common today (Heymann 2010). Knowledge of the circulation of air masses arose based on earlier observations and theories. In 1686, Edmond Halley explained that solar radiation differs for low and high latitudes, that heated tropical air is replaced by cooler air flowing equator-wards from the polar regions, thus causing a north–south circulation (Halley 1686). This circulation, as George Hadley pointed out a few decades hence, is deflected by the Earth's rotation (Hadley 1735). Because the speed of rotation differs at each point on the Earth's surface, as Heinrich Dove showed a century later, the deflection of air masses differs as well, causing a difference in rotational speed between moving air masses and the places to which these masses have moved (Dove 1837). These differences slowly change the direction of the currents: for instance, when they flow south from the North Pole, the deflection changes the direction from north, to northeast, to east. In 1858, William Ferrel rediscovered the Coriolis effect, applied it to the atmosphere and thus conceived an early global circulation model of the atmosphere (Ferrel 1858; see also Fleming 2002; Persson 2006; Gramelsberger 2017). This conceptual circulation model consists of three cells for each hemisphere: the polar cell, the midlatitude cell and the Hadley cell in the tropics.

2.2.2 Towards Computer Experiments

Climate from the late nineteenth century onwards was increasingly viewed as an abstract object of averaged meteorological variables on a regional and global level.

Local 'climates' were increasingly displaced by a physical and mechanical understanding of climate as a system (Heymann 2010 Coen 2018; for other overviews of the history of weather and climate science, see Nebeker 1995; Friedman 1993; Harper 2008; Edwards 2010; Fleming 2016). It was William Ferrel who not only conceived an early global circulation model but also outlined an advanced mathematics to describe such a model (Ferrel 1886). This modern understanding initiated the merging of climatology with weather science on the basis of the same medium – the atmosphere – and the same physical laws governing this medium, although it would be more than half a century before Ferrel's 'dynamic approach' was applied in climate science (Heymann 2010).

A modern climate model consists of a set of seven partial differential equations describing the hydro- and thermodynamic properties of the atmosphere like a heat engine. With these equations the seven main meteorological variables are computed – temperature, pressure, density, humidity and wind velocity in three directions. The heat engine is driven by solar radiation absorbed by the atmosphere and the Earth's surface, transformed into motion (wind) and energy (convection) or reflected and emitted back to space (see Edwards 2010). The greenhouse gases – the most important among them water vapour and carbon dioxide – keep energy in the system, acting like a shield that keeps the surface temperature of the Earth at a liveable average of around 15°C.

The main problem of meteorology is that a model consisting of seven partial differential equations is far too complex to derive an algebraically exact solution. Thus, numerically computing the equations for a discrete spatial grid and time step is the only way to achieve results. However, these 'numerical simulations' are only approximations of an unknown solution. They are never exact, and they require enormous amounts of computation depending on the resolution of the spatial grid. This is the reason atmospheric scientists, for climate as well as for weather predictions, have to make use of the fastest supercomputers available in order to compute weather development faster than the real weather evolves and to simulate, usually over the course of several days of computing, climatic changes over decades and centuries.

However, until the 1940s electronic, digital and free-programmable computers did not exist. Nevertheless, meteorologists like Ferrel, Max Margules and Felix Maria Exner tried to compute predictions by hand using mechanical desk calculators (Gramelsberger 2017). In 1922, the British scientist Lewis F. Richardson published an elaborate weather prediction computed by hand for a very coarse spatial grid. Although it took him over six weeks to complete the computations, his predictions proved to be somewhat off the mark – an error which was nonetheless later found to have been caused as much by unreliable data as by unreliable computation (Lynch 2006). Nevertheless, Richardson's 'simulation style' was quite ahead of its time. When in 1950 Jules Charney, John von Neumann and colleagues computed the very first weather model on an electronic computer, they essentially applied Richardson's practices. The same holds for Norman Phillips, who in 1955 conducted simulation experiments with the very first general circulation model – a simple model of the northern hemisphere that was able to represent the global circulation pattern which had been described by Ferrel. With these models, the era of computer simulation in meteorology had begun (Nebeker 1995; Harper 2008; Edwards 2010). The computer had become the dominant technology, embedded within a new set of practices and norms which would coalesce into distinctive cultures of prediction.

2.2.3 Model Building

Although the historical literature tends to refer to the Phillips model as the very first climate model, his model cannot really be seen as a 'climate model'. It was a simple 'barotropic' model that represented the global circulation patterns for a dry version of the atmosphere, neglecting thermodynamics, and it involved many other unrealistic abstractions in order to reduce the amount of computations.[1] Modelling in the 1950s aimed to provide a discrete version of the basic hydrodynamic equations and to numerically compute them. However, this changed over the next decades as climate researchers became aware that climate is not equivalent to the atmosphere but rather involves a whole suite of different processes and systems. While weather forecast models are concerned only with the circulation of the atmosphere, climate models, even in their simplest form, also need to take into account the behaviour of the ocean and sea-ice coverage (Manabe and Bryan 1969). The top few metres of the oceans hold more heat energy than the entire atmosphere. Thus, modelling 'climate' in the sense we understand it today truly started by steadily adding subsystems to the models, like the ocean, the cryosphere (sea-ice and the large ice shields and glaciers), the pedosphere and the marine and terrestrial biospheres. As all these subsystems influence each other, and as the climate system is shaped by many feedbacks between them, simulation models are the only reliable tool that can get us close to grasping this complexity (Gramelsberger and Feichter 2011; Bauer *et al.* 2015).

The barotropic models of the 1950s were replaced in the 1960s with global circulation models (GCMs) of the atmosphere and in the 1970s with coupled atmosphere and ocean global circulation models (AOGCMs). The first years of the twenty-first century saw the rise of 'Earth system' models (ESMs), which

---

[1] A barotropic model consists of a single atmospheric layer and simulates the circulation of the atmosphere as a function of advection.

include features such as the biosphere at ever-increasing levels of complexity. As the models have grown over the last few decades, scientists from various disciplines have become involved in the collective effort of modelling climate (Edwards 2011). However, while today's models include many more subsystems, processes and interactions, they can only represent climate to the extent of their spatial resolution. Every process and phenomenon which is smaller than the chosen computing grid resolution falls through the gaps. While global climate was computed with AOGCMs for a spatial resolution of about 500 km for the first Assessment Report of the Intergovernmental Panel on Climate Change (IPCC) in 1990, the recent fifth IPCC report is based on ESMs and resolutions of about 60 km (see IPCC 2014, especially section 1.5.2). However, a 60-km resolution is still too coarse for many climate-relevant processes, most notably the behaviour of clouds. Therefore, these unresolved processes have to be explicitly programmed into the climate models through 'subscale parametrisations'. Typical subscale parametrisations concern radiation transport, stratiform clouds, cumulus convection, subscale orographic effects, processes at the Earth's surface and horizontal diffusion (Gramelsberger 2010). Enormous research efforts have been devoted to improving these parametrisations, often in a 'gamble' involving great uncertainty as to whether the effort and resources will be matched by a marked improvement in model performance (Guillemot 2017a). Subscale parametrisations remain a major source of uncertainty for climate models, and thus for climate predictions. This is the other side of the coin of the problem that the seven foundational equations cannot be solved exactly but need numerical simulations. As one climate modeller stated: 'It is ironic that we cannot represent the effects of the small-scale processes by making direct use of the well-known equations that govern them' (Randall *et al.* 2003:1548).

In the 1990s, controversy raged within the climate science community over the norms governing the practice of 'flux adjustment'. This was essentially a subscale parametrisation which tried to account for the exchange of energy between atmosphere and ocean. This exchange could not be computed directly, so it had to be prescribed, or 'forced', in order to keep the simulation of climate realistic. For some modellers, this was anathema to the professional norms of climate simulation – the focus should be on perfecting the computation of the dynamic equations and letting the simulation run without interference. For others, flux adjustment was vital to producing 'realistic' simulations which could feed the growing desire for prediction to inform policy-making (Shackley *et al.* 1999). As we will explore below, these kinds of differences in the norms of climate modelling can be mapped onto distinct institutional cultures. By the early 2000s, flux adjustment was no longer necessary in most GCMs and ESMs. But parametrisations still play an important role in all models of the climate system, and conflict still reigns over the weight which should be given to either theoretical or empirical treatments of processes such as cloud formation (Guillemot 2017a). This conflict between the technological capabilities, computational practices and professional norms of climate science is part of a broader struggle to deal with the unavoidable uncertainties involved in understanding and predicting the climate system.

2.3 Domesticating Uncertainty

Predictive knowledge always involves uncertainty. Imperfect models, data limitations and restrictions of computer power generate significant uncertainties in climate model simulations, which often cannot be easily quantified. Modellers therefore have to make use of practices such as model validation, the comparison either of model and observational results or of the simulation results of many models, in order to determine the reliability and credibility of modelling practices. For climate projections, ensemble modelling – the simulation of future climate with multiple models – has become a standard technique, as has the practice of global model 'intercomparison'. The spread of model results is assumed to be an appropriate substitute measure for uncertainty margins, though this assumption could be challenged, because all models could be wrong in similar ways (Parker 2010). Furthermore, as Masson and Knutti (2011) point out, individual climate models are not independent entities – they often share identical or similar components, shared among developers or ported between institutions as scientists move around. Uncertainty persists at a range of levels, and strategies for its appropriate sampling, assessment and communication are contested. It potentially devalues predictive claims and divests them of their persuasive power and social effectiveness. This raises the question of how producers of predictive knowledge domesticate uncertainty in order to stabilise their claims and make them trustworthy and effective in society and politics. Obviously, modellers risk the immediate loss of political and public trust if their results turn out to be false and their models and practices unreliable.

There have been a number of moments in the history of the physical and statistical sciences when socially significant knowledge claims have been underpinned by newly introduced scientific practices – the experimental method, the development of population statistics and the rise of computer simulation, for example. The establishment of public trust in these practices and their resulting truth claims is therefore an important subject of inquiry (Shapin 1994; Porter 1995; MacKenzie 2001; Gooday 2004). Theodore M. Porter argues that trust in quantification and statistical procedures in the nineteenth century was constructed through standardised, procedural forms of accountability, which replaced personal trust relations. Porter calls these approaches 'technologies of distance' and a 'strategy of impersonality' (1995:ix, xi), which helped to make uncertain knowledge 'take the form of objectivity claims' and 'not depend too much on the particular individuals who author it' (ibid.:229). Computer modelling and simulation, it can be argued, has become such a technology of distance within a strategy of impersonality in climate science's efforts to gain public trust.

Public trust in predictive knowledge, however, is a complex phenomenon that cannot be reduced to the rigidity of scientific procedures or to the accomplishments of high technology. Domesticating uncertainty is a social process involving all actors invested in prediction: the scientists producing and framing it, customers interpreting and making use of it, competitors questioning it, interest groups challenging it and media distorting it (Lahsen 2005; Nowotny 2016). Public trust is a social construct, which becomes obvious when economic and climate modelling are compared. Forecasts about the development of economic indicators such as growth, investment and unemployment fail regularly without seriously compromising the status of economic knowledge and economists, who continue largely unimpeded to construct their projections (Mirowski 2013). Weather and climate prediction, in contrast, work under conditions of public scrutiny and continuously have to invest in domesticating uncertainty and in impression management to maintain scientific and public authority. The more contested knowledge claims and public authority are, the more has to be invested in the containment of uncertainty and in appropriate rhetorical devices and strategies. Public and political challenges to the claims of climate prediction pushed the IPCC, for example, to carefully define procedures and protocols and to devise new terminologies for the treatment of uncertainty – in turn creating a virtual science of uncertainty itself (Gramelsberger and Feichter 2011; Landström 2017).

The domestication of uncertainty is further complicated by the fact that uncertainty represents a resource that is mobilised for different ends by scientists, regulators, industry, media and other actors. Uncertainty can furnish powerful arguments for regulation to protect safety and health in the face of risks or conversely to reject regulation due to the lack of certain knowledge (Jasanoff 1990; Oreskes and Conway 2010). Hence, domesticating uncertainty is not simply a matter of inventing practices to produce and communicate robust and reliable knowledge. It is a matter of conflict, negotiation and boundary work and is intricately linked to the establishment of social credibility, legitimacy and authority of scientific claims and policy responses (Shapin 1995; Gieryn 1999). The domestication of uncertainty depends, first, on how scientists contain uncertainty and construct trust in models and, second, on how uncertainty is framed, communicated, negotiated and accommodated in the public sphere.

2.3.1 Trust in Models

The novelty of the methods and the epistemic problems involved in computer simulation raise the question of how confidence in climate models and simulations emerged in the first place (Heymann 2013). Model validation is an important part of constructing confidence and trust in models but is not sufficient to fully explain it. While atmospheric modellers usually posit model validation as the litmus test of model performance, it is rarely clear what exactly successful validation means and where its limits are. Historian and sociologist of science Hélène Guillemot has shown that no general protocol for the evaluation of climate models exists. Norms of validation have been shaped locally and differ across institutions and cultural contexts (Guillemot 2010). For example, whether or not a fit between simulated and observation-based data is considered 'good' is subject to local traditions and personal expertise. Evaluation proves to be a social process based on locally shared practices, norms and values, and on negotiation, agreement and compromise. Sociologist Gary Alan Fine calls it an 'interactional achievement' (2009:16).

An important source of confidence is the generation of so-called emergent features in simulations. It proves reassuring if elements of established physical theory in models (even though simplified in many regards) are able to produce patterns familiar from observation, even though the individual computation steps remain largely opaque to the modeller and s/he has no means to decide whether good results were achieved for the right or the wrong reasons. Historically, even if these patterns did not completely correspond with those in observations, they were considered important evidence that model development was on the right track. Norman Phillips' conclusions from his first general circulation experiment in 1955 provide a typical example. 'It is of course not possible to state definitively that this [...] is a complete representation of the principal energy changes occurring in the atmosphere, since our equations are so simplified', Phillips admitted, 'but the verisimilitude of the forecast flow patterns [with observed patterns] suggest quite strongly that it contains a fair element of truth' (Phillips 1956:154).

A second source of confidence derives from the multiplicity of models with similar behaviour. Climate scientist William Welch Kellogg, who strongly pushed for using climate models for the prediction of future climate change, suggested dealing pragmatically with the uncertainty of models and simulations. 'It can be seen, then, that there is an entire hierarchy of models of the climate system', Kellogg explained. 'It is reassuring to see that, when we compare the results of experiments with the same perturbations ... but using different models, the response is generally found to be either about the same or differs by an amount that can be rationalized in terms of recognized model differences or assumptions.' Kellogg adds a qualification right away. 'Of course, it is possible that all our models could be utterly wrong in the same way, giving a false sense of confidence', he concedes, 'but it seems highly unlikely that we would still be so completely ignorant about any dominant set of processes' (Kellogg 1977:9). In his argument, Kellogg applies a rhetorical figure that proved very typical in the attempt to domesticate uncertainty (see Heymann 2013): first, a statement of trust (multitude of models), second, a significant qualification of this statement (this argument could be wrong and trust may be unwarranted), which is followed, third, by a reinforcement of the power of the first statement (highly unlikely that all our models are wrong).

Paradoxically, a third source of confidence represents the invisibility of uncertainty. An argument made by climate modeller Stephen H. Schneider – using the same type of rhetorical figure as Kellogg – provides a telling example. In 1975, Schneider published an overview paper summarising predictive estimates of future climate by various scientists. In the abstract he made the following statement: 'Based on current understanding of climate theory and modelling it is concluded that a state-of-the-art order-of-magnitude estimate for the global surface temperature increase from a doubling of atmospheric CO_2 content is between 1.5 and 3 K, with an amplification of the global average increase in polar zones' (Schneider 1975:2060). Right after this statement he added an important qualification. 'It is pointed out, however, that this estimate may prove to be high or low by several-fold as a result of climatic feedback mechanisms not properly accounted for in state-of-the-art models' (ibid.). This sequence of statement and qualification is striking. While the statement appears to give a proper scientific result (the likely increase of temperature within certain ranges), the qualification suggests that this statement is so unreliable that it appears questionable altogether – and perhaps should not have been made at all.

Schneider shows that he is aware of the uncertainty and clearly acknowledges its importance. However, he cannot quantify it meaningfully. Instead, he decides to provide a range of temperature change based on the existing literature *without* any quantified uncertainty range (such as an expected temperature change of, e.g., 2 ± 6 K). Rhetorically, the way he puts his argument appears much more powerful. A clear number is out in the world (between 1.5 and 3 K) and likely to stick in the mind of the reader (cf. van der Sluijs *et al.* 1998), whereas the uncertainty, which is only suggested qualitatively, lacks the power of a clear quantitative expression and seems ineffective in devaluing the preceding number and less likely to stick with the reader. As a consequence, ignorance about the extent of uncertainty tended to produce an effective ignorance of uncertainty. It made uncertainties effectively less visible, if not invisible (Heymann 2013). Schneider would later reflect on the normative challenge of communicating uncertainty about climate change – he diagnosed a 'double ethical bind', by which the norms of scientific conduct were pitted against the rules of media discourse. Being an effective media operator, and spreading the word about the danger of climate change, might necessitate, Schneider suggested, periodically tempering one's scientific commitment to a full and honest treatment of uncertainty (see Russill 2010). Domesticating uncertainty, for Schneider, pitted the limits of computational practices and technologies against the norms which governed scientific conduct.

Finally, the invisibility of uncertainty reinforces a further psychological process. Myanna Lahsen has shown through close ethnographic study that modellers become closely accustomed to the uncertainties and the insurmountable limitations of computer model use. They have learned to live in a computer model world, have gotten to know the behaviour of their models after many years of experience and have over time developed increasing familiarity and trust. In everyday language in the simulation laboratories scientists talk about clouds, radiation, aerosols and the like while referring to *representations* of clouds, radiation and aerosols in computer models. Computer models and the virtual world they create take on a life of their own that is easily mingled (in language and thinking) with the real world. Models allow climate scientists to get what they consider as a feeling for climatic processes such as radiative transfer, convection, etc., even though it is a feeling of model behaviour rather than the real atmosphere. Lahsen (2005) suggests that models have a tendency to 'seduce' in their blurring of a virtual world with processes in the real world. Scientific, social and psychological factors all play their role in the domestication of uncertainty and the construction of trust in models. Uncertainty is pragmatically managed in the process of using models and in exploring both their scientific and political performance (Geden 2015). Below, we consider how institutional cultures shape these practices.

2.4 Institutional Forms

The rise of institutionalised predictive expertise should be understood against the backdrop of wider cultural or political anxieties, which called for new ways of making authoritative claims about the future (see below). The institutionalisation of weather prediction proceeded apace in the nineteenth century, as meteorologists sought to render the atmosphere more amenable to commerce, trade and imperial expansion (Anderson 2005; Mahony 2016). Where local or regional networks of (usually voluntary) observation, analysis and prediction existed, these were increasingly centralised into institutions like the United States Weather Bureau (Fleming 2000). These networks grew not just out of local curiosity about the weather but through the rise of what Baker (2018) calls 'meteorological government' – the co-production of climatic knowledge with state efforts to evaluate, calculate and monitor bodies and territories. Gradually, institutions of meteorological knowledge production became important 'centres of calculation' (Latour

1987), embedded within broader networks of human actors (meteorological observers, tabulating clerks) and non-human actors (self-recording instruments, punchcards, the weather itself). The internal structures of forecasting institutions could likewise be described with Thomas P. Hughes' concept of a socio-technical system – formed around a central technology, these systems develop both momentum and stability as they grow and become more complex. Computer models and simulation have helped to build a powerful socio-technical system of weather and climate prediction (Heymann *et al.* 2017b). Fine (2009:101) points to the role that technology plays in the 'institutional legitimation' of predictive claims, while Phaedra Daipha shows how technology disciplines practice. In the early 1990s for example, the US National Weather Service (NWS) introduced its Advanced Weather Processing System which 'truly anchored the NWS forecasting routine onto the computer' (Daipha 2015:34). This system, and its later updates, transformed forecasting from being based on geographical regions into a uniform grid system and replaced text-based efforts with a graphical process, which in turn automatically generated textual forecast statements. The system curtailed the freedom of the individual forecaster and harmonised procedures across the network of NWS offices. Resistance and subversion ensued, but this 'symbolic and material emphasis away from expert judgment and toward mechanical objectivity' helped strengthen the institutional authority of the NWS 'in an increasingly competitive weather market' (Daipha 2015:54, 35).

Despite the strong entanglement of weather science and the state, governmental weather offices have often had to compete with private operators for the public's attention and trust (Pietruska 2017). The performance of mechanised objectivity has been a key means of building trust and credibility in scientific forecasting and in distancing institutionalised meteorology and climatology from the 'subjectivity' and irrationality of its competitors (Henry 2015; Martin-Nielsen 2017). Yet technology alone does not determine the institutional cultures which in turn shape practices of prediction. Weather forecasting and climate prediction are forms of collective and distributed action. Institutions, such as the NWS or the UK Met Office, provide structures of formalised and routinised operation and define the rules and norms which generate a uniformity of practice. Although the global climate models which feed into IPCC assessment reports often have a shared ancestry, and are increasingly subject to questions and experimental designs standardised at the international level through the IPCC process, important variations exist in the institutional cultures within which they are situated. In 2001 Simon Shackley described an array of 'epistemic lifestyles' at major modelling centres in the United Kingdom and the United States, contrasting the policy-focused, hierarchical structure of the UK Met Office Hadley Centre with the greater intellectual freedom on offer amid the non-hierarchical arrangement of the National Center for Atmospheric Research (NCAR) in Boulder, Colorado (Shackley 2001). 'Climate seers' dominated at the Hadley Centre – those concerned with exploring and experimenting with the modelled climate system and its changes – in contrast to the 'model constructors' of NCAR, who take model building as an end in itself. And while the research priorities of the Hadley Centre were shaped by the demands of the IPCC and national policy-makers, modellers at the Max Planck Institute for Meteorology in Hamburg were largely 'aloof from politics' and thus concerned with asking rather different questions of their models (Krueck and Borchers 1999:123). Shackley *et al.* (1999) show how these very different institutional cultures produced very different attitudes to the practice of 'flux adjustment' described above, pitting 'purists' against 'pragmatists', with the latter being those who, as at the Hadley Centre, were primarily concerned with delivering usable results to policy-makers and other scientists.

Mahony and Hulme (2016) seek to take this analysis a step further, situating the culture of the Hadley Centre within a broader landscape of institutionalised ways-of-knowing. Noting Jasanoff's observation that in well-established institutions like scientific organisations 'societies have access to tried-and-true repertoires of problem-solving, including preferred forms of expertise, processes of inquiry, methods of securing credibility, and mechanisms for airing and managing dissent' (Jasanoff 2004:39–40), the Hadley Centre can be placed within a broader tradition of governmental knowledge-making in British politics. Jasanoff describes the 'civic epistemology' of British politics as one which prizes independent, pragmatic and trusted judgement, which exhibits an empiricist distrust of overextended modelling and prediction and which emphasises the political value of 'sound science' (Jasanoff 2005). In the quest to establish a modelling effort which would be 'independent' of the United States, which would build on an already established relationship of trust between institutions and which emphasised the pursuit of spatial realism in order to reconcile modelling with deep-lying empiricist tendencies in British civic epistemology, the establishment of the Hadley Centre illuminates how institutional cultures of prediction form in relationship to broader ways of making knowledge in and for politics. As the tools of climate simulation increasingly circulate around the globe (Mahony 2017), the comparative study of different institutional cultures of prediction can reveal important differences in the ways in which otherwise standardised technologies and practices of knowledge-making are put to work amid the diversity of norms involved in deliberating the science and politics of climate change.

Table 2.1 *Cultures of prediction: preliminary summary and points of comparison*

	Modes of computation	Domesticating uncertainty	Institutional forms
Technologies	• Computational capacity • 'Portability' of code and models (Mahony 2017) • 'Independence' of models from each other (Masson & Knutti 2011)	• Degree of standardisation of procedures for assessing uncertainty of climate model projections • Use of model intercomparison	• Standardising and disciplining role of technological systems (Daipha 2015)
Practices	• Spatial and temporal resolution • Dependency on parametrisations • Opacity of underlying code	• Degree of standardisation of procedures for assessing and communicating uncertainty • Rhetorical forms used to communicate uncertainty and develop trust (Heymann 2013) • Distribution of different perceptions of uncertainty in relation to 'distance' from model (Lahsen 2005)	• Relationship between modellers and policy-/decision-makers • Degree and nature of competition/cooperation with rival modelling centres or information brokers
Norms	• 'Epistemic lifestyles' (Shackley 2001): purists vs pragmatists • Priority given to either empirical or theoretical treatments of sub-grid scale phenomena (Guillemot 2017a)	• Norms of model validation • Degree of openness about uncertainties (Russill 2010)	• Institutionalised norms (e.g. 'epistemic lifestyles'– Shackley 2001) • Civic epistemologies (Jasanoff 2005)

2.5 Cultural Politics of the Future

In this final section, we expand the scope of the notion of 'cultures of prediction' further, to begin the task of setting the practices and institutional forms described above (Table 2.1) against the backdrop of a wider cultural politics of the future. In the previous section, we moved from the level of technology to questions of 'national' political culture in the shaping of cultures of prediction. But the rise of influential cultures of prediction is a transnational and transhistorical phenomenon and speaks to broader questions of human values and meaning-making. Cultural historians are starting to provide important insights into this interplay between predictive scientific technique and cultural values, which can help us round out our understanding of cultures of prediction (Heymann *et al.* 2017a).

All human existence is fundamentally uncertain. Predictive efforts are a fundamental way to contain this uncertainty and to alleviate the contingencies and risks of decision-making in the mastering of everyday life. Uncertainty in this more anthropological sense is a condition of life, and prediction a response to it. Lynda Walsh argues that scientists play a core role in the cultural response to uncertainty: 'polities call on science advisers to manufacture certainty for them, not uncertainty' (2013:196). Fine concurs: 'The dark heart of prediction is defining, controlling, and presenting uncertainty as confident knowledge' (2009:103). Prediction usually means relieving the challenge of decision-making under uncertainty by deferring to the authority and expertise of others. Oracles, priests, prophets, philosophers, fortune tellers and doctors have served the predictive needs of different societies, occupying privileged social positions justified by their access of esoteric knowledge of the future. Prediction, thus, is not a new phenomenon but a cultural necessity that has, in different forms, served all human societies.

Historically, times characterised by perceptions of rapid change or rising instability, crisis and uncertainty generated a felt need for prophecy and prediction. The rapid and deep transformations experienced in western Europe in the nineteenth century, for example, created new tensions and anxieties and increased the demand for prevision and reassurance. Georges Minois calls the nineteenth century 'a century of prediction' with an 'unusual abundance of popular prophecy' (Minois 1998:619, 614). Jamie Pietruska describes the period from the mid-nineteenth century to the First World War in the United States as an age of 'propheteering', in which a diversity of actors, often self-taught amateurs, served the predictive needs of farmers, traders and travellers (Pietruska 2009; see also Pietruska 2017). Industrialisation, urbanisation and social struggle in the nineteenth and twentieth centuries also increased demands for governmental provision and thus promoted a professionalisation of prevision, planning and prediction in domains such as demography, welfare policy, insurance systems, social hygiene, urban planning, agriculture and forestry. Experiences of deep crisis in the twentieth century, such as the World Wars and deep political and economic shocks, proved culturally unsettling and challenged state authorities to intensify state planning and intervention, of which science-based prediction became a crucial element (Andersson and Rindzevičiūtė 2015).

In the early Cold War, predictive efforts and 'planning euphoria' served the ambitions of technological innovation, military control and social progress. During the 1960s, however, this more optimistic frame of mind in parts of western culture began to collapse. New social movements increasingly questioned social conditions and institutional authorities. Strong environmental concerns contributed to creating an unsettled cultural state, which nurtured new interests in prediction and furnished it with new meaning and import. It sparked the 'rise of a movement – indeed virtually an industry' of prediction (Rescher 1998:29) led by institutions such as the Swiss think tank and consulting company PROGNOS and the newly founded International Institute for Applied Systems Analysis, IIASA. The publication of 'The Limits to Growth' report by the Club of Rome in 1972 is a powerful example. The simulation of future developments with a simple 'world model' showed that unabated population growth and resource use would lead to stagnation and collapse 'within the next century, at the latest' (Meadows et al. 1972:126). Launched with a highly effective marketing campaign, ultimately translated into thirty languages and selling over ten million copies worldwide by 1999, it boosted the idea of environmental and resource crisis and of an inherent danger of societal collapse sometime in the twenty-first century (Elichirigoity 1999).

The Report of the Club of Rome shows the two-faced character of prediction. It mirrored social interests and responded to cultural demands and, at the same time, produced new perceptions and culture-shaping knowledge. More recently, climate prediction has become one of the major preoccupations of both scientists and broader societies, and the IPCC assumes a similar role of responding to cultural demands and establishing certainty amid uncertainty. Lynda Walsh has argued that scientists such as those of the Club of Rome and of the IPCC fill the role of the prophets, a role which in other places and times might have been filled by oracles, priests or others who assumed access to the supernatural and to the future. Walsh traces a continuous thread linking the cultural roles of prophets from the Pythia, known as the oracle of Delphi, to Francis Bacon, Robert Oppenheimer, Carl Sagan and the IPCC. Studying the linguistic formations of contemporary scientists, she observes 'how similar their speech patterns were to the biblical prophecy' and how similarly they worked in both ancient and contemporary societies. In her rhetorical genealogy of prophetical *ethos* she argues that scientists have persistently assumed the cultural practices constituting prophetical *ethos* by constructing a 'persistent, recognizable cluster of rhetorical strategies'. Scientists, she contends, have served as the prophets of the modern world, empowered and authorised to manufacture 'political certainty' (Walsh 2013:iix, 2).

The cultural prevalence of predictive knowledge, and its determination of both the government policies and personal decisions which shape our everyday lives, means that social action is increasingly orientated around a future which is continuously present. The idea of the Anthropocene likewise telescopes deep, geologic timescales into our present moment, as it posits human society as the chief driver of environmental transformation, putting the post-war 'Great Acceleration' on a par with the ordinarily slower, more ponderous drivers of geological change (Davies 2016). This folding of past and future into the politics of now is an epistemological and ontological revolution which students of culture are only beginning to come to terms with. 'If in modernism time was seen to flow from the present to the future', in an endless struggle to escape the past, 'today we increasingly experience time coming toward us, from the future to the present' (Braun 2015:239). Returning to an earlier meaning of apocalypse as revelation, the Anthropocene for Bruno Latour is 'apocalyptic', 'in the sense of the revelation of things that are coming *toward* us' (Latour 2015:153). While Walter Benjamin's Angel of History may have flown into the future with its head turned towards the past, observing the catastrophes of human history piling up wreckage at his feet, in the Anthropocene, in Latour's reading,

the angel does an about turn, an angel of the passing modern, caught in future headlights, figuring out what is due to us: what we are culpable for, what we deserve, what is in store for us, what's coming to us, what's left to us, what is our legacy. And what might be due; about to arrive, coming just around the corner, on time or unavoidably delayed. (Matless 2017:11)

The scientific prediction of what's 'just around the corner' is therefore culturally significant not just for the power it exerts through the governmentalisation of environmental change (Fletcher 2017) but for its participation in the broader redrawing of the temporal and, indeed, spatial contours of human existence. A number of scholars have pointed to the co-evolution of new forms of global environmental monitoring with new forms of global consciousness, often focusing on the 1960s and the emergence of a new discourse of 'think global, act local', which sought to harness this new consciousness to shape an emerging environmental politics which was sceptical of capitalist consumption and state power and which advocated a renewed engagement with local ecologies and an insurgent 'grassroots' politics. The 'Earthrise' and 'whole Earth' photographs of the late 1960s and early 1970s are often taken as totemic of this new environmentalism, although the paradox of their status as products of the very military-industrial complex which many environmentalists sought to confront points to the ambiguities of environmentalism's relationship to 'science' (Cosgrove 2001; Grevsmühl 2014). While ecology has often functioned as something of an insurgent science, being used to challenge entrenched forms of power and their damaging effects, the sciences of environmental change more broadly are inseparable from state power and the military occupation (even weaponisation) of the spaces which environmentalists seek to protect – the atmosphere, the cryosphere, the oceans and so on (Masco 2009; Edwards 2012; Hamblin 2013). The sciences of environmental change are increasingly bound up within a matrix of power through which the government of human populations is conducted through a suite of new disciplinary and biopolitical technologies (Knox 2014). Nonetheless, despite the abstraction of a global climate, and the collapsing of human life into measurable and manageable 'populations', we can still assert the existence of a 'global sense of place' as a qualitatively new product of post-war science and politics in the West (Heise 2008). But while this may be a hopeful observation, some analysts have proposed that the global focus of the science and politics of environmental change creates a psychological distance from people's everyday lives which is hard to surmount and inimical to motivating political action (O'Neill and Hulme 2009). The abstract, statistically and computationally constructed space of a global climate is a long way from the embodied, visceral experience of weather, and the timescales of environmental prediction too far from the horizons of everyday decision-making, for meaningful everyday engagement with climate to be possible, or even desirable (Hulme 2010; Jasanoff 2010). A whole industry of 'climate change communication' has grown up in this supposed fissure between the everyday and what Timothy Morton (2013:58) calls the 'hyperobject' of climate change – something that is so vast, spatially and temporally, that it becomes 'almost impossible to hold in the mind'. This communication industry aims to translate the predictions of climate science for a variety of publics, bridging scales of space and time in a bid to manufacture 'engagement' and political will (Callison 2014). Yet as a number of scholars have argued (e.g. Hulme 2009; Latour 2015), it is futile to expect the facticity of scientific observations, or the power of scientific predictions, to motivate on their own a turn against the driving structural forces of global climate change. Jasanoff (2010:236) argues,

Representations of the natural world attain stability and persuasive power ... not through forcible detachment from context, but through constant, mutually sustaining interactions between our senses of the *is* and the *ought*: of how things are and how they should be.

The boundaries of 'science' and 'politics' need to be redrawn, the 'continual interchange between epistemic, social and ethical sense-making' acknowledged and better regard paid to 'the layered investments that societies have made in worlds as they wish them to be' (Jasanoff 2010:245, 236). Only by reconnecting descriptions of the *is* with the politics of the *ought*, by unashamedly bringing scientific predictions onto the terrain of the difficult politics of what kind of future we want to build, can the latent cultural power of scientific prediction be harnessed in creating a better Anthropocene.

2.6 Conclusion

In this chapter we have developed the notion of 'cultures of prediction' as a way to describe local and trans-local entanglements of scientific practices, technologies, rhetorical techniques, institutional forms and cultural anxieties. By zooming out from the code underlying climate models to the broader cultural politics of the future, we have argued that distinctive 'cultures of prediction' can be identified at a range of spatial scales. We have proposed a set of characteristics which may be used as vectors of comparison of the technologies, practices and norms which make up these cultures – modes of computation, the domestication of uncertainty, the degrees and character of institutionalisation and the broader cultural politics of the future. Further multidisciplinary research is required to develop a fuller picture of how distinct cultures of prediction are reshaping science, society and politics.

A number of important directions for new research might be summarised. Studying the geographical relocation of the technologies, practices and norms which make up cultures of prediction can shed important light on how certain cultures become dominant or hegemonic (Mahony and Hulme 2012). This is both a historical task and a question to ask of contemporary climate knowledge politics. Relatedly, more work is required to understand the comparative politics of predictive claims – to situate the production and application of predictive knowledge within broader political cultures and contexts. Historians and sociologists of science tend towards case study research, for clear reasons. Greater interaction with the comparative inclinations of political science, and with the spatially sensitive approaches of human geographers, would yield important new insights into how cultures of prediction differ not only at the level of the lab but at the level of the nation-state or the international institution. There is an increasingly recognised need for the insights of humanities and critical social science scholarship to engage with the institutions of global environmental knowledge-making (Castree *et al.* 2014; Lövbrand *et al.* 2015). As practitioners grapple with the challenges of domesticating and communicating uncertainty, and building institutions capable of constructing political authority across domains and constituencies, insights into the co-production of cultures of prediction with cultures of government and political decision-making have an important role to play.

Such insights are particularly important at a moment when climate science is both becoming ever-more tightly coupled with policy-making, with decision-makers seeking proposals and evaluations of policy performance (Dahan and Guillemot 2015; Guillemot 2017b), and when it is facing renewed political challenge from emboldened right-wing critics (De Pryck and Gemenne 2017; McGee 2017). Cultures of prediction are sites of both the subtle shaping of life and politics and more bellicose contestation over our collective rights and futures. Understanding their origins and their character has never been more urgent.

Acknowledgements

We would like to thank the editors for inviting this contribution and for their guidance in improving the chapter. We are also greatly indebted to members of the DFG-Funded network 'Atmosphere and Algorithms' and to the contributors to the volume *Cultures of Prediction in Atmospheric and Climate Science* whose work has inspired and informed this chapter.

References

Anderson, K. 2005. *Predicting the Weather: Victorians and the Science of Meteorology*. Chicago, IL: University of Chicago Press.

Andersson, J., and Rindzevičiūtė, E. (eds.) 2015. *The Struggle for the Long-Term in Transnational Science and Politics: Forging the Future*. London: Routledge.

Baker, Z. 2018. Meteorological frontiers: Climate knowledge, the west, and U.S. statecraft, 1800–1850. *Social Science History*, 42(4), 731–761.

Bauer, P., Thorpe, A., and Brunet, G. 2015. The quiet revolution of numerical weather prediction. *Nature*, 525(7567), 47–55.

Callison, C. 2014. *How Climate Change Comes to Matter: The Communal Life of Facts*. Durham, NC: Duke University Press.

Castree, N. *et al.* 2014. Changing the intellectual climate. *Nature Climate Change*, 4(9), 763–768.

Coen, Deborah R. 2018. *Climate in Motion: Science, Empire, and the Problem of Scale*. Chicago, IL: University of Chicago Press.

Cosgrove, D. 2001. *Apollo's Eye: A Cartographic Genealogy of the Earth in the Western Imagination*. London: John Hopkins University Press.

Dahan, A. and Guillemot, H. 2015. Les relations entre science et politique dans le régime climatique: à la recherche d'un nouveau modèle d'expertise?, *Natures Sciences Sociétés*, 23, 6–18.

Daipha, P. 2015. *Masters of Uncertainty: Weather Forecasters and the Quest for Ground Truth*. Chicago, IL: University of Chicago Press.

Daston, L. 2008. Unruly weather: Natural law confronts natural variability, in Daston. L. and Stolleis M. (eds.) *Natural Law and Law of Nature in Early Modern Europe: Jurisprudence, Theology, Moral and Natural Philosophy*. Aldershot: Ashgate, 233–248.

Davies, J. 2016. *The Birth of the Anthropocene*. Oakland, CA: University of California Press.

Dove, H. 1837. *Meteorologische Untersuchungen*. Berlin: Sander.

Edwards, P. N. 2010. *A Vast Machine: Computer Models, Climate Data, and the Politics of Global Warming*. Cambridge, MA: MIT Press.

Edwards, P. N. 2011. History of climate modelling. *Wiley Interdisciplinary Reviews: Climate Change*, 2(1), 128–139.

Edwards, P. N. 2012. Entangled histories: Climate science and nuclear weapons research. *Bulletin of the Atomic Scientists*, 68(4), 28–40.

Elichirigoity, F. 1999. *Planet Management: Limits to Growth, Computer Simulation, and the Emergence of Global Spaces*. Evanston, IL: Northwestern University Press.

Ferrel, W. 1858. The influence of the Earth's rotation upon the relative motion of bodies near its surface. *Astronomical Journal*, 109, 97–100.

Ferrel, W. 1886. *Recent Advances in Meteorology (Annual Report of the Chief Signal Officer)*. Washington, DC: US War Department.

Fine, G. A. 2009. *Authors of the Storm: Meteorologists and the Culture of Prediction*. Chicago, IL: University of Chicago Press.

Fleming, J. R. 2000. *Meteorology in America, 1800–1870*. Baltimore, MD: Johns Hopkins University Press.

Fleming, J. R. 2002. History of meteorology. In: Biagre, B. S. (ed.) *A History of Modern Science and Mathematics*, vol 3. New York: Scribner's, 184–217.

Fleming, J. R. 2016. *Inventing Atmospheric Science: Bjerknes, Rossby, Wexler, and the Foundations of Modern Meteorology*. Cambridge, MA: MIT Press.

Fletcher, R. 2017. Environmentality unbound: Multiple governmentalities in environmental politics, *Geoforum*, 85, 311–315.

Friedman, R. M. 1993. *Appropriating the Weather: Vilhelm Bjerknes and the Construction of a Modern Meteorology*. Ithaca, NY: Cornell University Press.

Geden, O. 2015. Policy: Climate advisers must maintain integrity. *Nature*, 521(7550), 27–28.

Gieryn, T. F. 1999. *Cultural Boundaries of Science: Credibility on the Line*. Chicago, IL: University of Chicago Press.

Gooday, G. 2004. *The Morals of Measurement: Accuracy, Irony, and Trust in Late Victorian Electrical Practice*. Cambridge: Cambridge University Press.

Gramelsberger, G. 2010. Conceiving processes in atmospheric models—General equations, subscale parameterizations, and 'superparameterizations'. *Studies in History and Philosophy of Science Part B*, 41(3), 233–241.

Gramelsberger, G. 2017. Calculating the weather: Emerging cultures of prediction in late nineteenth- and early twentieth-century Europe. In: Gramelsberger, G. and Mahony, M. (eds.) *Cultures of Prediction in Atmospheric and Climate Science*. London: Routledge, 45–67.

Gramelsberger, G. and Feichter, J. 2011. Modelling the Climate System: An overview. In: Gramelsberger, G., and Feichter, J. (eds.) *Climate Change and Policy: The Calculability of Climate Change and the Challenge of Uncertainty*. London: Springer, 9–90.

Grevsmühl, S. V. 2014. *La Terre vue d'en haut: L'invention de l'environnement global*. Paris: Seuil.

Grevsmühl, S. V. 2016. Images, imagination and the global environment: Towards an interdisciplinary research agenda on global environmental images, *Geo: Geography and Environment*, 3(2), 00020.

Guillemot, H. 2010. Connections between simulations and observation in climate computer modeling. Scientist's practices and 'bottom-up epistemology' lessons. *Studies in History and Philosophy of Science Part B: Studies in History and Philosophy of Modern Physics*, 41(3), 242–252.

Guillemot, H. 2017a. How to develop climate models? The 'gamble' of improving climate model parameterizations. In: Gramelsberger, G. and Mahony, M. (eds.) *Cultures of Prediction in Atmospheric and Climate Science*. London: Routledge, 120–136.

Guillemot, H. 2017b. The necessary and inaccessible 1.5°C objective: A turning point in the relations between climate science and politics? In: Aykut, S. C., Foyer, J., and Morena, E. (eds.) *Globalising the Climate: COP21 and the Climatisation of Global Debates*. London: Routledge, 39–56.

Hadley, G. 1735. The cause of the general trade-wind. *Philosophical Transactions of the Royal Society London*, 29, 58–62.

Halley, E. 1686. An historical account of the trade-winds and monsoons observable in the seas between and near the Tropicks, with an attempt to assign the physical cause of said winds. *Philosophical Transactions of the Royal Society London*, 16, 153–168.

Hamblin, J. D. 2013. *Arming Mother Nature: The Birth of Catastrophic Environmentalism*. Oxford: Oxford University Press.

Harper, K. 2008. *Weather by the Numbers*. Cambridge, MA: MIT Press.

Heise, U. K. 2008. *Sense of Place and Sense of Planet: The Environmental Imagination of the Global*. Oxford: Oxford University Press.

Henry, M. 2015. 'Inspired divination': Mapping the boundaries of meteorological credibility in New Zealand, 1920–1939. *Journal of Historical Geography*, 50, 66–75.

Heymann, M. 2010. The evolution of climate ideas and knowledge, *Wiley Interdisciplinary Reviews: Climate Change*, 1(4), 581–597.

Heymann, M. 2013. Constructing evidence and trust: How did climate scientists' trust in their models and simulations emerge? In: Hastrup, K. and Skrydstrup, M. (eds.) *The Social Life of Climate Change Models: Anticipating Nature*. Abingdon: Routledge, 203–224.

Heymann, M., Gramelsberger, G., and Mahony, M. (eds.). 2017a. *Cultures of Prediction in Atmospheric and Climate Science: Epistemic and Cultural Shifts in Computer-based Modelling and Simulation*. London: Routledge.

Heymann, M., Gramelsberger, G. and Mahony, M. 2017b. Key characteristics of cultures of prediction. In Gramelsberger, G. and Mahony, M. (eds.) *Cultures of Prediction in Atmospheric and Climate Science*. London: Routledge, 18–41.

Hulme, M. 2009. *Why We Disagree about Climate Change: Understanding Controversy, Inaction and Opportunity*. Cambridge: Cambridge University Press.

Hulme, M. 2010. Problems with making and governing global kinds of knowledge. *Global Environmental Change*, 20(4), 558–564.

Humboldt, A. v. 1817. Memoire sur les lignes iso-thermes'. *Annales de Chimie et de Physique*, 5, 102–111.

IPCC 2014. *Climate Change 2013: The Physical Science Basis*. Cambridge: Cambridge University Press.

Jasanoff, S. 1990. *The Fifth Branch: Science Advisers as Policymakers*. Cambridge, MA: Harvard University Press.

Jasanoff, S. 2004. Ordering knowledge, ordering society. In Jasanoff, S. (ed.) *States of Knowledge: The Co-Production of Science and Social Order*. London: Routledge, 13–45.

Jasanoff, S. 2005. *Designs on Nature: Science and Democracy in Europe and the United States*. Princeton, NJ: Princeton University Press.

Jasanoff, S. 2010. A new climate for society. *Theory, Culture and Society*, 27, 233–253.

Kellogg, W. W. 1977. *Effects of Human Activities on Global Climate*. WMO Technical Note No. 156. Geneva: World Meteorological Organization.

Knox, H. 2014. Footprints in the city: Models, materiality, and the cultural politics of climate change. *Anthropological Quarterly*, 87(2): 405–429.

Krueck, C. P. and Borchers, J. 1999. Science in politics: A comparison of climate modeling centres. *Minerva*, 37(2), 105–123.

Kyle McGee 2017. *Heathen Earth: Trumpism and Political Ecology*. New York: Punctum.

Lahsen, M. 2005. Seductive simulations? Uncertainty distribution around climate models. *Social Studies of Science*, 35(6), 895–922.

Landström, C. 2017. Tracing uncertainty management through four IPCC assessment reports and beyond. In: Gramelsberger, G. and Mahony, M. (eds.) *Cultures of Prediction in Atmospheric and Climate Science*. London: Routledge, 214–230.

Latour, B. 1987. *Science in Action: How to Follow Scientists and Engineers through Society*. Cambridge, MA: Harvard University Press.

Latour, B. 2015. Telling friends from foes in the time of the Anthropocene. In: Hamilton, C., Bonneuil, C., and Gemenne, F. (eds.) *The Anthropocene and the Global Environmental Crisis: Rethinking Modernity in a New Epoch*. London: Routledge, pp. 145–155.

Lövbrand, E. et al. 2015. Who speaks for the future of Earth? How critical social science can extend the conversation on the Anthropocene. *Global Environmental Change*, 32, 211–218.

Lynch, P. 2006. *The Emergence of Numerical Weather Prediction: Richardson's Dream*. Cambridge: Cambridge University Press.

MacKenzie, D. A. 2001. *Mechanizing Proof: Computing, Risk, and Trust*. Cambridge, MA: MIT Press.

Mahony, M. 2016. For an empire of 'all types of climate': Meteorology as an imperial science. *Journal of Historical Geography*, 51, 29–39.

Mahony, M. 2017. The (re)emergence of regional climate: Mobile models, regional visions and the government of climate change. In: Gramelsberger, G. and Mahony, M. (eds.) *Cultures of Prediction in Atmospheric and Climate Science*. London: Routledge, 139–158.

Mahony, M. and Hulme, M. 2012. Model migrations: Mobility and boundary crossings in regional climate prediction. *Transactions of the Institute of British Geographers*, 37(2), 197–211.

Mahony, M. and Hulme, M. 2016. Modelling and the nation: Institutionalising climate prediction in the UK, 1988–92. *Minerva*, 54(4), 445–470.

Manabe, S. and Bryan, K. 1969. Climate calculations with a combined ocean-atmosphere model. *Journal of the Atmospheric Sciences*, 26(4), 786–789.

Martin-Nielsen, J. 2017. Scientific forecasting? Performing objectivity at the UK's Meteorological Office, 1960s–1970s. *History of Meteorology*, 8, 202–221.

Masco, J. 2009. Bad weather: On planetary crisis. *Social Studies of Science*, 40 (1), 7–40.

Masson, D. and Knutti, R. 2011. Climate model genealogy. *Geophysical Research Letters*, 38(8), L08703.

Matless, D. 2017. The Anthroposcenic. *Transactions of the Institute of British Geographers*, 42(3), 363–376.

Meadows, D. H. 1972. *The Limits to Growth: A Report for the Club of Rome's Project on the Predicament of Mankind*. New York: Universe Books.

Middleton, W. 1969. *Invention of the Meteorological Instruments*. Baltimore, MD: John Hopkins University Press.

Minois, G. 1998. *Geschichte der Zukunft: Orakel, Prophezeiungen, Utopien, Prognosen*. Zurich: Artemis & Winkler.

Mirowski, P. 2013. *Never Let a Serious Crisis Go to Waste: How Neoliberalism Survived the Financial Meltdown*. London: Verso.

Morton, T. 2013. *Hyperobjects: Philosophy and Ecology After the End of the World*. Minneapolis, MN: University of Minnesota Press.

Nebeker, F. 1995. *Calculating the Weather: Meteorology in the 20th Century*. London: Academic Press.

Nowotny, H. 2016. *The Cunning of Uncertainty*. Cambridge: Polity.

Oldroyd, D. R. and Grapes, R. H. 2008. Contributions to the history of geomorphology and quaternary geology: An introduction. *Geological Society, London, Special Publications*, 301(1), 1–17.

O'Neill, S. J. and Hulme, M. 2009. An iconic approach for representing climate change. *Global Environmental Change*, 19(4), 402–410.

Oreskes, N. and Conway, E. M. 2010. *Merchants of Doubt: How a Handful of Scientists Obscured the Truth on Issues from Tobacco Smoke to Global Warming*. London: Bloomsbury.

Parker, W. S. 2010. Predicting weather and climate: Uncertainty, ensembles and probability. *Studies in History and Philosophy of Science Part B – Studies in History and Philosophy of Modern Physics*, 41(3), 263–272.

Persson, A. O. 2006. Hadley's principle: Understanding and misunderstanding the trade winds. *History of Meteorology*, 3, 17–42.

Phillips, N. A. 1956. The general circulation of the atmosphere: A numerical experiment. *Quarterly Journal of the Royal Meteorological Society*, 82(352), 123–164.

Pietruska, J. L. 2009. *Propheteering: A Cultural History of Prediction in the Guilded Age*. PhD thesis, MIT Program in Science, Technology and Society.

Pietruska, J. L. 2017. *Looking Forward: Prediction and Uncertainty in Modern America*. Chicago, IL: University of Chicago Press.

Porter, T. 1995. *Trust in Numbers: The Pursuit of Objectivity in Science and Public Life*. Princeton, NJ: Princeton University Press.

De Pryck, K. and Gemenne, F. 2017. The Denier-in-Chief: Climate change, science and the election of Donald J. Trump. *Law and Critique*, 28(2), 119–126.

Randall, D. *et al.* 2003. Breaking the cloud parameterization deadlock. *Bulletin of the American Meteorological Society*, 84(11), 1547–1564.

Rescher, N. 1998. *Predicting the Future: An Introduction to the Theory of Forecasting*. New York: SUNY Press.

Russill, C. 2010. Stephen Schneider and the 'double ethical bind' of climate change communication. *Bulletin of Science, Technology & Society*, 30(1), 60–69.

Schneider, B. and Nocke, T. (eds.). 2014. *Image Politics of Climate Change: Visualizations, Imaginations, Documentations*. Bielefeld: Transcript.

Schneider, S. H. 1975. On the carbon dioxide–climate confusion. *Journal of the Atmospheric Sciences*, 32(11), 2060–2066.

Shackley, S. *et al.* 1999. Adjusting to policy expectations in climate change modeling: An interdisciplinary study of flux adjustments in coupled atmosphere-ocean general circulation models. *Climatic Change*, 43(2), 413–454.

Shackley, S. 2001. Epistemic lifestyles in climate change modelling. In Miller, C. A., and Edwards, P. N. (eds.) *Changing the Atmosphere*. Cambridge, MA: MIT Press, 107–134.

Shapin, S. 1994. *A Social History of Truth: Civility and Science in Seventeenth-Century England*. Chicago, IL: University of Chicago Press.

Shapin, S. 1995. Cordelia's love: Credibility and the social studies of science. *Perspectives on Science*, 3(3), 255–275.

Van Der Sluijs, J. P., Van Eijndhoven, J., Shackley, S., & Wynne, B. 1998. Anchoring devices in science for policy: The case of consensus around climate sensitivity. *Social Studies of Science*, 28(2), 291–323.

Walsh, L. 2013. *Scientists as Prophets: A Rhetorical Genealogy*. Oxford: Oxford University Press.

3

Visualising Climate and Climate Change
A *Longue Durée* Perspective

SEBASTIAN VINCENT GREVSMÜHL

3.1 Introduction

This chapter explores the challenging terrain of images and scientific visualisations as a way of investigating and understanding climate and climate change within the context of society and culture. I retain the notions of images and visualisations here, as opposed to representation, in order to avoid the numerous metaphysical connotations the action of 're-presenting' may imply (Rheinberger 2001). Visualisation in the context of climate change is thus understood as a means of knowledge construction, and it necessarily implies imagination, political sensibilities, as well as material actions. My focus here will be on spatial visualisations of climate in the form of maps and in particular the history of climate classifications. The study draws mainly on a systematic investigation of several important scientific map collections, complemented by careful readings of several influential scientific treaties on meteorology and climate, as well as a critical survey of the available historical literature on climate and its cultural understandings.[1] Some of the questions I would like to address here are: What would a *longue durée* history of climate visualisations look like? What can graphs and maps tell us about a changing climate and its history? And more importantly, what can we learn from such a *longue durée* visual perspective, in particular within the contemporary context of anthropogenic climate change?

I argue that adopting a *longue durée* perspective can be fruitful because current debates on climate change resonate astonishingly well with many early questions

---

[1] Several important map collections have been partially digitised, the most important being the private Rumsey map collection (www.davidrumsey.com), with over 83,000 items available online (January 2018). A very useful resource on especially nineteenth-century geographical atlases can be found at www.altassen.info. Moreover, some thematic map exhibitions, in particular 'First X, Then Y, now Z' by John Delaney at Princeton University Library, were also very helpful.

philosophers and cartographers wished to address when they first mapped out 'climates'. Indeed, some of the major motivations were strikingly contemporary in nature – for instance, defining the role geographical regions play in harbouring favourable conditions for life, exploring the various ways in which climate may relate to population and migration, and finally reflecting on the kinds of world we can, or cannot, live in. To put it in other words: since the introduction of the notion of 'climate' and related vocabulary, fundamental explorations of our being and living in this world are at the very heart of climatic thinking.

As I will show below, approaching this set of questions from the visual realm can prove fruitful for several reasons. Visualisations, as the word indicates, fulfil an important role in rendering the invisible visible, and this is particularly true for the context of climate change where the vast majority of geophysical structures, processes and phenomena would otherwise stay out of reach of investigation. Climate visualisations form an important part of this interrogation. Maps in particular played (and still play) a leading role in shaping our understanding of the Earth's climates, regrouping important cultural resources of diverse nature, fuelled (in function of varying historical contexts) by reasoning of biblical, cosmological, geographical or geophysical nature. In the past, climate knowledge was, in other words, never stable, and the ways in which we think about climate has changed considerably, invoking at times highly disparate concerns related to medical, agricultural, geographical, economic and even racial questionings (Fleming and Jankovic 2011). 'Climate' should thus be read here in a broad sense, because of its unstable and shifting meaning over time, evolving constantly according to varying social and cultural contexts.

Indeed, as Heymann rightly points out in his genealogy of climate ideas, 'different understandings of climate were instrumental for the shaping of scientific interests and directions of research and the formation of scientific and public discourses about climate' (Heymann 2010:581). This should, however, not obscure the fact that climate, and in particular its changes, was hard to seize by science and, more importantly, that the view from science tells us of course only one small part of this story. As historians Locher and Fressoz have shown, François Arago, for instance, enlisted as France's climate expert during the first half of the nineteenth century, had to admit that in the absence of long-term meteorological records and in regard of the huge uncertainties attached to the climate question, a serious scientific investigation was an impossible task (Locher and Fressoz 2012). This did not mean that learnt discourse did not have important things to say about climate and its changes. Indeed, in order to adopt a *longue durée* perspective on climate and its changes, we rather have to engage in now-defunct understandings of 'climate theory where technique, political form, environment, and bodies all overlapped' (Locher and Fressoz 2012:581).

Some of the major shifts in past understandings of climate and its changes are captured by the rich iconography mobilised to visualise climate ideas and theories, others did not find entry into the visual realm and others again reveal historical transitions which sit squarely with received historical conceptions of climate change. So before considering the realm of contemporary climate change visualisations addressed at the end of this chapter, it is worthwhile asking which role images played in the forming of our understandings of climate in a *longue durée* perspective. This chapter thus speaks to a growing interest in understanding in more detail what 'climate' and 'climate change' mean culturally, revealing in particular how and why the dominant physical and statistical history of climate is closely interwoven with its cultural history (see also Hulme 2015).

3.2 Shifting Meanings in Climate Visualisations

Climatic theories and ideas were from the very beginning closely associated with images, in particular maps and diagrams. Amongst the earliest 'climate' maps historians have at their disposal today, one may count what contemporary map classification systems identify as medieval zonal maps.[2] These maps are all based upon or were inspired by Macrobius' *Commentary on the Dream of Scipio* from the early fifth century. Images were central to the *Commentary* because the text gives very precise instructions on how to draw four diagrams, and in particular a zonal diagram, showing the division of the Earth into five 'bands' or 'belts' made up of the 'frigid', 'temperate' and 'torrid' zones, with only the two temperate zones considered habitable, and only the northern one for sure. Many of these medieval zonal diagrams have a strikingly 'contemporary' look, not only because most of the hand-painted copies which were made throughout the Middle Ages mobilise a very familiar colour coding still in use today – with red systematically attributed to the torrid zone and blue often reserved to the frigid zones, as may be observed in Figure 3.1 showing a French example from around 1150 – but also because zonal division remains until today deeply enshrined in Western geographical imagination.

The importance of images is very clear here because all diagrams were explicitly conceived as pedagogical tools, allowing, according to Macrobius, to apprehend more easily the idea of zones (Hiatt 2007). The initial concept was introduced by Pythagoras, Parmenides and Aristotle not on observational but purely astronomical and mathematical grounds (Altmann and Schramm 2005). The idea was to express geographical differences in sunshine due to the inclination of the Sun relative to the Earth's surface. Thus, early twentieth-century historical commentators, such as climatologist Robert Ward, referred to the three types of zones as 'zones of solar

---

[2] Sometimes they are also referred to as 'hemispheric' maps because they always show only one hemisphere.

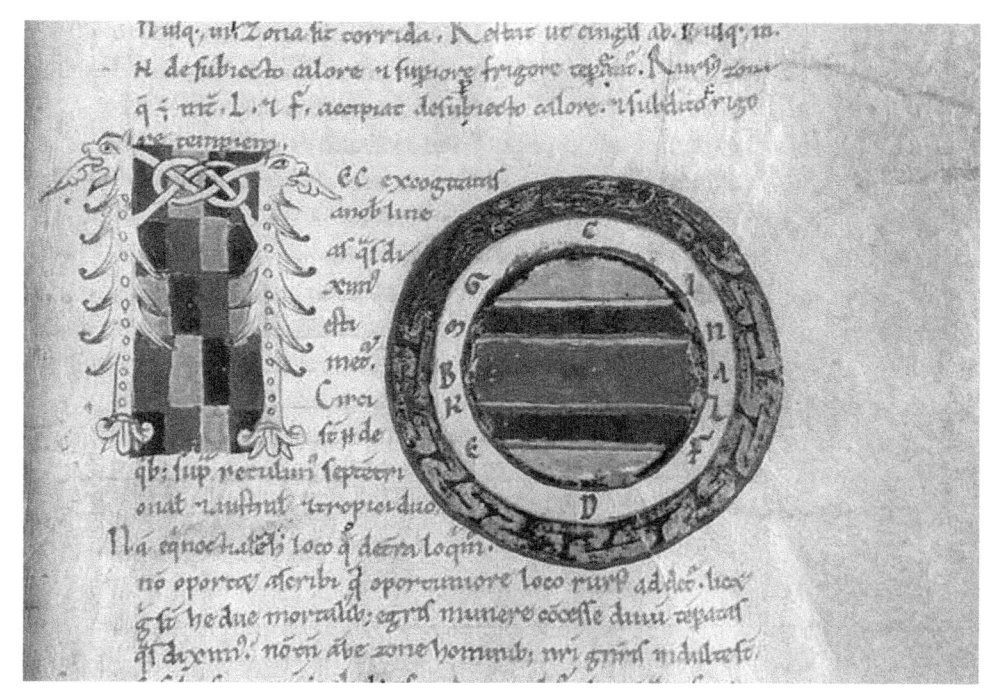

Figure 3.1 The five zones of the Earth according to Macrobius' commentary on Cicero's *Somnium Scipionis*, c. 1150 (Source: Wikimedia Commons)

climate' (Ward 1905:386; see also Ward 1908), remobilising a notion Humboldt had already popularised in 1813 in order to differentiate what he calls 'solar climate' and 'real climate' (von Humboldt 1813:471). The zonal diagrams were often accompanied by what is known as Macrobius' world map which included (in its standard version in copies of the text produced by scribes from the tenth century onwards) five zones, a schematic outline of the relationship of ocean to land in the northern hemisphere, an equatorial ocean and symbols indicating the direction of ocean flows moving from the equator to the poles (Hiatt 2007).

Curiously, however, the actual term 'climate' was not used for the zonal descriptions because it was reserved for a long time for another tradition that I also would like to briefly discuss here. Although at first sight these ideas might appear conceptually close - because they were visualised together on maps on many occasions particularly during the eighteenth century - the five zones depicted in zonal diagrams are not to be confused with the ancient Greek geographical concept of *klima*. The division into *klimata* was initially only applied to parts of the northern hemisphere, which were first divided into seven, and then later nine parallel, horizontal bands. Often named after the places through which they passed (Meroe, Syene, Alexandria and so on), these early *klimata* were distinguished from one another by the length of their solstitial day, and at first they were all located within a single zone, the temperate zone. This coincided at the same time with the *oikoumene*, believed to be the only inhabited part of the Earth (Corneille 1708; see also Honigmann 1929; Hiatt 2007).

For the Western context, it is important to note that this tradition is historically closely linked to Ptolemaic ideas. Ptolemy's reintroduction during the fifteenth century in the Western world was made possible through Islamic scholarship where ancient Greek natural philosophy was intensely conserved, studied and translated, and unsurprisingly, one of its most famous and influential examples, the Al-Idrisi map commissioned in the twelfth century by King Robert II, Norman ruler of Sicily, included seven climates in Ptolemaic tradition (see map reproduced in Cosgrove 2007:87; see also Honigmann 1929). In Europe, this cartographic tradition of 'climates' reaches back, for instance, to the 1480 atlas *Geographia* from Francesco Berlinghieri that contains a colourful world map depicting seven climates (Berlinghieri 1480), and this tradition was continued by many cartographers throughout the sixteenth and seventeenth centuries (e.g. Fries 1541; Ruscelli 1561). Other typical visual examples can be found in world maps redrawn after ancient Greek knowledge, for instance, the commentary of Strabo published in 1731 by Christoph Cellarius, showing a map of the northern hemisphere with seven parallel climates, numbered 1 through 7, running through Meroe, Syene and so forth, each line separated by half an hour difference in length of day (Cellarius 1731). As a distinct, separate cartographic tradition, this geographic climate category remained astonishingly stable well into the latter half of the eighteenth century. All of these visual examples show well that at its origin, the word 'climate' was not at all defined in terms of temperature or meteorological conditions but rather associated with the idea that a change of location, where the longest day is, say, half an hour or an hour longer than at the starting point, represents a change in climate.

3.3 The Slow Erosion of Fixed Divisions and Boundaries

The fifteenth and sixteenth centuries' voyages and discoveries, however, progressively made these divisions and boundaries more fluid, particularly in regard to the question of the extension of the habitable world in certain zones, but also in regard of the number of climates to be included. Indeed, the 'torrid zone' was not only inhabited in many places but it also could be traversed. Moreover, the voyages to the New World proved that the new lands had meteorological conditions that were not simple extrapolations of the known European ones. Historian Anthony Grafton, for instance, mentions the telling story of Jesuit scholar and traveller José de Acosta whose 1580 treatise on the New World describes his incursion into the 'torrid zone' as a rather chilling experience: he and his fellow travellers were not dying of heat – they were actually feeling cold (cited in Martin 2006:3). Also the seven traditional climates were successively extended, in particular because of early 'polar' exploration reaching far into the 'frigid zone'.

As a consequence of direct observation and the rapid expansion of the *oikoumene*, ancient Greek knowledge started to be questioned, at least in part, even though Aristotelian natural philosophy and climatic theories forged by Hippocrates proved astonishingly flexible and adaptable in the long run (Martin 2006). To be sure, the idea of the inhabitability of the 'torrid zone' was firmly rejected by most scholars from the sixteenth century onwards. However, causal climatology and climatic determinism remained well into the eighteenth-century influential climate doctrines, especially in medical climatology. As a philosophical doctrine, climatic determinism was frequently mobilised to explain differences in cultures and societies (Heymann 2010). Indeed, as Jim Fleming and other historians have shown, the second half of the eighteenth century was dominated by climatic concepts introduced by Abbé Du Bos, Montesquieu and Hume which forged the idea that cultures are determined or at least fundamentally shaped by climate and that, in turn, the collective actions of society shape the climate itself (Fleming 1998).

What counts here is that in world maps, zonal distribution (mostly five zones) and the 'modern' division of the globe into twenty-four or thirty 'climates' persisted side by side throughout the seventeenth and eighteenth centuries, yet without really interacting with each other (Mauelshagen 2016). This historical evolution is neatly summarised not in a map but in a hierarchical diagram, published by cartographer Nicolas Sanson's son Guillaume (Sanson and Sanson 1697). The diagram describes in detail which elements should be included in terrestrial globes and world maps, distinguishing in particular the two dominant, separate types of classification: 'zones' and 'climates'. According to Sanson, the 'zones' are made up of the classic three types ('one torrid, two temperate and two cold or frozen') as discussed above. 'Climates', on the other hand, underwent a historical transformation. Whereas the ancient Greek identified first seven, later nine 'climates' (running through well-known places and landmarks such as Meroe and Syene as discussed above), the 'moderns' defined thirty different 'climates', namely twenty-four for every half hour between the equator and the polar circle, and six for each month between the polar circle and the pole. Typical world maps would thus include both zonal and climate distribution in function of solstitial days.[3]

---

[3] The Rumsey map collection contains numerous examples illustrating this fact, such as the French 'Mappemonde géosphérique ou nouvelle carte idéale du globe terrestre' of Lattré from 1760, or the Italian 'Il mappamondo o sia descizione generale del globo' of Zatta/Pitteri from 1774. See: www.davidrumsey.com/ (accessed 5 December 2017).

However, not only the voyages to the New World made abundantly clear that neither day length nor latitude was a reliable or even useful indicator for meteorological conditions to be encountered when travelling around the globe. It was the rapidly growing popularity of weather instruments, especially during the eighteenth century, which brought about a quantitative approach to weather recording, even though this was a gradual process that once again does not easily fit into the idealised accounts of the Scientific Revolution. As historians Jankovic and Golinski have shown for the British case, meteorological observations were guided well into the eighteenth century by classical Aristotelian conceptions and remained often qualitative and descriptive (Jankovic 2000; Golinski 2007). Yet with such important inventions as gauges, barometers, manometers, thermometers, hygrometers and even machines designed to measure the speed of winds, the material basis was laid for empirical weather research which was at the origin of physical climatology, and this shift became increasingly clear from 1750 onwards.

The slow emergence of a modern understanding of climate is precisely captured by the famous *Encyclopédie* of Diderot and d'Alembert. The third volume from 1753 contains two 'climate' entries, a geographical one and a medical one. After a classic explanation of half-hour and monthly climates, the first geographical entry written by d'Alembert also states: 'One should not be misled in thinking that the temperature is exactly the same in countries located in the same climate: because an infinity of circumstances, such as the winds, the volcanoes, the proximity to the sea, the location of the mountains, become more complex under the sun's action, and often render the temperature very different in places situated at the same parallel' (Diderot and d'Alembert 1753:533, own translation). The second, a medical entry written by physician and chemist Gabriel François Venel, is also very explicit, placing temperature at the heart of the climate definition: 'Physicians consider climates only as a function of temperature or degree of heat ...: climate, in this sense, is even exactly synonymous to temperature' (Diderot and d'Alembert 1753:534, own translation).

D'Alembert's and Venel's climate definitions thus stand paradigmatically for the slow introduction of a complex, physical understanding of climate, driven by the instrumental revolution, a change that took place gradually from the second half of the eighteenth century onwards. Heymann notes that this is also the period when a disciplined, non-personal engagement in observation started to become a dominant trait of systematic weather observations, even though still at a very reduced level (Heymann 2010). The climate notion thus gradually shifted to a physical understanding, introducing an important complexification, guided by causal reasoning and a new, dynamic interpretation of climate (Mauelshagen 2016). Once hegemonic from the mid-nineteenth century onwards, with the installation of the first monitoring networks and the establishment of meteorological societies, the empirical approach to weather research radically changed inherited conceptions of climate, an aspect to which I will turn in the next section (Locher 2008).

For this early period, it is important to note that zonal and climate maps provide some important insights when considering a *longue durée* perspective on climate and climatic thinking. First, zonal maps make very clear that the supposed fundamental 'break' or 'rupture' between traditional natural philosophy and new knowledge that emerged with the so-called Scientific Revolution was by no means an abrupt one. Many ancient beliefs persisted well into the eighteenth century, and others regained importance at times, such as causal climatology in the Hippocratic tradition, which experienced many revivals in the form of medical climatology as a framework for explaining differences in human culture, or as a hygienist doctrine during the nineteenth century supporting explanations of racial superiority (Heymann 2010; Locher and Fressoz 2012). The visual approach reveals that zonal distribution persisted as a classification system for centuries, mainly because it could easily be adapted and integrated into different theoretical and moral frameworks. Well into the twentieth century it served as a visual guide in teaching the concept of climate zones, often accompanied, as Livingstone has shown, by moral considerations.[4] Moreover, this means that in this early period there is a dominant 'visual style' attached to the idea of climate that proved particularly robust throughout history, even though the meanings of 'climate' shifted of course substantially over time.

This visual continuity should, however, not obscure the fact that the word 'climate' was for a long time attached to a completely separate geographical cartographic tradition that is often completely overlooked, even though it had important repercussions for the history of climatology. Within cartography, both knowledge traditions started to merge during the middle of the eighteenth century, and as I will argue now, this merging led to a new climate classification at the beginning of the nineteenth century.

3.4 Visual Contributions to a Quantitative Understanding of Climate

Classical zonal distribution would not stay immutable throughout history. Indeed, the instrumental revolution brought about a new kind of 'zonality', based not on latitude but on temperature. And here again, the visual realm was fundamental in forging new ideas about climate, although it took a relatively long time to translate the growing

---

[4] Many geography textbooks continue well into the twentieth century to use zonal maps as pedagogical tools for teaching global climate distribution. For instance, in the 1907 edition of Matthew Fontaine Maury's classic textbook *New Elements of Geography*, one still finds on page 17 the idea of distinct zones defined according to the ancient Greek tradition. Well into the twentieth century, scientific accounts of atmospheric conditions are supplemented by moralistic pronouncements (Livingstone 2002).

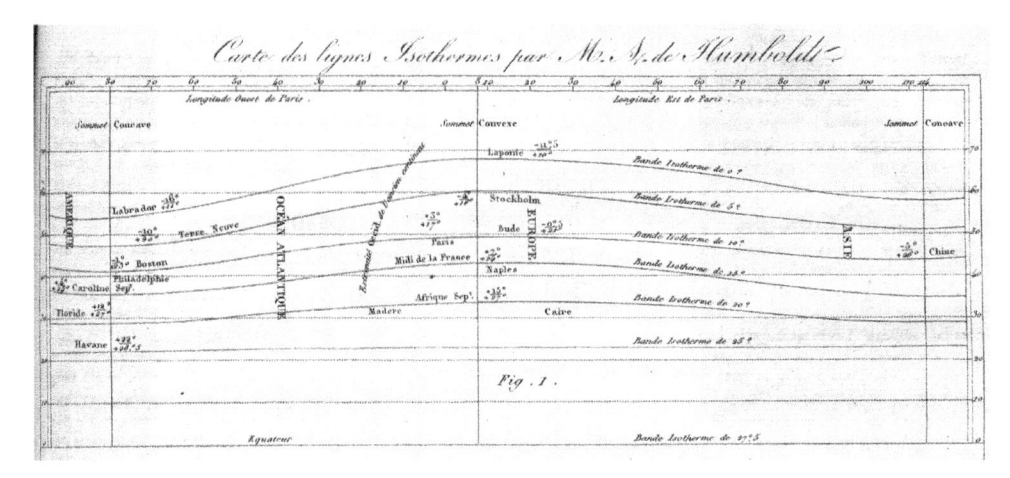

Figure 3.2 Isoline map as introduced by Alexander von Humboldt in 1817 showing mean temperature on a global scale (Source: Princeton Library Historic Maps Collection)

number of meteorological records into climate maps. One historical figure, as probably no other, is closely associated to this historical development: German naturalist and explorer Alexander von Humboldt. At the beginning of the nineteenth century, Humboldt laid important foundations for the nascent scientific discipline of climatology by popularising innovative visualisation techniques. One of the most famous visual tools he introduced to climatology can be found in a now well-known map showing so-called isothermal lines published in 1817 (Figure 3.2). This early climate map may be counted amongst the most influential examples of thematic mapping, even though its simplistic appearance does not render it immediately recognisable as a map. Humboldt's 'Carte des lignes isothermes' shows seven isolines (also known as contour lines), each connecting values of a selected average temperature, hence the notion 'isothermal line' coined by Humboldt. Here, the power of images becomes clearer than ever: the figure reveals a spatial phenomenon that if not mapped would not have any existence at all. Indeed, it is the masterful introduction of what I have called elsewhere a 'global imaginary' (Grevsmühl 2014), the visual creation of a geophysical phenomenon at a very important scale that nobody could or would ever see.

Moreover, in tracing the spatial distribution of average temperature, Humboldt took the crucial step from mere data collection (realised beforehand exclusively in the form of tables) to data analysis, resulting in careful scrutiny of the form of visual distribution, the numerous relationships between different phenomena and, most importantly, their underlying laws. The 'map of isothermal lines' is thus precisely what Regnauld calls a 'visual model', reaching well beyond mere illustration because it expresses a completely new theoretical concept (Regnauld 2016).

This fundamental shift, which marks in this sense (at least in the visual realm) the birth of modern climatology, is well known today; yet it remains too important a shift to be ignored. Several features of Humboldt's map are worth being noted here. First, reduced to an absolute minimum of information, the cartographic content is merely suggested here. The outlines of the continents are not shown but the graticule and the geographical inscriptions (as, for instance, 'America', 'Europe' and 'Asia') indicate that we are looking at a map showing large parts of the northern hemisphere, centred on the meridian of Paris. Yet the true radical novelty of this map is the introduction of a new type of 'zonality' based on seven isolines of average temperature and differentiated by mostly 5°C steps. The map shows at first in the North the 0°C isotherm passing through Lapland, moving in 5°C steps all the way down to the sixth isotherm of 25°C running through Havana, and finally the equator, which coincides with the isotherm of 27.5°C. Henceforth, climate was defined as a statistical representation of weather based on carefully selected field observations.[5]

Although there is an obvious visual resemblance with the geographical climate tradition – and as a visual style, isotherms helped in fact connect the visual tradition of ancient zonal theory to the geographical tradition – Humboldt himself sought inspiration not in Eratosthenes nor Parmenides but in a novel data visualisation technique of another global imaginary, first introduced (when considering a global scale) by Edmond Halley in geomagnetism in around 1700, showing magnetic declination at a truly global scale and based on actual measurements. Halley was already aware of the immense possibilities his innovative visualisation techniques could offer (in particular because he also drafted one of the first known meteorological charts), claiming that with his novel maps many natural phenomena 'may be better understood, than by any verbal description whatsoever' (Halley 1686:163; see also Wilford 1981).

Humboldt agreed of course with this view as he was deeply convinced of the usefulness of visualisation techniques for global data analysis, stating that '[t]he use of the graphic method will help clarify phenomena which are of the highest interest for agriculture and for the social state of the inhabitants. If instead of geographical maps, we only possessed tables covering latitude, longitude and altitude, a great number of curious relationships, offered by the configuration and unequal surface qualities of the continents, would have stayed for ever lost' (von Humboldt 1813:510–511). Citing Halley's preliminary work on isolines, Humboldt further wrote: 'I traced on a map isothermal lines analogous to lines of

---

[5] Although field observations are crucial to the map, isolines create the illusion of continuous measurement, effectively hiding the number of actual measurements used in the construction of the map (see Grevsmühl 2014). For instance, despite its global appearance, Humboldt retained only fifty-eight places to construct his isothermal map (see Schneider 2016).

magnetic inclination and declination' (von Humboldt 1813:488), finding that on many occasions 'they are neither parallel to the equator, nor parallel to each other' (von Humboldt 1813:511).

Humboldt's groundbreaking contribution to the nascent field of climatology thus consisted mainly in successfully disconnecting climate from the rigid parallels, separating temperature from latitude with the help of a powerful visualisation tool, the isoline, that he helped popularise during the first half of the nineteenth century. Henceforth, neither latitude nor day length would be deemed appropriate criteria for climate description. Temperature, and in particular average temperature, would be retained as one of the most important guiding criteria. This fundamental choice is still at the basis of our current definition of climate, which, according to the third Intergovernmental Panel on Climate Change (IPCC) report, is referred to in common parlance as 'the average weather' and involves in scientific terms 'the statistical description in terms of the mean and variability of relevant quantities [temperature, precipitation, etc.] over a period of time ranging from months to thousands or millions of years' (IPCC 2001).

3.5 Humboldt's Lasting Legacy

The first to adopt Humboldt's novel visualisation technique in America was a Christian educator and school teacher, William Channing Woodbridge. His map titled 'Isothermal Chart, or View of Climates & Production Drawn from the accounts of Humboldt & others' from 1823 (Figure 3.3) shows for the first time the relationship of mean annual temperatures to crop growth and world climates. The carefully hand-coloured map visually merges the well-known seven climates with zonal theory by introducing seven 'climate regions', delimited now not by parallels but isotherms, which would become more and more 'undulated' and complex in nature with the inclusion of a rapidly growing number of global observations throughout the nineteenth century.

Faithful to the tradition of Humboldt's early map, Woodbridge's isothermal chart convinces with a simplistic design, reducing visual information to seven climate regions (delimited by isothermal lines and differentiated in function of colour), the geographical limits of agricultural production in function of temperature and the continental outlines, including only a strict minimum of topographical features. Traditional topographic information one would usually expect in particular in a school atlas was deliberately left out in order to direct the reader's attention to the essential information Woodbridge wished to convey. His chart also picks up a similar colour coding as already encountered in the medieval zonal maps discussed above, with strong red colour reserved to the 'torrid region', light green to the 'temperate' regions and blue tones to the 'wintery' and 'frozen' ones. These

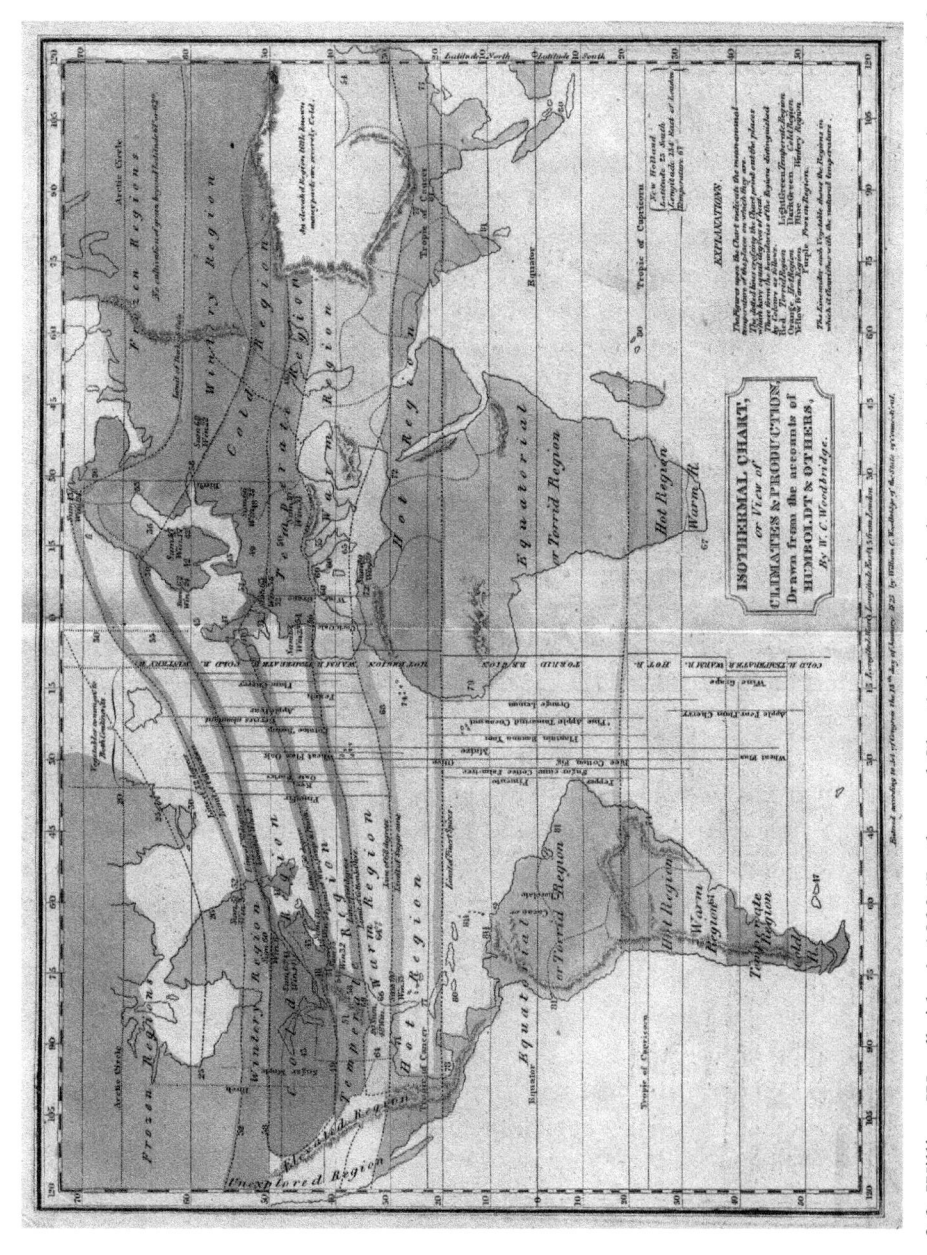

Figure 3.3 William Woodbridge's 1823 'Isothermal Chart' showing agricultural production in function of climatic zones defined by isolines (Source: HIST 1952, Harvard University)

elements make Woodbridge's map an early forerunner of modern climate classification systems, although his work is mostly ignored by general classification histories.[6] Finally, Woodbridge also believed in the superiority of visual media in geographical education, stating 'it is only by this process of [visual] comparison, that the great objects of geography – the expansion of the mind, and the discipline of the reflective powers – can be attained' (cited in Delaney 2012:62). This visual comparison of climate and agricultural production was made possible with two major achievements, a new climate definition, which was since Humboldt completely disconnected from the parallels, and the introduction of maps as analytical tools, allowing the spatialisation of all sorts of information for comparative and global analysis. Climate and its changes were central to this new visual, comparative approach, because it allowed the tackling of many pressing questions.

In particular, the link with agricultural production, spatially theorised in plant geography, had far reaching cultural, economic, scientific and political implications throughout the nineteenth century, of which only a few can be mentioned here (Schulten 2010). For instance, agricultural production was central to colonial discourse from the mid-eighteenth century onwards in the French and British contexts, where it had become increasingly clear that insular climates were highly dependent on forestry questions and precipitation regimes (Grove 1995). In Algeria, in the eyes of the French colonists, it was their duty to rehabilitate and amend deleterious climate by, for example, planting eucalyptus trees in large numbers and introducing massive irrigation infrastructures. Back in Europe, regional climate changes were also noted, for instance in 1832 by Austrian astronomer and director of the Vienna observatory Joseph Johann Littrow, who thought it is very likely that climate in Middle and Northern Europe must have changed (it was probably rising according to him) due to the important draining of marshes and massive deforestation. General climate change, as discussed by many at the time, in particular, after large parts of Europe experienced severe weather events (resulting from the 1815 eruption of Tambora in Indonesia), should however, according to Littrow, not be inferred from temperature records (Littrow 1832). The observation of general climate change tendencies was in other words a highly controversial topic, and whereas some authors claimed to have proof for historical climate changes, Danish geographer and botanist Schouw noted, for instance, only a few years before Littrow that there were no indications for a changing climate in any of the meteorological and other records one could find in Denmark and Scandinavia (Schouw 1827).

---

[6] No mention, for instance, in the Glossary for Meteorology of the American Meteorological Society: http://glossary.ametsoc.org/wiki/Climatic_classification (accessed 5 December 2017) nor in the classic textbooks, such as Owen Box (2016).

3.6 Historical Trajectories towards Contemporary Climate Change Icons

The link established between plant distribution, temperature and its underlying laws was absolutely crucial for the debates that followed because it had important repercussions on climate change theory as it resurfaced during the Cold War, thus shaping at least in part many of the famous climate change icons we know today. For instance, the field of historical climatology was born precisely out of this new interconnected field of investigation, infused by the widely circulated climate visualisations of Humboldt and Schouw, and thereafter picked up, during the second half of the nineteenth century, in many influential thematic maps produced by Woodbridge, Berghaus, Petermann, Johnston, Cartee, Bromme, Guyot and others. Although Woodbridge's visual simplicity would not persist as a dominant visualisation mode (indeed, more and more information was included in maps especially from the 1850s onwards), the visual style of Woodbridge's maps dominated weather and climate visualisations in atlases well into the 1870s. More importantly, these maps spawned research on historical vegetation data – pushed in France by astronomer François Arago, based for instance on the dates of grape harvest or the geographical distribution of date palms – in order to allow the reconstruction of climates of the past few thousand years (Arago 1858; see also Locher and Fressoz 2012).

Although not directly related to our visual climate history, one must nonetheless mention here that the timescales of historical climatology were even further expanded by the rise of the theory of ice ages during the second half of the nineteenth century, born once again out of fears of a changing climate. It dominated the climate research agenda for almost 100 years until the rapid expansion of the Earth Sciences at the beginning of the Cold War and the subsequent introduction of numerical modelling in climatology.[7] The existence of ice ages in former times required addressing timescales that were much vaster than any timescales considered beforehand, demanding also novel explanations for large changes in mean temperature. Thus, especially two research fields had a lasting impact on the following climate change debate: the modification of the Earth's orbital configuration and the chemical composition of the atmosphere, the latter including of course the carbon dioxide–induced greenhouse effect. Yet contrary to common historical belief, many of the so-called precursors, generally identified as the initiators of the theory of climate change (John Tyndall, Svante Arrhenius or Thomas Chamberlin),

---

[7] Born out of the Swiss context of glacier research and fears of a gradual cooling in Europe, Swiss engineer Ignace Venetz proposed in 1821 (published in 1833) that in the dim past, enormous glaciers covered the alpine regions (Venetz 1833).

were not so much worried about human-induced climate change but were rather seeking to explain the ice ages (Brönnimann 2002, Locher and Fressoz 2012).

Finally, and this is crucial for our *longue durée* visual perspective, visualisations of geographical plant distribution (in function of mean temperature and precipitation) gave rise to the most influential climate classification still in use today: the Köppen–Geiger climate classification. Picking up on Humboldt's and Woodbridge's important work, and further popularised by Berghaus, Dove, Supan, Grisebach and others, the Russian-German meteorologist Wladimir Köppen first proposed in 1884 a draft of a new climate classification system, transformed in 1900 into a mathematical system of climate classification, dividing the world into various climate types, a system he would subsequently refine throughout his career until his death in 1940 (Köppen 1884). His three-letter system consisted in classifying climate according to five climate categories (first level of description: tropical, dry, temperate, boreal and polar climate, labelled A to E, derived maybe from Swiss-French botanist De Candolle's vegetation groups), taking further into account seasonal variations in precipitation (second level of description) and monthly mean temperature (third level). The system's main aim was to represent areas with similar climatic conditions whose statistical properties would yield similar vegetation types, thus revealing the geographical complexity of climates at a regional level. After Köppen's death, German climatologist Rudolf Geiger – with whom Köppen collaborated to author the multiple-volume work *Handbuch der Klimatologie* (Handbook of Climatology) – published in 1961 the last version of his climate classification as a large wall map for use in teaching (Geiger 1961). Today, it remains the most popular and most influential climate classification, with applications in a broad diversity of fields (Rubel and Kottek 2011).

And since modern climatology is more concerned with prediction than description (see Mahony *et al.*, this volume), the Köppen climate classification did not just remain a useful resource for teaching, but it also became an important tool of climate change research. For instance, the Köppen classification was used in 1975 by pioneers of numerical climate modelling Manabe and Holloway as one of the first benchmarks to verify the output of global climate models (Manabe and Holloway 1975). Many other influential applications followed (as, for instance, its use in model intercomparison) so that today the Köppen classification is considered 'a standard part of future efforts to develop and evaluate AOGCMs' (atmosphere-ocean general circulation models) (Rubel and Kottek 2011:363; see also Jylhä *et al.* 2010). Indeed, since the publication of digital world maps for an extended period (covering now 1901 to 2100), the visualisation of global trends and projected shifts of climate zones according to different IPCC scenarios has become an influential tool for climate change research. Thus, since the 1970s, climate classifications and their visualisations are once again firmly embedded in climate change discussions, with the naturalistic, quantitative approach now clearly established as the dominant way to describe climate and its changes.

Two types of visualisations illustrate this historical development very clearly. The first are global maps and computer animations produced by an interdisciplinary team of climate researchers, showing on world maps for the twenty-first century shifting Köppen climate zones in function of different IPCC greenhouse gas scenarios, with expected shifts of certain climate classes for some scenarios estimated by the end of the century at almost 5 per cent (see Figure 3.4) (Rubel and Kottek 2010).[8] Although isolines are now absent from these maps because of a more regional approach to climate classes and the inclusion of precipitation data, there is still a striking resemblance with Woodbridge's isothermal chart from 1823, in particular the colour coding which remains almost the same.

In a very similar way, but on a more regional scale, scientists from the Finish Meteorological Institute produced maps showing shifting Köppen climates on the scale of Europe. Backed up by a questionnaire, their study provides some evidence that maps can be an effective way of visualising and disseminating information concerning regional climate change, especially to non-experts, policy-makers and the general public (Jylhä *et al.* 2010).

A second, closely related type of visualisation is that of 'spatial analogues' which consists mostly in looking 'for regions where the present-day climate resembles the anticipated future climate of the study area' (Jylhä *et al.* 2010:162). Used since the 1980s in particular to inform agricultural studies on climate change (Parry 1988), climate analogue studies have remained, especially with the rise of numerical modelling since the 1990s, an influential approach, despite numerous drawbacks such as 'distorting the geographical configurations between locations' (Jylhä *et al.* 2010:162; see also Hallegatte *et al.* 2007). Today, with the help of web tools such as the Climate Analogues online platform, anybody with an Internet access can search for and explore possible climate analogues for almost any place on Earth.[9]

3.7 Conclusion: Moving towards Mobile Climates

The two previous examples show very well that global climate change now confronts us with *mobile* climates where fixed categories no longer exist. Climates within the contemporary climate change regime, and this is a key insight of our *longue durée* visual perspective, have in other words become migrant – along with crops, animals,

---

[8] Animated maps can be accessed here: http://koeppen-geiger.vu-wien.ac.at/ (Accessed 5 December 2017).
[9] See: www.ccafs-analogues.org/tool/ (accessed 5 December 2017).

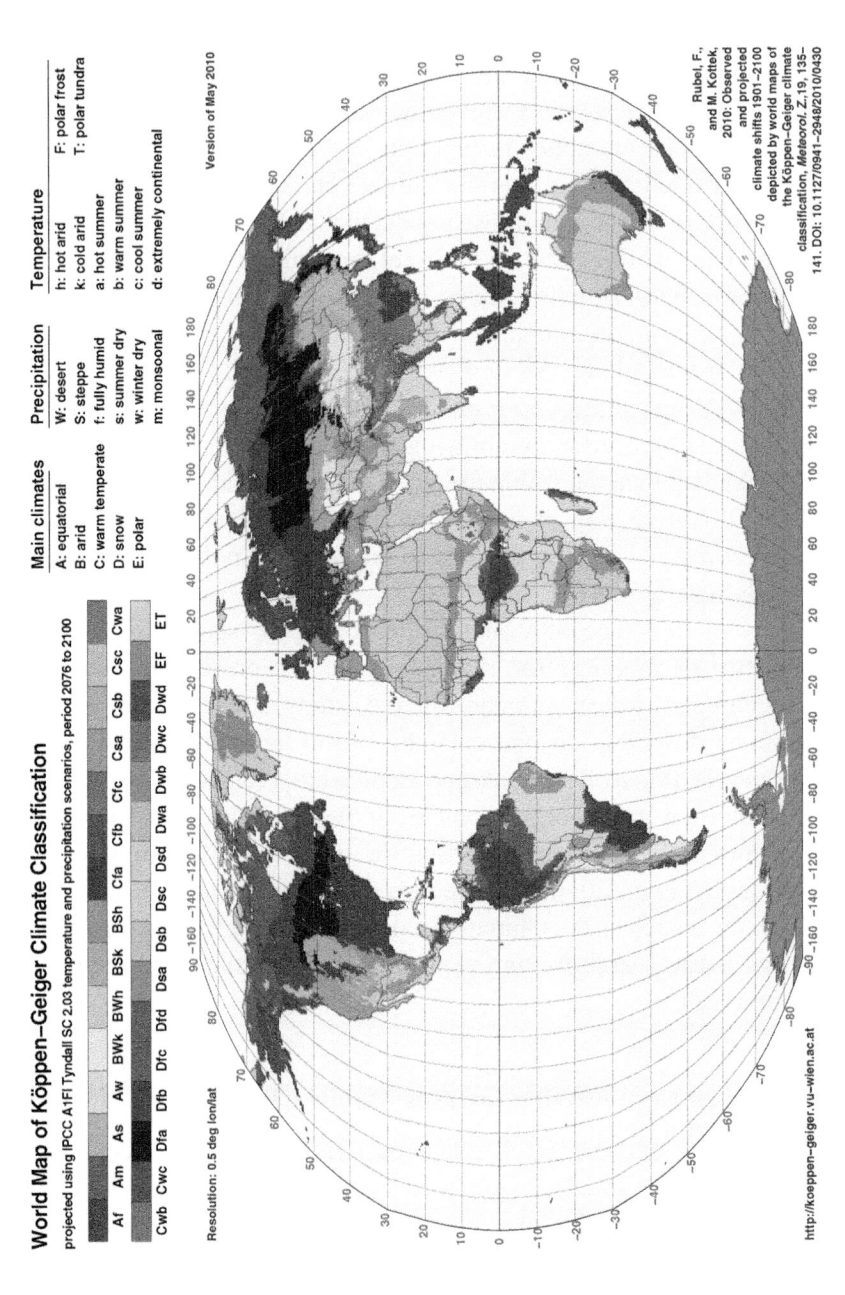

Figure 3.4 World Map of Köppen–Geiger climate classification projection for the period 2076–2100 using the A1FI emission scenario (rapid economic growth with use of fossil intensive energy sources) visualising shifting climate types (Source: Meteorologische Zeitschrift)

diseases and people. It is no longer the colonial administrator who confronts himself to the dangers of the so-called torrid zone, or the explorer who ventures into the forbidding climes of the polar regions, equipped each time with ideas of stable climates and distinct zones, each attached to specific views on nature and society. Today, the situation is somewhat inversed, with the 'tropics' inviting themselves to 'us', as evidenced, for instance, in climate change exhibitions such as *Postcards from the Future* (London 2010) or in the recurrent use of the motif of glaciers and palm trees in the Swiss publishing context (Brönnimann 2002; Mahony 2016). Both examples reveal that mapping is only one of many visual strategies deployed, with montage being another key visual strategy used to visualise the mobility of climates.

There are, however, also important limits to these approaches. Indeed, as Mahony has convincingly argued, the fear of tropical invasion, as promoted in *Postcards from the Future*, comes with a highly typified view of the tropics 'neglecting a fuller politics of human mobility, urbanisation and the global interconnections through which human populations are jointly engaged in the ongoing composition of a common, climate-changed world' (Mahony 2016:14).

Yet, despite these limitations to which scholars must remain of course attentive, visually exploring the mobility of climates can provide nonetheless a means for conveying (at least to some degree) the highly dynamic nature of climate change, especially when addressing the lay public. The above-mentioned climate analogue studies may be helpful here in particular when it comes to communicating the observed rapidity of changing climates. In the past, most geographical climate analogue studies explored analogies between current and *future* climates, but it is also possible to take a look at the past. For instance, in a recent study, Beniston proposed 'to assess the manner in which climatic conditions in European cities in the last decade compare with those that prevailed in the 1950s in other locations generally located well to the south' (Beniston 2014:1839). The advantage of this study and of the resulting maps is that they give an idea of how rapidly our climates are changing, with isotherms moving in some cases between geographical locations as fast as 14 kilometres per year in a northward direction. This means that some places are now experiencing temperatures that were recorded already over half a century ago at places located up to 500 kilometres to the south.

To be sure, as a purely statistical approach this type of analogue study comes once again with major drawbacks, mainly because local geographic conditions are voluntarily neglected and temperature and precipitation remain the sole important variables considered. However, as a *historical* approach, it opens an important field of investigation which can prove fruitful in the future. Indeed, the historical perspective – just as attempted by this *longue durée* visual perspective on climate and its changes – invites us to think more deeply about *climatic difference* and to trace the changing values and functions attached to climates, the geographical imaginations and moral assumptions mobilised when referring to them, as well as the cultural associations that go with them. As we have seen, the visual perspective can be crucial here, revealing largely ignored geographical traditions, showing also that the dominant geophysical approach grew gradually out of far broader investigations into climate and its changes and finally that moral assumptions were often closely attached to climate classifications and that these considerations are not likely to go away in current heated debates on the refugee crisis and the idea that the 'tropics' are inviting themselves to 'us'.

Moreover, and this is a second, closely connected key insight, the *longue durée* visual perspective reveals that throughout history, climate visualisations were accompanied by a fundamental aesthetic and educational desire making the eye the primary organ for knowledge construction and pedagogical reasoning. As we have seen above, this tradition reaches from Macrobius to Woodbridge and Humboldt and stretches even to the most recent climate classifications and their projections. This begs of course the important question about the role that the visual can and should play in the construction and communication of climate knowledge and its imaginings.

The pessimistic observer may claim that after all, climate science has already brought about some of the most robust scientific climate change icons we are all familiar with today, such as the Keeling curve, showing ever-rising carbon dioxide levels since the International Geophysical Year (1957–1958). Yet even the Keeling curve – made up of its two powerful storylines of a 'breathing Earth' and an atmosphere that is becoming less and less respirable – failed to induce long-term political action, just as most of the other famous climate change icons failed too (Howe 2015).

There is of course no simple answer to this dilemma, nor is there a single explanation that could account for the past failures (but see Grevsmühl 2017). However, as I have argued elsewhere, some of these issues are at least partly due to the fact that most global environmental icons struggle to become meaningful in everyday living (Grevsmühl 2016). They remain 'psychologically sterile' because they promote a very specific worldview that is clearly not shared by everyone, or to put it in other words: the 'view from everywhere' (Hulme 2010) rarely coincides with the 'view from everyone'. One major bias is that there are privileged sites for detecting global environmental change and not all individuals, institutions or states have the same access to the tools and infrastructures that allow to create global environmental knowledge, nor do they consume this knowledge in the same way. In particular in hindsight of recent political developments and a general crisis of trust, accompanied by calls for more democratic participation processes, the pessimistic observer may therefore conclude that the importance of the visual is overstated and that one urgently should look for other alternatives based on more inclusive knowledge-construction processes.

Yet recent developments show that this pessimism is perhaps unfounded. Indeed, many of the recent initiatives within climate change research are well aware of at least some of the above-mentioned shortcomings. As a potential first step in the right direction, the Finnish example discussed above of shifting Köppen climates mapped at a European level shows that the visual can contribute to making climate change meaningful for local populations by taking into account at least to some degree local geography – which is of course all about the complex, local interactions between societies and nature. Bringing climate change so to speak 'back to the ground', with the help of, for instance, more regional or local mapping approaches, is therefore probably one of the greatest challenges but also one of the most promising contributions the visual domain will offer in the immediate future. Yet the full potential of the visual will most certainly only be attained by including a historical perspective that invites challenges to general cultural assumptions associated with changing climates. Thus, the predictive tools of scientific and political analysis will only become truly meaningful if equipped with fundamental reflections on how the past confronted the climate question.

After all, in light of recent discussions concerning the notion of the Anthropocene, of the need to connect both the natural history of humans as a species and their cultural history as a more differentiated historical, economic and thus political account of humanity (Chakrabarty 2009), climate is, as observed above through the eye of the visual, a unique opportunity to rethink from a *longue durée* perspective the relationship not only of humankind and nature but also of culture and climate.

Acknowledgements

The author would like to thank his colleagues of the environmental history research group at EHESS (GRHEN) for thoughtful critiques, comments and ideas as well as the editors for their valuable support, guidance and suggestions.

References

Altmann, A. and Schramm, L. 2005. Judah Halevi's theory of climates. *Historical Studies in Science and Judaism*, 5, 215–246.

Arago, F. 1858. *Sur l'état thermométrique du globe terrestre. Œuvres complètes, tome 8*, Paris: Gide, 184–646.

Beniston, M. 2014. European isotherms move northwards by up to 15 km year^{-1}: Using climate analogues for awareness-raising. *International Journal of Climatology*, 34, 1838–1844.

Berlinghieri, F. 1480. *Geographia*, Firenze: Nicolo Todescho.
Owen Box, E. 2016. *Vegetation Structure and Function at Multiple Spatial, Temporal and Conceptual Scales*, New York: Springer.
Brönnimann, S. 2002. Picturing climate change. *Climate Research*, 22, 87–95.
Cellarius, C. 1731. Veteris Orbis Climate ex Strabone, map 1:50 million, Leipzig.
Chakrabarty, D. 2009. The climate of history: Four theses. *Critical Inquiry*, 35(2), 197–222.
Corneille, T. 1708. *Dictionnaire universel, géographique et historique*, tome 1, Paris.
Cosgrove, D. 2007. Mapping the world. In Akerman, J., and Karrow Jr., R. (eds.), *Maps: Finding Our Place in the World*, Chicago: University of Chicago Press, 65–116.
Delaney, J. 2012. *First X, Then Y, Now Z: An Introduction to Landmark Thematic Maps*, Princeton: Princeton University Library.
Diderot and d'Alembert. 1753. *Encyclopédie ou dictionnaire raisonné des sciences, des arts et des métiers*, vol.3, Paris.
Fleming, J. 1998. *Historical Perspectives on Climate Change*. Oxford: Oxford University Press.
Fleming, J. and Jankovic, V. 2011. Revisiting *Klima*. *Osiris*, 26(1), 1–15.
Fries, L. 1541. *Typus Orbis Descriptione Ptolemaei*, Vienna: Gaspard Trechsel.
Geiger, R. 1961. *Köppen-Geiger, Klima der Erde*, map 1:16 million, Gotha: Klett-Perthes.
Golinski, J. 2007. *British Weather and the Climate of Enlightenment*, Chicago: University of Chicago Press.
Grevsmühl, S. V. 2017. A visual history of the ozone hole: A journey to the heart of science, technology and the global environment. *History and Technology*, 33(3), 333–344.
Grevsmühl, S. V. 2016. Images, imagination and the global environment: Towards an interdisciplinary research agenda on global environmental images. *Geo: Geography and Environment*, 3(2), e00020.
Grevsmühl, S. V. 2014. The creation of global imaginaries: The Antarctic ozone hole and the isoline tradition in the atmospheric sciences. In: Schneider, B., and Nocke, T. (eds.), *Image Politics of Climate Change*. Bielefeld: Transcript Verlag, 29–53.
Grove, R. 1995. *Green Imperialism: Colonial Expansion, Tropical Islands, Edens, and the Origins of Environmentalism, 1600–1860*, Cambridge: Cambridge University Press.
Hallegatte, S., Hourcade, J.-C., and Ambrosi, P. 2007. Using climate analogues for assessing climate change economic impacts in urban areas. *Climatic Change*, 82, 47–60.
Halley, E. 1686. An historical account of the trade winds.... *Philosophical Transactions of the Royal Society of London*, 16, 153–168.
Heymann, M. 2010. The evolution of climate ideas and knowledge. *WIREs Climate Change*, 1(4), 581–597.
Hiatt, A. 2007. The map of Macrobius before 1100. *Imago Mundi*, 59(2), 149–176.
Honigmann, E. 1929. *Die sieben Klimata*, Heidelberg: Carl Winter's Universitätsbuchhandlung.
Howe, J. 2015. This is nature, this is un-nature: Reading the Keeling curve. *Environmental History*, 20(2), 286–293.
Hulme, M. 2015. Climate and its changes: A cultural appraisal. *Geo: Geography and Environment*, 2(1), 1–11.
Hulme, M. 2010. Problems with making and governing global kinds of knowledge. *Global Environmental Change*, 20(4), 558–564.
von Humboldt, A. 1813. 'Des lignes isothermes et de la distribution de la chaleur sur le globe', *Mémoires de physique et de chimie de la Société d'Arcueil*, vol.3.
IPCC 2001. *Climate Change 2001: The Scientific Basis. Contribution of Working Group I to the Third Assessment Report of the Intergovernmental Panel on Climate Change*, Cambridge: Cambridge University Press.

Jankovic, V. 2000. *Reading the Skies: A Cultural History of English Weather*, Manchester: Manchester University Press.

Jylhä, K., Tuomenvirta, H., Ruosteenoja, K., Niemi-Hugaerts, H., Keisu, K., and Karhu, J. A. 2010. Observed and projected future shifts of climatic zones in Europe and their use to visualize climate change information. *Weather, Climate and Society*, 2, 148–167.

Köppen, W. 1984. Die Wärmezonen der Erde. *Meteorologische Zeitschrift*, 1, 215–226.

Littrow, J. J. 1832. *Über den gefürchteten Kometen des gegenwärtigen Jahres 1832 und über Kometen überhaupt*, Wien: Carl Gerold.

Livingstone, D. N. 2002. Race, space and moral climatology: Notes toward a genealogy. *Journal of Historical Geography*, 28(2), 159–180.

Locher, F. 2008. *Le savant et la tempête: Etudier l'atmosphère et prévoir le temps au XIXe siècle*, Rennes: Presses Universitaires de Rennes.

Locher F. and Fressoz, J.-B. 2012. Modernity's frail climate: A climate history of environmental reflexivity. *Critical Inquiry*, 38(3), 579–598.

Mahony, M. 2016. Picturing the future-conditional: Montage and the global geographies of climate change. *GEO: Geography and Environment*, 3(2), e00019.

Manabe S. and Holloway, J. L. 1975. The seasonal variation of the hydrologic cycle as simulated by a global model of the atmosphere. *Journal of Geophysical Research*, 80, 1617–1649.

Martin, C. 2006. Experience of the new world and Aristotelian revisions of the earth's climates during the renaissance. *History of Meteorology*, 3, 1–15.

Mauelshagen, F. 2016. Ein neues Klima im 18. Jahrhundert. *Zeitschrift für Klimatologie*, 1, 39–57.

Maury, M. F. 1907. *New Elements of Geography*, New York: American Book Company.

Regnauld, H. 2016. Coastal landscape as part of a global ocean: Two shifts. *Geo: Geography and Environment*, 3(2), e00019.

Rheinberger, H.-J. 2001. Objekt und Repräsentation. In: Heinz, B., and Huber, J. (eds.), *Mit dem Auge denken*, Zürich: Edition Voldemeer, 55–61.

Rubel, F. and Kottek, M. 2011. Comments on: 'The thermal zones of the Earth' by Wladimir Köppen (1884). *Meteorologische Zeitschrift*, 20(3), 361–365.

Rubel, F. and Kottek, M. 2010. Observed and projected climate shifts 1901–2100 depicted by world maps of the Köppen-Geiger climate classification. *Meteorologische Zeitschrift*, 19(2), 135–141.

Ruscelli, G. 1561. *Ptolemaei Typus*, Venice.

Sanson, N. and Sanson, G. 1697. *Cartes et tables de la géographie ancienne et nouvelle*, Paris.

Schneider, B. 2016. Der 'Totaleindruck einer Gegend': Alexander von Humboldts synoptische Visualisierungen des Klimas. In: Ette, O. and Drews, J. (eds.), *Horizonte der Humboldt-Forschung*, Hildesheim: Georg Olms Verlag, 53–78.

Schouw, J. F. 1827. Über die vermeintliche Veränderung der klimatischen Verhältnisse Dänemarks und der benachbarten Länder und über die Periodicität dieser Verhältnisse. *Hertha, Zeitschrift für Erd-, Völker-, und Staatenkunde*, 10, 307–353.

Schulten, S. 2010. *Mapping the Nation: History and Cartography in Nineteenth-Century America*, Chicago: University of Chicago Press.

Venetz, I. 1833. Mémoire sur les variations de la température dans les Alpes de la Suisse. *Mémoires de la Société Helvétique des Sciences Naturelles*, 1(2), 1–38.

Ward, R. 1908. *Climate, Considered Especially in Relation to Man*, London: John Murray.

Ward, R. 1905. The climatic zones and their subdivisions. *Bulletin of the American Geographical Society*, 37(7), 385–396.

Wilford, J. N. 1981. *The Mapmakers*, New York: Knopf.

4

Indigenous Knowledge Regarding Climate in Colombia
Articulations and Complementarities among Different Knowledges

ASTRID ULLOA

4.1 Introduction

Global climate change and subsequent mitigation and adaptation processes are issues that transcend local contexts and bring together different knowledges, ideas and practices related to territory, landscapes, ecosystems and the environment, both human and non-human. Climate in particular brings together different cultural conceptions and different ways of knowing, interpreting, perceiving, representing, acting and reacting in relation to the weather, climatic phenomena and climatic variability. These notions respond to practices, daily experiences and relationships with beings that exist in the material and symbolic environment. They are also related to the ways in which diverse cultures interact with nature, which implies that there are multiple notions of nature that coexist in relations of confrontation, complementarity or inequality. Global climate change and its causes, effects and adaptation mechanisms are closely interconnected with culture (Crate 2008, 2011; Crate and Nuttall 2009a, 2009b; Heyd 2011; Ulloa 2011a, 2011b).

However, global actions against climate change have been backed by a single type of knowledge. The notions that support the global policies of climate change are based on the concepts of the Intergovernmental Panel on Climate Change (IPCC), which interprets climate based on scientific constructions of natural processes (Rossbach 2011). Likewise, the IPCC responds to specific ways of producing knowledge that are reflected in the implementation of programmes and actions based on this knowledge (Baucon and Omerlsky 2017; Castree 2017). This global knowledge not only has political implications but also makes a local impact because it is implemented via a global policy that does not include local worldviews and knowledges. Climate change global discourse could be considered, as Heikkinen *et al.* (2016) state, as an ontological politics that refers 'to global discourses as human artifacts that instrumentalize the social and political way of problem framing by including certain epistemic assumptions and standardized methods for knowledge production' (2016:214).

Local inhabitants (indigenous, Afro-descendant and peasants) have their own views on weather and other non-human elements of the environment. They demonstrate the intricate articulations of these conceptions, as they relate to climate, through music, territory, health, body, time, agriculture and dreams, among others. Indigenous peoples' knowledge of meteorological phenomena and climate has allowed them to manage its impact through many generations. This knowledge is also closely related to gender, age, specialisation, location and environmental transformations, including climate change, and has occurred historically. Indigenous peoples, like other societies, have generated strategies to manage and confront environmental transformations. For example, in times of drastic changes in climate, indigenous knowledges related to practices of agricultural diversity have allowed the management of food production, the generation of alternative crops in diverse ecosystems and the creation of exchange networks with other peoples. Indigenous peoples have the smallest ecological footprint, but the effects of climate change are, and will be, more evident in their territories due to different circumstances, such as: their close interrelation with the non-human environment; the fragility, biodiversity and endemism of many ecosystems where they live; the loss of territories by forced displacement and violence; the refusal to recognise their rights and the little influence and participation in political decision-making arenas regarding global and national policies of climate change that affect their territories and their lives.

Approaches to local knowledge related to climate management and prediction strategies, as well as discussions related to vulnerability, are basic to understanding the cultural dimensions of climate (Crate 2008, 2011; Ulloa 2008; Crate and Nuttall 2009a, 2009b). However, there is a lack of knowledge about the ways of knowing of indigenous peoples and the climate variability management strategies that they use. To account for these strategies, it is necessary to start from local knowledge situated in both specific and sacred places. These are spaces of meeting and memory; they intersect with everyday experience and are embedded in practices that reinforce indigenous peoples' relationships with the non-human elements in their lives.

Even though proposals have been made for the articulation of various knowledges to address climate change, attempts at intercultural and interdisciplinary agreements continue to face problems regarding comparisons with specialised science in terms of indicators, scales and ways of systematising knowledge. There also exist methodological differences surrounding the variables to be considered, especially in relation to prediction processes, as in the case of climate scenarios and the local ways of predicting their impact. Likewise, the dialogue between specialised scientific and indigenous knowledges is presented in terms of expert knowledge based on a particular ontology, which cannot be homologous to other ontologies, such as that of indigenous peoples. However, indigenous peoples question the imposition of a geopolitics of knowledge which does not include them and at the same time demand acknowledgement of their own position in terms of their knowledges. In a similar way, indigenous peoples consider that their exclusion from the global strategies agenda is a problem of climate injustice. These situations reveal that the existing processes of articulation between knowledges have not allowed for the positioning of different ontologies and have maintained the expert scientific tradition.

In this context, I argue that local knowledges, given their ontologies, epistemologies and gender differences, require a different understanding based on territorial relationships and ways of producing knowledge. Therefore, it is necessary to recognise other cultural ways of approaching knowledge. To achieve this, I propose that a new way of analysing climate should be created based on local knowledges and territorialised knowledge and highlighting local peoples' ways of producing knowledge, including a gender perspective based on their ontologies and relational epistemologies.

This analysis is based on two phases of fieldwork, which were conducted with a team of indigenous and non-indigenous researchers. The investigations were developed in urban, rural and indigenous contexts in Colombia between 2005 and 2013. However, I focus here on the perspectives of indigenous peoples, which emerged from collaborative research with indigenous researchers in their territories. Their perspectives are presented in sections of this text to show their voices (from ethnographic fieldwork, interviews and representations), the complexity of cultural perspectives and their practices in relation to the notions of meteorological change and climate in specific contexts.

This chapter is divided into four sections. The first section presents a discussion on how indigenous knowledge has been approached regarding climate. The second section presents the dialogic strategies that exist between different knowledges, the surrounding ways of approaching and understanding climate variability, the predictions, the joint strategies used to face vulnerability and the problems and disagreements that have emerged. The third section explores a proposal for understanding indigenous knowledges based on their territorial relationships and gender differences, highlighting the perspective of indigenous women, drawing on Pasto[1] and Kamëntsá Biyá.[2] The final section

[1] Specifically, Pasto people who live principally in Panan, in the Department of Nariño, Colombia.
[2] Specifically, Kamëntsá Biyá people who live in the Valle de Sibundoy, in the Department of Putumayo, Colombia.

discusses the scope of a climate perspective that arises from the recognition of other ontologies and epistemologies and the implications that this might have for indigenous peoples.

4.2 Indigenous Knowledge on Weather and Climate

Since ancient times, indigenous peoples have developed systems of knowledge and interaction with nature. For example, rainbows, different kinds of precipitation, atmospheric transformations such as frost and climatic cycles such as winters and summers are considered living beings and perceived and interpreted according to territorial and cultural particularities. The conceptions about weather and climate are culturally constructed; in some cultures there are no considerations of weather or climate other than as living beings. These conceptions are related to knowledges that are not transmitted as rules or formulas; they are affected by everyday experience, immediate perception, accumulated observation, beliefs or generalised opinions in the social environment which are transferred through oral communication and the relationship of their specific practices with the climatic conditions at particular times. Weather and climate are culturally assigned values, which allow analysis of everyday experiences and environmental transformations.

Multiple analyses have been done on indigenous peoples and their relation to climate, highlighting the conceptions, ritual practices and symbolism associated with specific meteorological phenomena. For example, rain, hailstorms, clouds or rainbows and their effects on health and disease are recurrent themes in various cultures (Goloubinoff *et al.* 1997). In this sense, Lammel (1997) explains that the most important meteorological phenomena for the Totonac people of Mexico are rain, storms and thunder, which are associated with the supernatural world, while the rainbow is attributed to a negative entity. Sierra (2011) explores how relationships between the Kaggaba, U'wa and Misak peoples of Colombia and the climate primarily revolve around the rains, the rainbow and the thunder, which come from the complex relations these peoples have with the water and their specific cosmovisions. These expressions and actions of cultural beings are affected by climate change, which has repercussions for the cultural practices associated with notions of territory.

The relationships between the manifestations of climate and culture are evident in practices such as annual predictions (e.g. the cabañuelas observation of the atmosphere during January to predict the weather for the year), the elaboration of local calendars, which establish temporal and spatial relationships with climatic effects, and local predictions based on observations. In relation to predictions, there is the case of the peasants of Peru, who at the end of June observe the departure of the Pleiades star cluster to predict the weather and the amount of rainfall (Orlove *et al.* 2004). Likewise, there are indigenous peoples that associate the climate with astronomical phenomena, such as the presence or absence of stars or planets at specific times. For example, among the Coras of the Sierra Nayarit, the deities that represent cold weather are associated with the planet Venus (Sprajc 1997).

The relationships between culture and environmental management in conditions of climatic variability or in extreme ecosystems (e.g. deserts) are evidenced in the links with water, specific species or meteorological changes. Among the Wayuu people in Colombia, the moon, which is a masculine being, and its changes allow the prediction of weather variations. If the moon is hidden at, for instance, seven o'clock at night towards the southwest, it means that it will not rain during the spring (Iguarán 2011).

Similarly, in the face of minor changes in ecosystems, there are cultural knowledges that show the complexities and subtleties of the relationship with the environment, as is the case in the African societies (such as the Gabaya, Komas and Tikar peoples) analysed by Dounias (2011), where their relation to subtle climatic changes and bioindicators, such as insects, reveals a deep understanding of the fragility of ecosystem balances. Dounias states, 'Insects are particularly relevant biotemporal indicators, since they are capable of reacting to minute alterations of climatic conditions, in thresholds that the human being is incapable of feeling' (Dounias 2011:237).

Rituals and symbols are another key aspect of the relationship between climate and culture. Weather and its changes are perceived and represented in diverse images that imply specific practices (performed by specialists or the whole community), for example, representations related to the abundance or scarcity of resources, as well as the power of non-humans, such as the wind or rain. Guaraldo (1997) reports that in Mexico there are mortuary representations that are related to the devastating and lethal power of rolling winds or to the perception of strong and excessive winds. Villela (1997) describes how the Nahua, Mixteca and Tlapaneca indigenous peoples of Mexico develop rituals around supernatural beings associated with the elements, such as rain, to promote agricultural production. These rituals and representations allow us to understand these conceptions and evidence the complex network of meanings between climate and culture.

It is important to highlight the work on cultural bioindicators, specifically in terms of fauna, which allows the prediction of climatic variations. In Mexico, the Nahua predict rainfall based on the behaviour of ants and swallows (Villela 1997). In Bolivia, Yana (2008) describes how, by reading animal behaviours, periods of drought or rainfall can be predicted. Other works relate climatic transformations with religious practices. In Mexico, Hémond and Goloubinoff (1997) analyse climate changes and the relationship between agricultural cycles and religious ceremonies.

In Colombia, studies on climate and weather are not so numerous. However, some researchers have reported indigenous cosmovisions and their relation to lunar cycles, planetary movements, star positions or weather and annual cycles linked to seasonal climate changes (see, for example, for Colombia, Arias de Greiff and Reichel 1987; Pardo 1987; Ortiz 1987; Correa 1987; Ulloa 2011a, 2011b; Ulloa and Prieto 2013). Likewise, there are authors who have investigated, with indigenous researchers, the perspective and interpretation of indigenous views on the meanings of water and the rainbow, for example, among the Guambianos or Misak in Colombia (Aranda *et al.* 1998).

Moreover, studies have been initiated that account for the knowledge of indigenous peoples in general and of women specifically with regard to climate change (Ulloa *et al.* 2008). Works such as that of Yana (2008), a Bolivian Aymara indigenous researcher, have begun to analyse and systematise climate bioindicators and indigenous perceptions according to gender. In her analysis she finds that there are differences in practices in relation to climate variability. Men react by changing their management responses to agriculture, while women create symbolic strategies to maintain the balance between humans and non-humans. For example, in terms of prevention strategies against drought, men say: 'when it snows in a dry place or in the mountain range, then we predict that we have to sow in humid places and sow little in dry places, because we know that we are going to lose', while the women say: 'we make a dish (offering of sweets, herbs, colored wool, etc.) [to offer to beings related to climate] ... with that we only defend ourselves' (Yana 2008:50).

Among the works from an indigenous perspective are a number of texts by Zonia Puenayán (2009, 2011, 2013), who analyses the effects of climate change on agriculture and the climate-related cosmovisions of the Pasto people. Also, Catherine Ramos, along with indigenous researchers Ana Tenorio and Fabio Muñoz (2011), investigated the conceptions of the Nasa in relation to climate changes, agricultural calendars and their relationship with the perceptions of cold and hot: 'The composition of the world in cold and hot is then the main feature of the Nasa worldview, which is reflected in its classification of the world of plants, diseases, soils and the body' (Ramos *et al.* 2011:254).

In terms of the effects of climate change, indigenous knowledge is also considered in relation to specific places and their historical transformations. For example, Tupaz and Guzmán (2011) show how, in their Pasto culture, the climate is formed by ancestral elders (sun, rain, rainbow, snow, frost) and their signals are expressed in their territory. In this way, the Pasto territory is related to memory and transformations that allow us to understand the behaviour of the ancestral elders.

In general, the examples presented above describe how indigenous peoples have conceptions and practices related to weather and climate. However, these examples do not present the way that indigenous peoples produce knowledges and why it is necessary to consider other ontologies to understand the diversity of relationships among humans and non-humans, including those related to climate.

The diversity of cultural relations with climate evidences multiple responses and proposals in the face of climate change. At the same time, cultural diversity and associated knowledges, perceptions, representations and practices are related to the visions of national and global discourses and policies on climate change, and this implies the need for dialogue with intent to establish intercultural relationships.

4.3 Dialogue between Different Climate Knowledges

The debates led by some social researchers about climate change, and the proposals of indigenous peoples to be part of the discussions and policies related to it, have opened new scenarios in which global and national policies of climate change try to include indigenous knowledges. In this process, experts and disciplinary knowledge began to include indigenous knowledge in order to establish relationships and comparisons in search of complementarity and integration and to contribute to the understanding of environmental and climatic problems (Nazarea 2006; Ulloa 2008). In these contexts, there have been several processes of interrelation, dialogue or complementarity of knowledges. These processes have been implied for indigenous peoples' different situations. For example, indigenous peoples have positioned their knowledges, and their demands for the recognition of these knowledges, in local proposals for climate variability management, which implies their inclusion. Indigenous peoples have managed to position their demands and knowledge using global terms and concepts, such as climate change and adaptation, which could be understood as hybrid natures or climate syncretism, as analysed by Rossbach (2011) and Cantor (2015). Indigenous peoples, in collaboration with researchers from different disciplines, have generated interdisciplinary and intercultural analyses that have allowed knowledge exchange related to local ways of predicting seasonal changes, rainfall or climatic variability. These strategies of interdisciplinary and intercultural dialogues have allowed a better approach to local processes related to climate (Ulloa 2013).

However, intercultural and interdisciplinary proposals present general problems, such as decontextualisation of knowledge, since they are converted into data that are not analysed in relation to specific contexts. This decontextualisation of knowledge is reflected in the processes of interrelation and/or comparison between indigenous knowledge and expert knowledge (Rossbach 2011; Reyes-García *et al.* 2016), which entail the following problems:

- Difficulty in comparisons in terms of indicators, scales and ways of systematising knowledge
- Methodological differences around the variables to be considered in relation to prediction processes, which is the case of climate scenarios and local ways of prediction
- Diverse temporal and spatial dimensions
- Prioritisation of expert knowledge based on a particular ontology
- Non-inclusion of gender and cultural differences
- Predominance of a single way of understanding the relationship between nature/culture
- Impositions of notions of risk or vulnerability focused on humans
- Inclusion of some practices and knowledges as requirements without actually being the basis from which climate change policies are developed

On the other hand, global policies and programs on climate change have intended to include indigenous knowledges. However, in practice, global policies privilege a single vision and knowledge and their application privileges expert knowledge in order to implement programs into local processes, in particular for indigenous peoples. This perspective disregards indigenous knowledges given that indigenous relations with the non-humans as living beings are not taken into account.

The aforementioned processes of articulation between knowledges have three fundamental problems: the notions of nature, i.e. what kind of ontological relationship is established between humans and non-humans; the understanding of what knowledges are and how they are produced from the specific ontology in which they are immersed and the lack of recognition of differences among different cultures related to age, localisation and especially gender differences.

Faced with the first problem, it has been established in the social sciences that there are diverse notions of natures (interrelated, separate, coproduced, among others) but that one has become hegemonic, the dual conceptualisation of nature/culture which is a reflection of Western thought or modern ontology (Escobar 2015; Ulloa 2016). This supremacy of Western thought has not allowed the positioning of other ontologies, nor the clear understanding of cultural diversity and the categories related to nature or climate. This leads to an inability to establish commensurability and equality between different categories and variables in the analysis of climate change.

Even though these problems are current, there have been conceptual streams, like the ones centred in the Anthropocene, which have opened new discussions and positioned the problem of climate change under a reconceptualisation of the duality of nature/culture in the social sciences. This has also led to a conceptual, methodological and political rethinking of these dualities. This new wave allows for academic knowledge to influence national/global decision-making processes and has opened the way to new ways of producing knowledge (Chakrabarty 2009; Latour 2013, 2014; Emmet and Lekan 2016).

However, unique geopolitics of knowledge remain present. These view climate change as a global problem that requires global responses and which erases historical relations of power and inequalities that have led to said transformations. The current process surrounding climate change reflects the environmentalism of the 1970s and 1980s that, even if it gave rise to different positions, tendencies and conceptions, in the end created a unified response, an ideal vision and a global proposal for sustainable development (Leff *et al.* 2003; Leis and Viola 2003). Now, in the face of climate change, the global solution becomes the responsibility of all citizens of the planet, centred on a unique vision of nature (reconfigured) and its management based on expert knowledge (Liverman 2015; Ulloa 2017). However, this whole process, which is global, gives rise to a series of actions and solutions focused on specific actors at an international level (COPs, IPCC) involving centralisation and, therefore, control over the production of global knowledge about climate change (Watts 2015; Heikkinen *et al.* 2016). It is a repositioning of an Anglo-Eurocentric episteme grouped in the 'rethinking' of duality which supposes to generate a new configuration (but based on dual categories) of the geopolitics of knowledge production. However, modern thought appears not only as the centre of the cause but also as the solution, and this implies that other ontologies are not included (Ulloa 2017).

The effect of this is the consolidation of a vision that emerges from the expert knowledge found in the universities of countries that lead the centres of academic production in Europe and the United States. These centres establish rules ranging from what should be done and how to solve the problems of climate change, or the Anthropocene, to what suitable political options are possible. In this sense, the geopolitics of knowledge are reconfigured as they become centred on the production and legitimacy of the experts. Therefore, the global discussions related to the Anthropocene do not include other forms of knowledge production related to climate change, as is the case of indigenous, Afro-descendent and peasant perspectives in Latin America.

Moreover, although climate change is presented as the result of human activities on the entire planet, its causes are not completely broken down, leaving aside the diversity of options and knowledge, as well as the unequal power relations. This leads to a lack of recognition of other ontologies and epistemologies.

The second problem implies that what we understand as knowledge differs when we think of indigenous peoples. For these groups, knowledge is interrelated among beings and is produced when beings are embedded in territory which is, at the same time, alive (Ingold 2000; Escobar 2015; Ulloa 2015, 2016). Generally, in climate change policies there is an absence of local knowledge and territorial situatedness, given that subjectivities, identities and practices around nature and its specific transformations, and in particular places, are not given equal status with expert knowledge. In other words, these policies do not include other ways of thinking about the relationships between humans and non-humans, for example, those which are mediated by diverse social, political and economic relations that, although articulated to global processes, respond to other ontologies. These other ontologies allow us to rethink political relations and actions associated with climate change and to position other points of view and strategies for their confrontation.

Finally, there is a lack of knowledge regarding gender inequalities, given that analyses on climate change and knowledge reflect very little on gender relations and do not recognise a deep differentiated knowledge between men and women, even in scientific knowledge. This is evident in the production of knowledge, as well as in the absence of local knowledge in global climate change policies. Moreover, global policies of climate change affect indigenous men and women differently, but the general polices are implemented without a gender perspective. For example, differentiated knowledge between men and women is not included, nor are the ways in which women manage climate variability. These processes have been questioned in international spaces, which in turn has allowed climate change policies, and results attributed to those policies and programs, to partially include women, although without a direct focus on indigenous women. Furthermore, access to information is minimal for various peoples who speak other languages, or who cannot access or understand the scientific knowledge produced. These exclusions are more evident for women (Iniesta-Arandia *et al.* 2016).

Given that global and even national policies on climate change do not contain gender differences, international organisations have initiated programs aimed at the inclusion of women. These range from participation processes to the visibility of the impacts of climate change and the search for funds to develop programs that respond to specific implications of climate change for women (Aguilar *et al.* 2009; FMICA 2010; Jungehülsing 2012; Davis *et al.* 2015). Based on the criticism of the absence of gender analysis in climate change research, various approaches have been developed that call for overcoming stereotypes or the mechanical inclusion of women in order to compensate for gender inequalities (Röhr 2007). However, in the scenarios of climate change, participation of indigenous women is very limited, as is their access to information, so they demand their inclusion in the discussions and in climate justice, in addition to the demands for the reversal of the causes of climate change.

The politics of climate change and the approaches of social sciences that centre on the Anthropocene have not changed, nor will they change, the unequal relationships in terms of the production of knowledges or gender inequalities. Therefore, indigenous peoples' knowledges, and their perspectives on the non-human, have become the centre of a permanent struggle to propose another way of rethinking climate change.

The previous analyses show the need to create new ways of approaching indigenous perspectives and their particular way of producing knowledges on climate change as a starting point. A dialogue needs to be established, not only among knowledges, but in terms of understanding the different ontologies and epistemologies that arise to really face the processes of global climate change, as well as the gendered process of the production of knowledges in specific cultural contexts.

4.4 Processes of Production of Knowledge among Indigenous Peoples

The previous conclusions lead to the need for a new proposal that articulates knowledges and recognises cultural and knowledge diversity regarding climate. To do this, we must first fully understand indigenous knowledge production processes. This proposal will approach local knowledges (indigenous knowledge or traditional environmental knowledge) by understanding them in a contextualised way and from their inherent worldview and dynamics, where the producers of such knowledge are seen from their capacity for action and according to gender. This leads to a positioning of other ontologies and relational epistemologies around climate.

Indigenous peoples maintain relations with the non-human; these relate to territories and cultural practices that respond to specific and localised notions of nature. The relationships established can be reciprocal between the human and the non-human (the latter understood as a category that can include diverse beings as related) and mediated by the use, access, control, rights (of all beings) and collective or individual decision-making that articulates diverse cultural and territorial dimensions. In specific places, these processes are evidenced by representations, classifications and practices according to gender. Places are those in which being and doing relate to each other according to the cultural and territorial places occupied by different living beings. Likewise, all these practices imply other notions of care given that, for indigenous peoples, the care of life includes the human and non-human, which in turn is part of indigenous knowledge fluidity and which responds to cultural principles of being and doing in specific places. This is called ancestral law, or laws of origin; these are the norms that relate with the non-human.

Ancestral law structures indigenous thinking and establishes forms of political and cultural control over the human and non-human through cultural governability. Indigenous peoples, therefore, demand their autonomy, which is related to territorial order and control, environmental management and food sovereignty, i.e. environmental self-determination. This cultural governability is based on cultural practices in accordance with each indigenous context.

For example, for the Pasto people, the territory is seen 'as a living being, a being that, according to the law of origin, feels, listens, produces sounds, gets sick and is restored' (Puenayán 2013:276). In the law of origin, the territory is sacred and a person must act in accordance with the norms of the said law in order to guarantee autonomy and governability of his or her territories.

By acting in accordance with their ancestral laws, indigenous peoples are seen as environmental authorities that take care and control of their territories/natures and food sovereignty. This leads to autonomous environmental management as a basic axis in the current demands of indigenous peoples, responding to the great territorial pressures and imposition of models of relationships with nature that have fragmented the territories and affected the knowledge related to ecological and food practices.

The principles that underlie the law of origin, or ancestral law, are part of the elements that articulate the ways of thinking and practices of indigenous peoples. Like the Pasto people, several indigenous peoples, such as the Kamëntsá Biyá people, demand recognition of their law of origin, given that actions such as those associated with climate cannot be imposed from outside but should take into account the cultural processes related to the ways of producing knowledge and the places where such knowledge is produced.

Under these notions of ancestral law, climate is often not a category, or something 'natural', but is the result of relations between humans and non-humans. Moreover, knowledge processes differ between men and women. In general, practices of indigenous women are not very well documented. Therefore, production of knowledge related to climate has to be contextualised historically and culturally.

For indigenous peoples, knowledge is differentiated between men and women. Here, the focus is on the knowledge of women. For them, care and continuity of life, expressed through the physical, the social and the territorial, as well as the non-human, are very closely related to the daily changes in weather. I follow the categories that I have developed in other texts in relation to the process that implies the production of knowledges among indigenous peoples. These involve, from situated and historical perspectives, ways of knowing; places of knowledge; spaces of power; words of power; life practices and local methodologies for research from their own viewpoint (Ulloa 2016, 2018a, 2018b). I focus specifically on Pasto and

Kamëtsä Biyá women in order to show how the production of knowledge related to non-humans is gendered.

4.5 Perspectives of the Pasto and Kamëtsä Biyá Women

The indigenous women of the Pasto and Kamëntsá Biyá peoples have a particular way of relating to the non-human, based on ancestral law, which implies a process of closeness and knowledge of specific places and beings and which allows for ways of caring that involve the body, the non-human and the territory. These complex networks of territory and the beings that inhabit them mean any transformation and degradation of their territories/environments, from internal or external causes, are perceived and felt by the women in a more immediate way.

In particular, for the Pasto people, environmental changes are affecting biodiversity, which implies a loss of knowledge. This leads to transformations of the practices of both men and women and the social networks that are established around territory/nature. For Pasto men, environmental scarcity has implied greater mobility as they search for new economic options, and this affects the processes of care of the human and the non-human. In the face of these environmental and sociocultural situations, women have created alternatives related to the consolidation of their networks in the recovery of knowledge, particularly those concerning seed and care practices, which have led to new gender relations in the communities (Puenayán 2013). Pasto women have organised themselves into collective activities called *mingas de pensamiento* (to think collectively) in order to strengthen the cultural and ancestral practices related to recovery of seeds and to exchange them as a way to maintain biodiversity and food sovereignty.

For the Kaméntsá Biyá people, environmental transformations and climate change are 'making the earth sick', leading to loss of practices (such as planting in spirals and having polycultures), crops and seeds (Juajibioy and Cantor 2013). They also point out that cultural change and its effects have transformed the territory, e.g. changes in traditional farming practices caused by the new surge of monocrops. In the same way, they point out that generational changes and the migration of younger generations also cause transformations in their territories. The women's response has been to return to practices like the *chagra* (area of cultivation) because they consider that, in a *chagra*, life relationships are learned among mothers and children, as Concepción Juajibioy states:

The chagra is like a family, some plants help each other and protect each other, to maintain balance all plants must be diverse, in the family they all help each other, the same happens with the plants, they take care of each other, each plant has its place next to the others, plants stick and grow in one place and not in others, they have their specific place, in the chagra one becomes closer to the children, that's where we learn the main values of respect and care towards others. (Jaujibioy cited in Cantor and Juajibioy 2013:163)

In the face of these transformations, the Pasto and Kamëntsá Biyá women have generated alternatives linked to the processes that involve the production of knowledge, which are reflected in daily practices related to food sovereignty.

In these contexts, food sovereignty, or food autonomy as it is also named, is directly linked to life in a territory and with the process of permanently maintaining the circulation of life since it is associated with all beings. It is also related to social processes and relationships between men and women, both individually and collectively, which establishes a social order. These processes involve not only production, reproduction and circulation of food but also knowledge that is reflected in the territory and in everyday and symbolic practices.

To understand these dynamics, it is important to explain how indigenous women produce the knowledge which circulates in specific places through words and life practices.

4.5.1 Ways of Knowing

Once cultural relations are established with entities or non-human beings related to the climate, there are ways to perform readings that indicate their stability or change. Conceptions, perceptions and local representations about weather and climate are based on indigenous knowledge of the relationships between meteorological and climatic conditions regarding flora, fauna and cultural aspects. These interrelationships allow us to talk about local knowledge systems based on experiences and perceptions of weather and climate, given that local knowledge is related to specific places in which relationships are established with the non-human. Within these approaches, cultural indicators and how they are understood are presented as basic to the relationships between climate and culture and as inputs for local strategies in the face of environmental transformations.

Knowledge about nature and climate is expressed in the management of indicators, i.e. in the reading of signals that are found in various aspects of environmental elements. These can relate to size and colour and to changes in seasons which affect appearance or transformations, among others. Likewise, the cultural knowledge of weather and climate is based on practices, experiences and perceptions, which manifest in the reading of the indicators. The indicators are complex and focus on aspects such as behaviours or reactions of humans and non-humans, and they can be of various types: astronomical, meteorological, biological, physical-geographical, forms of objects, body sensations, sensitive responses or symbolic signals (Ulloa 2014).

For the Pasto people, some of the indicators respond to symbolic elements that are presented through dreams and that express the relationships that must be considered in regard to environmental changes, e.g. as they relate to water scarcity:

[Pasto people consider that] each one of the natural resources is protected by the spirits of nature. [Narcisa tells that] she lived this experience of meeting the spirit of water, which revealed to her in her dreams that the community should protect and take care of the water in the Larga swamp, over the Paramo of the *resguardo* (reserve). (Puenayán 2013:302).

Dreams are very important for Pasto people; through them the ancestral beings related to climate show their annoyance to different people as a result of human practices which affect the water or the forests. Therefore, the person who had the dream has to share the situation with the community to begin the ritual and ask for permission from any ancestral being that is related with the cultural practices that affect them.

4.5.2 Places of Knowledge

Among indigenous peoples, knowledge production processes are associated with specific places within which concepts flow and practices associated with such knowledge are established. Among these places are the cultivation areas in which knowledge is transmitted and circulated following socially established relationships between knowledge and differentiated practices and between men and women. In these spaces, there are interactions between complex processes of transmission between generations and new knowledge that is spatialised.

As an example, I focus on the cultivation sites among the Kamëntsá Biyá people. The *Jajañ*, the *chagra* or space for cultivation, is a place of vital importance and can be understood as a central space in life. It is a place where a great variety of species are planted, guaranteeing food and traditional medicine. The *Jajañ* is the place where the family strengthens its bonds and where social and cultural practices are transmitted; it is the place where the individual reproduces their culture and learns to relate to the non-human and to others. It is also the place where the people grow their own food and medicinal plants, and they cultivate their thoughts and the culture of the Kamëntsá Biyá man and woman (Cantor and Juajibioy 2013). These spaces are an example of the inscribed knowledge in the territory; they are spaces that midwives, for example, can access, allowing for the circulation of their particular knowledge in their daily practices.

In particular, climate change has brought about transformations in these cultivation places, which has caused the loss of seeds and plant varieties, as well as associated practices and knowledge. As Clementina Muchachasoy puts it, 'If you lose the Kamëntsá Biyá thinking, you lose the feeling, the language and the practice, the cultivation of the chagra' (Cantor and Juajibioy 2013:172).

Faced with these dynamics, women have recovered cropping practices, such as spiral cultivation, that use little space, allow the combination of a diversity of species and represent the interrelations between the various worlds of the territory. Thus, the women have recovered ancestral knowledge related to biodiversity. Parallel to this, they have created networks of women to exchange experiences, seeds and strategies and to maintain exchange of knowledge.

4.5.3 Spaces of Power

Sacred sites are spaces that are appropriated and socially constructed according to each culture. They have specific places of decision-making or of symbolic and cultural importance. These sacred sites are places of power and they become increasingly important in environmental processes since the decision-making surrounding the relations between humans and non-humans is carried out in those places.

For Pasto people the territory is a living being that feels, listens, talks, gets sick and recovers. The ancestral beings have inscribed in the territory the history of the Pasto people and structured it into three worlds (top, middle and bottom). The sacred sites are 'where passive and active energies are united, these cosmic energies, the sun and the moon, the spirits and all the other components of nature are interacting in the territory to manifest life' (Puenayán 2013:276). Thus, life manifests in the lagoons, in the spiritual places where men and women meet with ancestral beings. They are spaces of power for memory and knowledge, as well as for relationships with the non-human, and they are reactivated through rituals (Puenayán 2013).

4.5.4 Words of Power

All places and knowledge are associated with the idea of words of power, which implies specialised and precise knowledge about specific relationships with the territory or the non-human. These words of power are linked to various representations associated with these relationships. Words of power representations bring together knowledge and beings that inhabit territories with specific characteristics and qualities; they are named with words that generate action and incidence of said beings in symbolic or everyday processes.

For the Kamëntsá Biyá people, words such as sun, wind, rain or thunder link not only the graphic, sound or tactile representations associated with these concepts but also the symbolic representations that imply power. The word links the knowledge of these beings and generates concrete actions, such as in the territories associated with environmental changes and transformations (Cantor and Juajibioy 2013).

For the Pasto people, traveling through the territory is associated with the word and the circulation of thoughts:

The territory and its natural components are full of ancestral wisdom; the mythical roads and places are interwoven with the words of the elderly through orality in the mingas [collective processes] of thought, mingas of work, and especially in storytelling. (Puenayán 2013, 290)

4.5.5 Life Practices

The above aspects are understood by indigenous peoples, men and women, surrounding ways and practices of life, which are the ways in which local cultures construct possible worlds based on diverse relationships among humans and between humans and the non-human.

For the Pasto people, one of the practices of life has to do with the *mingas*, or collective processes, which seek to strengthen ancestral knowledge and that allow putting into practice the laws of origin in the use and ordering of the territory in accordance with local planning and cultural governability. These processes are also related to the practices of knowing and walking the territory as a process of knowledge (Ulloa 2018a, 2018b).

4.5.6 Methodological Strategies for Knowing

The aforementioned practices can be combined with methodological strategies to enable indigenous peoples to focus on ways of perceiving and acting in accordance with indigenous ways of knowing and suggest cultural methodological tools to understand the environment. These ways of knowing involve knowledge transmitted from generation to generation according to gender and to bodily and cognitive dispositions to be, feel, do and act in relation to all non-humans in the territory. That is the case of Pasto men and women who follow specific criteria for doing research within their territories (Table 4.1).

As evidenced, there are different methodological ways to approach the relationships between knowledge and climate. In summary, we can say that the knowledge related to climate responds to conceptions that each culture has

Table 4.1 *Conceptual criteria of indigenous Pasto research*

Observe	Look at the daily events, look in detail at how I am doing a job or experience, observe the cycles, the times.
Asechar (Wait cautiously for some purpose)	Have the ability to catch a concept, an idea, a strategy.
Alert, prevention	Be aware of something, predict the action (A preventive person is worth two).
Fixate	Concentrate the senses on something you want to know, noticing is different to just seeing.
Attention	Have your senses fixed on the subject, words or work, be very focused.
Contemplate	Part of meditation and seeing the nature and the cosmos.
Be Curious	Part of investigating what draws the attention of the sacred or the mysterious.
Chapar (watch carefully)	Be vigilant of someone who wants to enter a sacred place and, in this case, their ideas or precepts.
Explore	Walk the worlds of mystery or of the unknown.
Try	Make an effort, an idea, a precept, a work, an action, experiment or sow.
Save	Have your knowledge in order, like the sacred archives of the soul.
Inspect	Monitor how jobs and processes are followed, surround, see, plan activities.
Interrogate	Ask: Why? Where? When? Who? What for? Find out. Converse and *minguiar* (act collectively).
Sharpen your hearing	Hear the word of the council, transcribe the message of the voice of conscience. Hear the voice of silence.
Meditate	Place the senses in the silence, recapitulate the past, the present, the premonition, reflect.
Be aware	Be very alert when the seeds are born, when the cycles begin and end.
Take the pulse	Measure the strength of spirit of knowledge and energies, the strength of the willpower with the conscience.
Recognise	Know the three states of water, recognise the natural processes that occur in the sacred territory.
Register- Mark- Signal	Indicate the measurement, mark, *amojonar* (to put a physical signal, usually a stone) when it starts and when it ends, ask how to advance, what to change.

Table 4.1 (*cont.*)

Memorise	Have the ability to keep the precepts, memories, ideas, secrets, mysteries.
Explore	Have precepts, criteria, opinions of others, surveys.
Keep track of	Be aware of something curious.
Touch	Experience the sense of knowing whether something is hard, soft, cold, hot.
Taste	Know if something is sour, sweet, bitter, salty, a friendship, a memory, a job.
Select, pick	Experience the beautiful, the sad, the memory, the idea, the fantastic, the mysterious.

Source: Asociación Shaquiñan (2008)

about the human and the non-human; beings or entities that make up the non-human; characteristics of the entities; related places; gender differences; associated practices and ways of knowing.

4.6 Final Reflections: Situated/Globalised Knowledge with a Gender Perspective

Indigenous knowledges are located and embedded in the territory and interrelated with different dynamics for each activity. These knowledges have been related to different conceptions of time and space, which implies a particular way of understanding the said dimensions of each practice. Likewise, these knowledges are differentiated according to gender, which implies ways of being, doing and feeling in a territory, and with non-humans, in a differentiated way for both men and women. At the same time, these knowledges are crossed by multi-scale processes that permanently link them to dynamics and policies based on other knowledges. Therefore, it is necessary to propose strategies of dialogue that include diverse ways of co-producing knowledges and conceptual and methodological changes that allow such dialogue. These processes allow us not only to understand knowledge in a contextualised way, but to make contributions visible in local practices and policies.

To establish this dialogue, I will review the advances that have been made from relational ontologies and feminist perspectives in relation to climate and climate change, since they allow us to address part of the problems previously raised.

These perspectives contribute to create a new way of analysing the ontologies and epistemologies that sustain worldviews of indigenous knowledge. The perspective of relational ontologies allows us to understand how 'human and non-human (the organic, the non-organic, and the supernatural or spiritual) are an integral part of these worlds in their multiple interrelations' (Escobar 2015:98). Previous debates and discussions have allowed us to understand and position the diversity of ontologies and the inequalities that exist between them in terms of power in decision-making in climate change discussions, especially indigenous ontologies, which exist under unequal conditions of recognition. These ontologies are based on diverse notions of nature, which leads to multiple epistemologies. Also, it is important to understand that knowledge is produced in specific local-territorialised contexts, which are in turn differentiated by gender, location, age or specialisation. Finally, we need to raise our awareness to understand that the non-human forms part of the environmental processes as this leads to the recognition of their rights.

Given that knowledge production processes and global geopolitics have generated inequalities, it is necessary to recognise indigenous peoples' demands for the inclusion of their knowledge according to gender differences in local contexts and their articulation of global-local transformations. Likewise, it is important to rethink environmental and climate geopolitics: knowledge; representations and relationships with the non-human. The feminist perspective on climate change highlights the inequalities between ways of knowing and the power relations that have been established from the geopolitics of expert knowledge, and which generate processes of production, circulation and localisation of knowledge in a differentiated way among indigenous peoples, and between men and women (Iniesta-Arandia *et al.* 2016). These perspectives demand that climate change policies and associated knowledge include ontological and epistemological plurality for an intercultural and interdisciplinary dialogue when analysing climate and climate change, as well as a gender perspective as central in all processes.

Discussions on climate change should recognise other alternatives and strategies based on cultural perspectives. From this point of view, it is necessary to take into account other cultural relationships between human and non-human beings. This implies the need to focus more on the causes and consequences of climate change and the unequal power relations around the production of knowledge between the Global North and South, and between these and indigenous peoples (Ulloa 2013, 2018a, 2018b).

These knowledge debates are linked to the demands of climate justice led by indigenous peoples, given that it has been the mechanism to initiate the debate on the need for a new understanding of knowledge geopolitics. Therefore, in the reconfiguration of environmental justice perspectives, four dimensions are proposed for rethinking the processes associated with both geopolitics of knowledge and inequalities that are not included in debates on climate change: reversing inequalities based on the dual notions of culture and nature; rethinking global environmental and climate policies; reconfiguring the legal and rights issues recognised in local/national and international contexts and including cultural demands and multiple perspectives.

In this way, indigenous peoples are putting their knowledge within the world's political fields, based on relational ontologies, as a basic premise to understand the reciprocity, complementarity and connections between all beings. This perspective also implies another conception of rights to consider: the rights of non-humans. This implies notions of being, existing and feeling, which expand the ideas of recognition and participation and include non-humans and territory as living beings with feelings and emotions and as political agents.

Acknowledgements

Translation from Spanish by Naira Bonilla. I want to thank the indigenous researchers and participants who participated in the research, especially Zonia Puenayán, Lucy Juajibioy and Luisa Cantor.

References

Aguilar, L, *et al.* 2009. *Manual de capacitación en género y cambio climático*, San José, Costa Rica: Unión Internacional para la Conservación de la Naturaleza (UICN) y el Programa de las Naciones Unidas para el Desarrollo (PNUD) en colaboración con la Alianza Género y Agua (GWA), la Red Internacional sobre Género y Energía Sustentable (ENERGIA), la Organización de las Naciones Unidas para la Educación, la Ciencia y la Cultura (UNESCO), la Organización de las Naciones Unidas para la Agricultura y la Alimentación (FAO) y la Organización de Mujeres para el Medio Ambiente y el Desarrollo (WEDO) como parte de la Alianza Mundial de Género y Cambio Climático (GGCA).

Aranda, M., Dagua, A., and Vasco, L. G. 1998. *Guambianos, hijos del aroiris y del aroiris y del agua*, Bogotá: Cerec, Los Cuatro Elementos, Fundación Alejandro Ángel Escobar, Fondo de Promoción de la Cultura del Banco Popular.

Arias de Greiff, J. and Reichel, E. 1987. *Etnoastronomías americanas*, Bogotá: Universidad Nacional de Colombia.

Asociación Shaquiñán. 2008. Cartilla recreando la memoria desde la casa de la realeza, Asociación de Cabildos y Autoridades Tradicionales del Nudo de los Pastos. Cumbal: Pasto.

Baucon, I. and Omerlsky, M. 2017. Knowledge in the age of climate change. *South Atlantic Quarterly*, 116(1), 1–18.

Cantor, L. and Juajibioy, L. A. 2013. Mujeres indígenas del pueblo kamëntsá biyá y sus conocimientos en torno al clima y el jajañ (municipiode Sibundoy. Veredas: Llano Grande, San Félix, Sagrado Corazón, La Menta, La Cocha, El Ejido). In: Ulloa, A. *et al.* unpublished. Informe Final Proyecto Perspectivas Culturales y Locales sobre el Clima en Colombia. Bogotá: Universidad Nacional de Colombia-Colciencias, 156–180.

Cantor, L. 2015. ¿Adaptación al Cambio Climático? Diálogos y estrategias entre lo global y lo local: Banco Mundial y Pueblos Indígenas, Putumayo – Colombia, unpublished thesis, Magister en Medio Ambiente y Desarrollo. Universidad Nacional de Colombia.

Castree, N. 2017. Global change research and the 'people disciplines': Towards a new dispensation. *South Atlantic Quarterly*, 116(1), 55–67.

Correa, F., 1987. Tiempo y espacio en la cosmología de los Cubeos. In: Arias de Greiff, J. and Reichel, E. (eds.). *Etnoastronomías Americanas*, Bogotá: Universidad Nacional de Colombia, 137–168.

Crate, S. 2008. Gone the bull of winter: Grappling with the cultural implications of and anthropology's role(s) in global climate change. *Current Anthropology*, 49(4), 569–595.

Crate, S. 2011. Climate and culture: Anthropology in the era of contemporary climate change. *Annual Review of Anthropology*, 40(1), 175–194.

Crate, S. and Nuttall, M. 2009a. Introduction: Anthropology and climate change. In: Crate, S. and Nuttall, M. (eds.). *Anthropology and Climate Change: From Encounters to Actions*. Walnut Creek, CA: Left Coast Press, 9–36.

Crate, S. and Nuttall, M. 2009b. Epilogue: Anthropology, science and climate. In: Crate, S. and Nuttall, M. (eds.). *Anthropology and Climate Change: From Encounters to Actions*. Walnut Creek, CA: Left Coast Press, 394–404.

Chakrabarty, D. 2009. The climate of history: Four theses. *Critical Inquiry*, 35(2), 197–222.

Davis, A., Roper, L., and Miniszewski, U. 2015. *Justicia climática y derechos de las mujeres: Una guía para apoyar la acción comunitaria de mujeres*. Boulder, CO: Fundación Ford, el Fondo Global Greengrants, el Fondo Global Wallace, Red Internacional de Fondos de Mujeres y Alianza de Fondos.

Dounias, E. 2011. Escuchando a los insectos: acercamiento etnoentomológico al cambio climático entre pueblos indígenas africanos de bosques húmedos tropicales. In: Ulloa, A. (ed.). *Perspectivas culturales del clima*. Bogotá: Universidad Nacional-ILSA, 223–245.

Emmet, R. and Lekan, T. (eds.). 2016. Whose Anthropocene? Revisiting Dipesh Chakrabarty's 'Four Theses'. RCC Perspectives, 2.

Escobar, A. 2015. Territorios de diferencia: la ontología política de los 'derechos al territorio. *Desenvolvimento e Meio Ambiente* 35, 89–100.

FMICA (Foro de Mujeres para la Integración Centroamericana). 2010. *Género y cambio climático. Aportes desde las mujeres de Centroamérica a las políticas regionales sobre cambio climático*, San José, Costa Rica: FMICA.

Jungehülsing, J. 2012. *Gender Relations and Women's Vulnerability to Climate Change. Contribution from an Adaptation Policy in the State of Tabasco Toward Greater Gender Equality: The Reconstruction and Reactivation Program to Transform Tabasco*. Mexico: Heinrich Böll Stiftung.

Goloubinoff, M., Katz, E., and Lammel, A. (eds.). 1997. *Antropología del clima en el mundo hispanoamericano*. Vols. I y II. Quito: Abya-Yala.

Guaraldo, A. 1997. Imágenes antropomorfas de aires rodantes en culturas prehispánicas del Golfo de México. Un problema abierto. In: Goloubinoff, M., Katz, E., and Lammel, A. (eds.). *Antropología del clima en el mundo hispanoamericano*. Quito: Abya-Yala, 157–180.

Heikkinen, H.I., Acosta García, N., Sarkki, S., and Lépy, E. (2016) Context-sensitive political ecology to consolidate local realities under global discourses: A view for tourism studies. In: Nepaland, S. and Saarinen, J. (eds.). *Political Ecology and Tourism*. New York: Routledge, 211–224.

Hémond, A. and Goloubinoff, M. 1997. El camino de la cruz del agua: Clima, calendario agrícola y religioso entre los nahua de Guerrero (México). In: Goloubinoff, M., Katz, E., and Lammel, A. (eds.). *Antropología del clima en el mundo hispanoamericano*. Quito: Abya-Yala, 237–261.

Heyd, T. 2011. Pensar la relación entre cultura y cambio climático. In: Ulloa, A. (ed.). *Perspectivas culturales del clima*. Bogotá. Universidad Nacional-ILSA, 17–30.

Ingold, T. 2000. *The Perception of the Environment: Essays in Livehood, Dwelling and Skill*. London: Routledge.

Iniesta-Arandia, I., Ravera, F., Buechler, S., et al. 2016. A synthesis of convergent reflections, tensions and silences in linking gender and global environmental change research. *Ambio*, 45 (Suppl. 3), 383–393.

Iguarán, M. T. 2011. *La mujer Wayuu frente al reto de la crisis del agua, paper presented at Quinto Seminario Internacional Mujeres Indígenas y Agua*. Bogotá, Colombia.

Juajibioy, L. and Cantor, L. 2013. *Jenan Bëngbe Vid. Sembradoras de vida*. Bogotá: Universidad Nacional de Colombia-Colciencias.

Lammel, A. 1997. Los colores del viento y la voz del arco iris. Representaciones del clima entre los Totonacas (México). In: Goloubinoff, M., Katz, E., and Lammel, A. (eds.). *Antropología del clima en el mundo hispanoamericano*. Quito: Abya-Yala, 153–175.

Latour, B. 2013. Telling Friends from Foes at the Time of the Anthropocene. Paper presented at the Symposium 'Thinking the Anthropocene', Paris, 14–15 November, 2013, EHESS-Centre Koyré- Sciences Po.

Latour, B. 2014. Anthropology at the Time of the Anthropocene – A Personal View of What Is to Be Studied. Paper presented at the conference of the American Anthropological Association, Washington.

Leff, E., Argueta, A., Boege, E., and Porto, C. W. 2003. Más allá del desarrollo sostenible: la construcción de una racionalidad ambiental para la sustentabilidad: una visión desde América Latina. *Medio Ambiente y Urbanización*, 59(1), 65–108.

Leis, R. and Viola, E. 2003. Gobernabilidad global posutópica, medio ambiente y cambio climático. *Nueva Sociedad*, 185, 34–49.

Liverman, D. 2015. Reading climate change and climate governance as political ecologies. In: Perreault, T., Bridge, G., and McCarthy, J. (eds.). *The Routledge Handbook of Political Ecology*. London: Routledge, 303–319.

Nazarea, V. D. 2006. Local knowledge and memory in biodiversity conservation. *Annual Review of Anthropology*, 35(1), 317–335.

Orlove, B., Chiang, J., and Cane, M. 2004. Etnoclimatología de los Andes. *Investigación y Ciencia*, 330, 77–85.

Ortiz, F. 1987. Etnoastronomía de los grupos arawak de los llanos (Colombia). In: de Greiff J. A. and Reichel, E. (eds.). *Etnoastronomías americanas*. Bogotá: Universidad Nacional de Colombia, 91–110.

Pardo, M. 1987. Términos y conceptos cosmológicos de los indios emberá. In: de Greiff J. A. and Reichel, E. (eds.). *Etnoastronomías americanas*. Bogotá: Universidad Nacional de Colombia, 69–90.

Puenayán, Z. 2009. Percepción indígena de los pastos sobre cambio climático. Resguardo Panán municipio Cumbal -Nariño- Colombia, unpublished undergraduate thesis, Universidad Nacional de Colombia, Bogotá.

Puenayán, Z. 2011. Percepción del cambio climático para los pastos del resguardo Panán, Nariño, Colombia. In: Ulloa, A. (ed.). *Perspectivas culturales del clima*. Bogotá: Universidad Nacional-ILSA, 275–313.

Puenayán, Z. 2013. Mingambis: minga de percepciones y concepciones propias de los indígenas pastos, sobre tiempo y clima, resguardo Panan, Cumbal (Nariño, Colombia). In:

Ulloa, A. and Prieto-Rozo, A. (eds.). *Culturas, conocimientos, políticas y ciudadanías en torno al cambio climático*. Bogotá: Universidad Nacional de Colombia-Colciencias, 273–316.

Ramos, C., Tenorio A., and Muñoz, F. 2011. Ciclos naturales, ciclos culturales: percepción y conocimientos tradicionales de los nasa frente al cambio climático en Toribio, Cauca, Colombia. In: Ulloa, A. (ed.). *Perspectivas culturales del clima*. Bogotá. Universidad Nacional-ILSA, 247–273.

Reyes-García, V., Fernández-Llamazares, Á., Guèze, M., Garcés, A., Mallo, M., Vila-Gómez, M., and Vilaseca, M. 2016. Local indicators of climate change: The potential contribution of local knowledge to climate research. *WIREs Climate Change*, 7(1), 109–124.

Rossbach, L. 2011. Del monólogo científico a las pluralidades culturales: dimensiones y contextos del cambio climático desde una perspectiva antropológica. In: Ulloa, A. (ed.). *Perspectivas culturales del clima*. Bogotá: Universidad Nacional-ILSA, 55–82.

Röhr, U. 2007. *Gender, Climate Change and Adaptation: Introduction to the Gender Dimensions*. Background Paper Prepared for the Both Ends Briefing Paper: Adapting to Climate Change: How Local Experiences Can Shape the Debate.

Sierra, E. 2011. Las lluvias, el arco iris y el trueno: representaciones simbólicas del paisaje y el sentido de lugar de los pueblos kággaba, u'wa y misak, Colombia. In: Ulloa, A. (ed.). *Perspectivas culturales del clima*. Bogotá: Universidad Nacional-ILSA, 329–365.

Sprajc, I. 1997. Observación de extremos de Venus en Mesoamérica. Astronomía, clima y cosmovisión. In: Goloubinoff, M., Katz, E., and Lammel, A. (eds.). *Antropología del clima en el mundo hispanoamericano*. Quito: Abya-Yala, 129–155.

Tupaz, D. and Guzmán, N. 2011. Tiempo y clima en la visión andina del pueblo de los pastos, Colombia y Ecuador. In: Ulloa, A. (ed.), *Perspectivas culturales del clima*. Bogotá: Universidad Nacional-ILSA, 315–328.

Ulloa, A., Escobar, E. M., Donato, L. M., and Escobar, P. (eds.). 2008. *Mujeres indígenas y cambio climático. Perspectivas latinoamericanas*. Bogotá: Universidad Nacional de Colombia, Fundación Natura, UNODC.

Ulloa, A. 2008. Implicaciones ambientales y culturales del cambio climático para los pueblos indígenas. In: Ulloa, A., Escobar, E. M., Donato, L. M., and Escobar, P. (eds.). *Mujeres indígenas y cambio climático. Perspectivas latinoamericanas*. Bogotá: Universidad Nacional de Colombia, Fundación Natura, UNODC, 17–34.

Ulloa, A. 2011a. Construcciones culturales sobre el clima. In: Ulloa, A. (ed.). *Perspectivas culturales del clima*. Bogotá: Universidad Nacional-ILSA, 33–53.

Ulloa, A (ed.). 2011b. *Perspectivas culturales del clima*. Bogotá: Universidad Nacional-ILSA.

Ulloa, A. 2013. Controlando la naturaleza: ambientalismo transnacional y negociaciones locales en torno al cambio climático en territorios indígenas, Colombia. *Iberoamericana*, 13(49), 117–133.

Ulloa, A. 2014. Estrategias culturales y políticas de manejo de las transformaciones ambientales y climáticas en Colombia. In: Lara, R., and Vides-Almonacid, R. (eds.). *Sabiduría y Adaptación: El Valor del Conocimiento Tradicional en la Adaptación al Cambio Climático en América del Sur*. Quito: UICN, 155–175.

Ulloa, A. 2015. Environment and development: Reflections from Latin America. In: Perreault, T., Bridge, G., and McCarthy, J. (eds.). *The Routledge Handbook of Political Ecology*. London: Routledge, 320–331.

Ulloa, A. 2016. Cuidado y defensa de los territorios-naturalezas: mujeres indígenas y soberanía alimentaria en Colombia. In Rauchecker M. and Chan, J. (eds.). *Sustentabilidad desde abajo: luchas desde el género y la etnicidad*, Berlin: Lateinamerika-Institut der Freien Universität Berlin-CLACSO, 123–142.

Ulloa, A. 2017. Geopolitics of carbonized nature and the zero carbon citizen. *South Atlantic Quarterly*, 116(1), 111–120.

Ulloa, A. 2018a. La confrontation d'un citoyen zéro carbone déterritorialisé au sein d'une nature carbonée locale-mondiale. In: Beau, R. and Larrère, C. (eds.). *Penser l'anthropocène*, Paris: Presses de Sciences Po, 283–300.

Ulloa, A. 2018b. Re-configuring climate change adaptation policy: Cultural strategies and policies for managing environmental transformations in Colombia. In: Klepp, S. and Chavez-Rodriguez, L. (eds.). *A Critical Approach to Climate Change Adaptation: Discourses, Policies and Practices*. London: Routledge Advances in Climate Change Research, 222–237.

Ulloa, A. and A. Prieto (eds). 2013. *Culturas, conocimientos, políticas y ciudadanías en torno al cambio climático*. Bogotá. Universidad Nacional de Colombia-Colciencias.

Villela, S. 1997. Vientos, lluvias, arcoíris: Simbolización de los elementos naturales en el ritual agrícola de la montaña de Guerrero (México). In: Goloubinoff, M., Katz, E., and Lammel, A. (eds.). *Antropología del clima en el mundo hispanoamericano*. Quito: Abya-Yala, 225–235.

Watts, M. 2015. Now and then: the origins of political ecology and the rebirth of adaptation as a form of thought. In: Perreault, T., Bridge, G., and McCarthy, J. (eds.). *The Routledge Handbook of Political Ecology*. London: Routledge, 19–50.

Yana, O. 2008. Diferencias de género en las percepciones sobre indicadores climáticos y el impacto de riesgos climáticos en el altiplano boliviano: estudio de caso en los municipios de Umala y Ancoraimes, departamento de la Paz. In: Ulloa, A., Escobar, E. M., Donato, L. M., and Escobar, P. (eds.). *Mujeres indígenas y cambio climático. Perspectivas latinoamericanas*. Bogotá: Universidad Nacional de Colombia, Fundación Natura, UNODC, 43–54.

5

Thin Place
New Modes of Environmental Knowing through Contemporary Curatorial Practice

CIARA HEALY-MUSSON

5.1 Introduction

The idea of a thin place comes from folklore and describes a marginal, liminal realm, beyond everyday human experience and perception, where mortals can pass into Otherworlds. In Irish folklore, Tír na nÓg, a legendary Irish Otherworld of youth and abundance, could only be entered through a 'thin place', such as an ancient burial mound or cave, or by voyaging across the sea. Similarly, 'Annwn', the name given to the Welsh Otherworld in the Mabinogion, was believed in some instances to be hidden behind waterfalls or located on an invisible island on the west coast. Like Tír na nÓg, it could only be accessed through a particular anomalous 'portal' in a landscape. Such portals and spaces between worlds were defined by anthropologist van Gennep (1960:111) as 'liminal' after the Latin word *limen*, meaning gateway or threshold.

This chapter is about a practice-based, interdisciplinary, curatorial research project installed at Oriel Myrddin, a contemporary art gallery in Carmarthen, Wales, in 2015. The project was titled Thin Place and was concerned with developing a new curatorial approach. It involved the careful selection and presentation of artworks in a gallery. The artworks acted as portals or gateways between bodies of knowledge and were further contextualised and supported by a symposium, an education programme and a publication. Together they enriched the breadth and depth of societal engagements with place in a changing climate. Thin places and worlds were extended beyond folkloric narratives in this research so that they might also stand for the points at which different belief systems, creative traditions, philosophies, narratives, experiences, scientific worldviews and historical perspectives overlap – all of which helped to create an expanded sense of place and new modes of environmental knowing.

The aim of my research was to show, through a particular form of curatorial practice – which I called 'thin curating' – that my actions as a curator could generate deeper engagements between human, non-human and ancestral worlds in particular places.

I believe it is possible to place even the most abstract of comprehensions in visual terms, to come towards a gradual understanding of how they begin to signify, or how they resonate, in unexpected ways. During this exhibition I invited many unlikely elements to coincide. This sense of strange and incomplete commonality invited both participants and contributors to this project to treat non-humans – animals, plants, earth, even artefacts and commodities – more carefully. To develop this capacity I revisited discredited philosophies of nature, risking what philosopher Jane Bennett calls 'the taint of superstition; animism, vitalism, of anthropomorphism and other premodern attitudes' (Bennett 2010:18).

These philosophies were exactly what Bennett embraced in her book *Vibrant Matter*. Bennett describes herself as a vital materialist, as someone who lingers in those moments during which she finds herself fascinated by objects, taking them as clues to the material vitality that she shares with them. I believe that this vital materialism sets us in the right direction to attend to thin places. It is a philosophy that builds a discourse around plurality and simultaneity, giving agency to organic and inorganic things. It gives us the ability to ascertain some, if not all, of the potentials that relational interactions can bring about.

To do this, I looked at the concept of animism and considered the French philosopher Felix Guattari's argument that those from animist societies understood subjectivity as being 'the core of the real' (cited in Melitopoulos and Lazzarato 2012:48).

As a result, 'everything breathes, and everything conspires in a global breath' (2012:49). This project therefore became primarily concerned with resisting the impoverishment of subjectivity through creative arts practices. In so doing, I hoped to redress and reorientate curatorial practice from a metanarrative approach derived from Enlightenment-based discourse, to include interventions that use archaic ideas. This was the process through which I critically modelled an alternative means of addressing climate change through the visual arts.

5.2 Curating: A Brief Definition

The term 'curation' is derived from the Latin word *curare*, meaning to take care. Before the eighteenth century, it was used in relation to those responsible for the care of the soul, such as priests. However, the term was later applied to those responsible for caring for objects and collections of art.

The urge to collect and to organise objects into systems of display in order to make sense of our place in the world is an intrinsic part of the human condition; from carved Neolithic bones to Victorian keepsakes and mementos of dead loved ones, the desire to organise and present material objects for personal or public display reveals how we relate to our time and place, past and present.

The 'Wunderkammer' was a term given to collections of objects, artworks and artefacts in the fifteenth century that attempted to represent the world in one room or cabinet. Collections were often eclectic accumulations of memories and personal experiences presented in random groupings rather than specific calculated structures. The display cabinets used to house Wunderkammer collections were often adapted from vitrines used by the church for preserving and venerating the relics of saints. Placed in these cabinets, these often secular and personal objects were transformed into sacred and untouchable treasures.

The diverse format these collections took was eventually dispersed during the period of the Enlightenment in the seventeenth century, when an interest in classification took place. This heralded the beginning of the curatorial interest in organising cultural artefacts into symbolic constructs and hierarchical displays which could range from the achievements of a nation, or a supposed hierarchy of humans and animals, or simply a differentiation between disciplines such as painting, printmaking and sculpture.

This system of classifying objects and artefacts into a specific order became increasingly popular during the nineteenth century when a hunger for education and public benefit made museums and galleries very popular. A rational, hierarchical, scientific means of representing the world was established by curators using clear systems of classification, good labelling, isolating objects and artworks from their context and from each other as well as providing a suitable background and a position in which the work can be readily and distinctly seen.

In the nineteenth century, the belief that virtue came from beauty prevailed. Sunday visits to museums and galleries would expose the general public to carefully classified objects and artefacts, which in turn, it was believed, would transform their behaviour. Museums and galleries became secular places of worship as well as a crucial means of developing a very specific way of seeing the world. The curator, in this context, became the priest, consciously and explicitly representing one particular paradigm of knowledge as a truth through the specific organisation and display of artworks and objects (Geoghegan 2010; Geoghegan and Hess 2015).

Curation throughout the nineteenth and early twentieth centuries continued to provide the public with a monotheistic understanding of their place in the world – that is, until the gallery itself began to assert itself as an aesthetic object in the mid-twentieth century. For minimalist sculptors such as Donald Judd in the 1960s, the removal of all contexts from a gallery space was liberating – thus emerged the modern white cube, which became inseparable from the modern artworks exhibited inside it. The gallery and the work were one.

By the 1970s however, critics such as Brian O' Doherty (2000) argued that the gallery space is not a neutral container but a historical construct. He claimed that the white cube not only conditions but also overpowers the artworks themselves in its shift from placing content *within* a context to making the context *itself* the content.

Today, curators have become increasingly aware of the history of the museum and gallery space and of the importance of maintaining a connection with the context of an artwork. They carry a great responsibility when they attempt to create representations of the world.

Hans Ulrich Obrist, director of the Serpentine Gallery in London for example, feels that it is important to challenge the viewer by creating a space of 'interlocking zones' (Obrist 2004). As a curator he feels that exhibitions and museum spaces need to be transformed into an 'archipelago' (ibid), where different time zones can be represented together and where an idea is produced through the act of curating rather than simply illustrating an idea that already exists.

Obrist's curatorial policies are not dissimilar to Nicholas Serota, former director of Tate Modern London. In his book *Experience or Interpretation*, Serota puts forward his curatorial proposal stating that exhibitions must have 'zones of influence' (2000:48). This means an open-plan exhibition space divided only by corners and half-length walls. This type of space means it is always possible to view an artwork in relation to other works in the exhibition. Serota argues that this approach is particularly useful when curating an exhibition of historical and contemporary references because it enables audiences to experience how the exhibition works as a whole. Serota's rejection of exhibiting artworks in chronological order emerged from his disillusionment with a curatorial approach that presented one metanarrative of a time and place, which he described as 'the conveyor belt of history' (2000:62). This was a system of representation most commonly used in what the cultural theorist James Clifford (1991) called 'majority museums'.

In his analysis of the representation of Native American artworks in museums in the early 1990s, Clifford (1991) critiqued the visible repression of cultures that hold an animistic perception of the world. This repression, he argued, was conducted through specific curatorial strategies, such as the forced sale of Native American artworks or, as happened more frequently, the omission of Native artworks from contextual displays in museums – if they were displayed at all. Clifford found that a significant number of Native artworks had been kept in storage from the time when they were first included in the museums' collections. All of these events facilitated, he argued, a decline in the presence and understanding of Native

American culture and values in the North American psyche. It was as if this vibrant culture only existed in the past.

While majority museums attempted to acknowledge or include Native American perspectives in their displays, Clifford felt that they did not accurately reflect the culture from which the work came. The most significant differences between majority museums and what Clifford referred to as tribal museums were to be found in the ways in which power and ownership were mediated. Majority museums sought to own a collection of objects, and when presenting these objects they sought to separate fine art from ethnographic culture. The structure, layout and curatorial policies of the majority museums therefore articulated the hierarchical ideals of the Enlightenment. Tribal museums, on the other hand, were perceived by Clifford to have a more cooperative agenda because they challenged the notion that history can only be represented in a museum as a metanarrative. Local and oral testimonies, subjective memories and personal experiences were presented alongside existing national and 'objective' histories in a tribal museum. Thus, a more holistic relationship between place, narrative, experience, spirituality and material was formed.

Central to my research then is the curator – not as an individuated practitioner representing a single (hierarchical) paradigm, but as a mediator, facilitator and nurturer of participation and contribution from those who inhabit many different worlds.

In this chapter, I give an account of how 'thin curating' – a term I developed during this research project in Wales – evolved. I use descriptions and reflections on the process of developing the exhibition, education programme, publication and symposium. I also make a case for the benefits of employing 'thin curating' as a methodology when working on interdisciplinary projects that aim to address climate change.

5.3 Thin Curating

Thin curating describes the process of engaging a community in place-based interdisciplinary thinking. It involves interweaving ancient and contemporary worldviews and constitutes an example of what cultural theorist Declan Kiberd (2004) calls 'the archaic avant-garde'. Kiberd (2004) used the phrase 'archaic avant-garde' to describe a way of rereading visual culture from the past in order to rewrite the present. He applied the term initially to artworks and artefacts created in Ireland during the Celtic Revival of the nineteenth century when motifs and symbols from ancient Ireland began to appear in arts and crafts as well as in industrial and architectural design. What began as an interest in ancient native culture by second-generation Anglo-Irish gentry and Victorian anthropologists later provided the visual tools to support a nationalist uprising. While the first example of the presence of the archaic avant-garde in Ireland might be seen in the poetry and artwork of the Celtic Revival of the nineteenth century, Kiberd also felt the term could be applied to contemporary society. This is because, he argued, the archaic avant-garde is a precursor to societal change and often manifests as new forms of creative practices that look to the past for inspiration during economic, social and, in our current time, environmental crisis. This argument was particularly relevant to this research because one of the key concerns for this project was to find out if thin curating could play a role in supporting a community to develop a more enriched understanding of place in a time of climate change.

5.4 Making a Case for Thin Curating

My practice as a 'thin curator' was informed by the impact archaic, sacred and animistic perceptions of the world have on different disciplines. It was also inspired by the ways in which emotional experiences can be triggered by specific places, especially places that were once seen as sacred in the archaeological record. I felt that a thin curatorial approach was a useful model of addressing climate change through the visual arts because it is animist thinking put into *action*. These actions were designed to temper the perception that all living things exist merely as a resource for human consumption. The recovery of, and dissemination of, animistic perceptions of the world in the twenty-first century coincides with discoveries in the sciences, where the traditional dualities of subject/object are no longer as clearly defined as had previously been considered (Adams and Green 2014). If that is the case, then I believed that the arts, sciences and humanities might be able to collectively consider new ways of addressing the current ecological issues we face and call into question the secular monotheism and the monocultures of what Donna Haraway calls 'Big Energy and Big Capital' (2016:160).

Like Haraway, I saw animism not as a New Age or neo-colonial fantasy but as a proposition for rethinking relationality. Creating less oppositional categories and processes requires, she argues, a deeper entanglement with our animal, plant and human worlds. Making kin in these interspecies knots enables a more intertwined relationship with place, where modern, traditional, past and present worlds become inextricably interconnected.

To engender this, Haraway created a literary proposal for a future filled with what she calls 'symbiogenetic' linkages (2016:159–160) where some of the genetic material of an endangered species is combined with the genes of a human child so that both human and non-human become forever linked in a deeper genetic and ontological understanding of each other's needs.

As sci-fi as this proposal may sound, it is not so very different from the ways in which human and non-human entanglements were perceived in the Western psyche in the past. Emma Wilby's book on magic, shamanism and witchcraft in seventeenth-century Scotland (Wilby 2010), for example, offers an analysis of the legal testimony of a young Scottish peasant called Isobel Gowdie, who was put on trial for witchcraft in 1662. Her testimony highlights the legacy of animistic and polytheistic thinking in Britain and offers an historical account of how the boundaries of perception and multi-species entanglement evolved. In her confession, Gowdie revealed how she, along with her peasant neighbours, inhabited a reality that was profoundly influenced by the natural rhythms of the land, of the body, the transition of stars and planets across the sky and their impact on her fate. Her testimony also reveals the magical and mystical powers (seen and unseen), which controlled both her physical and emotional well-being. At times, this experience of existence was frightening, complex, overwhelming and violent, yet always animate. By 1662, Isobel Gowdie's social superiors had, according to artist and cultural geographer Iain Biggs, largely 'internalised a strictly dualistic Calvinism' (2013). However, Gowdie and her neighbours had not. Their world remained shaped by a multitude of uncertain conflicting forces. These included the dead, wise women and their familiars, revenants, wizards, old popular Catholic sites and 'the good neighbours' – the elves or fairies. These entanglements between human and non-human were polyverse – and share many similarities with the ecological and socio-political climate in which we live today, at least in their structural complexity and radical uncertainty. How we negotiate our relationship with, and our responsibility towards, this ever-unfolding, ensouled world requires a much greater permeability and porosity between disciplines.

This is something that is not yet in evidence in most sectors, particularly in education, where disciplines continue to exist in separate silos, making it increasingly difficult to grasp the immensity and complexity of climate change and what we can do collectively to address it.

My methodological approach was therefore multi-dimensional and broadly speaking engaged in theoretically informed action research. It had an orientation consistent with established *anthropological* (Turner 1986; van Gennep 1960), *archaeological* (Mulk and Bayliss-Smith 2007), *philosophical* (Guattari 2008; James 1985), *social* (Heelas and Woodhead 2005), *educational* (Edwards et al. 1998) and *curatorial* (Obrist 2004; Serota 2000; Bennett 1995; Clifford 1991) practices.

As my thin curatorial practice related to issues around relational modes of being, I turned to sociologists of religion Heelas and Woodhead, who coined the term 'holistic milieu' (2005:15) during their socio-logical study of the sacred landscape of the town of Kendal in the north of England.

They argued that western Europe has been experiencing an ongoing shift away from more or less authoritarian ideologies of faith – what they call the 'congregational domain'[1] – towards a more inclusive, subjective and relational experience of spirituality, which they call the 'holistic milieu'.[2] They define this as a 'spiritual revolution' (2005:3). The primary concern of the Kendal study was to examine what Heelas and Woodhead describe as the heartlands of religious and spiritual life.

By investigating the perceptions and activities of the congregational domain and the holistic milieu, they could distinguish between associational forms of the sacred. It is worth noting that their study took place in a small, predominantly white town like Carmarthen (the locale of my practice), rather than a large city with a substantive migrant population. What they did find, however, was that the congregational domain tended to live what they titled a 'life-as' existence, which was generally characterised by duty, roles and moral judgement, where language was dominated by 'should' or 'ought', where church or chapel offered a sense of order, meaningfulness and security: 'Individuals are told what to do by a higher authority, rather than being encouraged to look to their own inner resources to decide for themselves' (2005:16). 'Subjective-life' was generally characterised by the pervasive use of 'holistic' language. These included words such as harmony, balance, flow, integration, interaction and connect. Above all, the holistic milieu is interested in growth, in moving beyond what Heelas and Woodhead (2005:3) describe as 'barriers, blocks, patterns or habits by making new connections'. It is concerned with healing the dynamic of the whole.

According to the 'spiritual revolution' (2005:3) claim, subjective-life spirituality is beginning to eclipse life-as religions. Heelas and Woodhead predict that, overall, the congregational domain will continue to decline for the next twenty-five to thirty years. They attribute the decline of the congregational domain, and subsequent growth of the holistic milieu, to a single process, what Taylor (1991 cited in Heelas and Woodhead 2005:2) calls 'the massive subjective turn of modern culture'. This subjective turn presents a challenge to the worldview that truth is arrived at objectively, empirically and methodically.

What I hoped to show through my thin curatorial practice was that patterns do exist in and across meanings, and that these would qualify Heelas and Woodhead's (2005) suggestion that the presence of the holistic milieu is a growing force and therefore cannot be isolated from larger cultural or educational issues. My hypothesis was that the living world possesses its own agency and sentience, the presence of which can be found everywhere. However, this agency can be given a voice through curation informed by animism, the archaic avant-garde and the holistic milieu.

Thin curating is also pragmatic in as much as it makes interventions *into* the world rather than simply revealing it. Because the Enlightenment project was, as

---

[1] For example, the public visible activities of church and chapel.
[2] The 'invisible' activities of what is often called alternative or New Age spirituality.

cultural theorist Owain Jones (2008:1601) suggests, 'a vast body with great diversity and momentum', its current destruction involves the 'breaking of many things, over large tracts of thought, space and time'. Making interventions into that destruction is one of the ways in which thin curating can salvage and reconstruct something from the wreckage, embedding what the nineteenth-century pragmatist William James articulated as a turning away from

abstraction and insufficiency, from verbal solutions, from bad a priori reasons, from fixed principles, from closed systems ... towards action and power. (James 2000 [1907]:28)

While this statement might at first appear to resonate with logical positivism, James' proposal is relativist, or a form of 'qualified relativism' (Thayer-Bacon 2002 cited in Jones 2008:1602). It claims that creativity, pluralism and experimentalism are the necessary means by which to understand the world. As a creative praxis then, pragmatists might view creative methodologies like thin curating as 'ethical act[s]' (Thrift 1996 cited in Jones 2008:1604).

Thin curating also owed its development to Guattari's concept of ecosophy as much as it did to theories concerned with animism and pragmatism. This is because ecosophy values the multi-faceted nature of subjectivity, yet simultaneously stresses the importance of collectivity. Guattari emphasised the notion that life is polymorphous and multi-dimensional. Social, political and environmental conflicts are taking place under what he calls a 'multipolar system' (2008:22), and yet they are still addressed through the lens of the traditional dualist oppositions that have guided social thought and geo-political cartographies since the Enlightenment. He proposes a 'group Eros principle of social ecology' (2008:67) as an alternative way of looking at discreet, separate categories and benchmarks.

While pragmatism, the concept of the archaic avant-garde, animism, ecosophy and the holistic milieu theoretically underpinned my methodology, my approach for thin place also developed through my interactions with, and correspondence with, members of staff at Oriel Myrddin, with the artists, writers and speakers I worked with, and with the information I gathered from questionnaires issued to visitors to the exhibition. It also developed through workshops and through my engagement with children in three schools in Carmarthenshire.

Part of what drove the development of this curatorial research project was a wish to re-engage viewers with the place in which they live, in the hope that it might restore in them a deeper emotional attachment, sense of responsibility towards and ownership of the physical, metaphysical and social geography they inhabited, particularly in a time of climate change. The wish that thin place would transform how place is perceived and valued impelled this research, but to prove that it had made a difference to the way in which the community felt about their place is a hard claim to substantiate empirically. However, the hope was to facilitate the possibility.

Carmarthen was an ideal locale for the exhibition in that respect. Around the time this research took place, it was one of the most economically deprived regions of Wales (and therefore of the United Kingdom). In the 2008 survey compiled by the National Assembly for Wales for the Office of National Statistics for example, 6 per cent of Carmarthenshire fell within 10 per cent of the most deprived areas in Wales. The majority of its areas were more deprived than the Wales average. Most significantly, the emotional well-being of the population in Carmarthenshire was described in the 2012/13 survey by the Office of National Statistics, *Personal Well-Being across the UK*, as having the highest percentage of low personal well-being in the whole of the United Kingdom (11 per cent). A greater proportion of people in Wales in the 2012/13 survey rated their life satisfaction and sense of feeling worthwhile as very low, the lowest in the United Kingdom as a whole (Office for National Statistics 2013; National Assembly for Wales Commission 2008).

Carmarthenshire's rich historical relationship with animism and sacred forms of place-based worship was another compelling reason for situating the research there. In the twelfth century, for example, it was associated with Merlin (Welsh *Myrddin*) from the Arthurian legends. Merlin's presence was recorded in the *Black Book of Carmarthen*, written by a scribe from the Priory of St John the Evangelist and Teulyddog before 1250, along with stories and poems about other legendary Welsh heroes. While the stories in the *Black Book of Carmarthen* are one example of a symbiotic relationship between an ancient past and an early Christian presence, the historian and psycho-geographer Graham Robb (2013:116–120) identifies another when he notes that, in pre-Roman times, certain points along the western coasts of Wales and Ireland were believed to be significant entrances into Otherworlds because they were on solstice paths. Summer and winter solstice angles were crucial points of reference for ancient civilisations. These paths were carefully measured using Pythagorean mathematics and thus combined what, today, we separate into the categories of science and religion. Because the Sun's light fades in the west, the solstice line system had a certain psycho-geographical logic. It was a physical reminder that the west was where the soul departed this world after death. In Carmarthen, a significant number of religious orders developed, including an Augustinian priory and a double-cloistered Franciscan friary on pre-existing sacred sites. This merging of two forms of worship in one place is common in the British solar network. The location of both the friary and the priory in Carmarthen therefore played a major role in the cultural life of Wales from a very early period.

The virtual erasure of these places of worship from Carmarthen following the Reformation led the way for many archaeological and architectural calamities.

Today, for example, the ancient Franciscan friary is buried under a large supermarket and a car park/shopping mall crudely named 'Merlin's Walk' leads the way to the priory, which is under yet another large supermarket carpark. Many ancient sacred pre-Roman burial places connected to Merlin were poorly excavated prior to redevelopment.

5.5 Case Study: Thin Place

The realisation of the Thin Place exhibition, publication, education programme and symposium was fully funded by the Arts Council Wales and supported by Oriel Myrddin, Carmarthen, Wales. The five exhibiting artists and the many other contributors to this curatorial project produced work that was concerned with, or responded to, two particular locations: west Wales and the west of Ireland. These were locations where, it was once believed, souls could easily enter Otherworlds.

My approach to the education programme was pedagogically underpinned by educational theorists Freire (1993, 1973), Edwards *et al.* (1998) and Robinson (2006). I was interested in developing a dynamic ethos in the gallery space through the use of enquiry, discovery, experimentation, problem-solving and expression (Jeffrey and Craft 2004:80). This idea is based on the Reggio Emilia co-participative approach (Edwards *et al.* 1998), which promotes holistic thinking and creativity. The Italian primary schools of Reggio Emilia incorporate into their practice strands of European and American progressive education, Piagetian constructivism and Vygotskian socio-historical psychologies, as well as participatory and democratic European postmodern philosophy. In other words, storytelling, drawing, critical thinking and animistic thought experiments are blended together and become the primary means by which children in these schools integrate a conscious and unconscious relationship to place. Edwards *et al.* (1998) argue that children who are educated in this way are much more likely to grow up to become people who will care for the future of a place. This alone seemed a relevant and significant reason to employ their methodologies in my educational practice.

As a thin curator, I wanted to develop an exhibition that evolved through dialogue. Rather than selecting artists in a formalised way within a tight deadline, the exhibition took over three years to come to fruition. The artists selected to exhibit were Adam Buick, Jonathan Anderson, Flora Parrott, Ailbhe Ní Bhriain and Christine Mackey. All of the artists produced new work for the exhibition. The work emerged in their studios as a result of our dialogues, correspondences, studio visits, exchanges of texts and essays. At no point was I prescriptive about what the work should look like or how it should evolve. I was interested in seeing what would emerge through ideas and discussion on what a thin place was.

Figure 5.1 Jonathan Anderson, *Pylon Totems*, 2014. Installation images from 'Thin Place', Oriel Myrddin, 2015

For example, Adam Buick developed a body of work that created an animist dialogue and exchange with the landscape. During a series of ritualised walks, which took place over several months, Buick placed hundreds of tiny votive jars made from unfired porcelain in particular locations along parts of the Pembrokeshire coast which, he felt, corresponded to the idea of a thin place. Photographic documentation of this work was then installed in the gallery. One small jar was placed over the main doorway of the gallery entrance to signify the crossing of a threshold.

Jonathan Anderson on the other hand developed a collection of sculptures using bitumen and rags. These sculptural objects ranged in shape and height from 5 cm to approximately 5 foot and also had a votive element. They made pan-cultural references that alluded to Crucifixes, Asian Buddhist statues, South American and African voodoo dolls. Yet their inspiration also came from the many hundreds of electricity pylons that criss-cross the Welsh Landscape. Assembled on a tiered platform, Anderson's sticky, black-coated oil and tar effigies raised ecological questions about the direction of Western worship whilst at the same time were a reminder of the political and ancient uses of these materials (Figure 5.1).

Ailbhe Ní Bhriain developed a two-screen adaptation of a body of film works titled 'Great Good Places' which she had originally developed in 2012–2013. The

Figure 5.2 Flora Parrott, *Fixed Position*, 2015, and Ailbhe Ní Bhriain, *Great Good Places*, 2011–2015. Installation images from 'Thin Place', Oriel Myrddin, 2015

two extracts from this larger body of work were adapted for Oriel Myrddin and included images of the coast of her native Co. Clare, Ireland. Seeping into these landscape images were tax offices, abandoned airports and famine villages. The confluence of these places, all presented simultaneously as permeable and dissolving layers on a screen, possessed an acknowledgement of brokenness – of the cracks in the realities we have constructed and hidden behind. The work addressed issues affecting Ireland following the collapse of the 1990s economic boom and questioned what remains possible in the aftermath and in the afterlife of both an image and a place (Figure 5.2).

Christine Mackey presented examples of her explorations of hidden histories, real and imagined, as well as ecological formations using diverse graphic sources – from town planning literature to ancient topographical drawings. This work began to evolve during a visit to her studio in Leitrim, Ireland, in 2013. It was out of these conversations that Mackey produced a body of work, which involved walking, mapping, mandala making and collecting. She repeated walks in certain parts of Ireland collecting seasonal wildflowers as she went. Each flower was labelled and recorded in a logbook before its dye was extracted. The dyes were then stored in vials and each walk was painted as a pseudo pie chart or mandala, documenting the flowers found on a particular day and in a particular place.

Flora Parrott, a London-based artist whose practice at the time of the development of the exhibition drew inspiration from west Wales, was also selected, as her work was both conceptually and aesthetically archaic avant-garde. Over three years we exchanged letters, texts and documents on geology, archaeology, physics and spirituality. The sculptures she constructed in response to this dialogue referred to ancient, shamanic-like practices or fragments from an archaeological discovery (Figure 5.2).

The absence of labels beside the artworks made the exhibition appear more cohesive and facilitated more direct and experiential engagement. There was a discretely presented map in the main foyer of the gallery which identified which artist was responsible for each work, but it was not possible to see who had made each work when in the gallery space itself.

In a review published in *This Is Tomorrow*, critic Rowan Lear (2015) immediately identified a relationship between the location of the exhibition, its history, spirituality and curatorial premise when she wrote:

Carmarthen is one of Wales' oldest towns, and myth names it the birthplace of Merlin, wizard of Arthurian legend. Oriel Myrddin itself is sited on Church Lane, a breath away from St Peter's Parish Church opposite. Presenting artists with Welsh, Irish and Scottish heritage and thus Celtic cultural memory, in a place already imbued with mystery and prehistory, Healy draws us a map of thin places and thin practices, suggesting connections between the two.

In considering how the concept of Thin Place is experienced, Lear posits that:

Thin Place is not intended as a clarification on boundaries, but a dissolving of distinctions, a way marker that complicates the route.

Finally, her review indicated that the works on show encouraged a more enriched and holistic experience of place:

On encountering a thin place, rarely does a visual sign alert its presence. Instead there's a peculiar, instinctual, bodily sense: a prickling awareness that cannot be described but is there nonetheless ... The most powerful artworks ... incite a 'gut' reaction ... Leaving the exhibition, and on the journey home, that odd feeling stayed with me, long after I left. (Lear 2015)

This review went some way to confirm that place could be perceived through more than one register, and that place might, in some way, have an agency of its own. It did not, however, indicate that this heightened perception of place might compel someone to address climate change. But Lear's review showed that the exhibition did elicit a strong reaction, an awareness of some form of encounter with the numinous, with a more mystical understanding of place.

William James argued that the 'opinion opposed to mysticism in philosophy is sometimes spoken of as rationalism' (1985:73). As a result, 'vague impressions of something indefinable have no place in the rationalistic system'. Our philosophies, he suggests, are the fruits of rationalism, as are the physical sciences. However, even though rationalism has the prestige, he claimed that if you have any intuitions at all they come from a deeper level of your nature:

Your whole subconscious life, your impulses, your faiths, your needs, your divinations, have prepared the premises, of which your consciousness now feels the weight of the result; and something in you absolutely *knows* that that result must be truer than any logic-chopping rationalistic talk, however clever, that may contradict it.

The duality between rationalism and mysticism James describes echoed the tension many people felt when they experienced the Thin Place exhibition, education programme and publication. Many were troubled to discover that their worldview was inherently dualistic. This was certainly the case with the symposium, as I will discuss shortly, but less so with the children's education programme.

5.5.1 *The Education Programme*

The children's education programme was designed as two interconnected workshops, which took place a week apart. The first workshop took place in three local schools; the second workshop took place in the gallery itself. The schools selected for the education programme were situated within the 10 per cent most deprived areas in Wales.

The first workshop included a series of storytelling events. Place-based folk tales from Welsh, Irish and English language books were used and children were encouraged to interpret the narratives through guided meditations and creative drawing and bookmaking activities. The stories focused predominantly on ideas of Otherworlds, the afterlife and overlaps between worlds, many of which bore a connection to their locale. The fact that many stories were based within their own region helped support the development of cultural and environmental literacy amongst the children.

Participating children were given time to reflect on the stories in their classrooms and to learn any new terminology or vocabulary through organised play. The exhibition catalogue was also used as a resource. From this, a richer vocabulary of place, animism and spirituality developed. This was then applied to writing practice following a visit to the exhibition. A critical writing competition was open to all children across Carmarthen.

The workshops in the schools cultivated an imaginative and poetic view of place. This opening up to the imaginal realm equipped the children with a new, intuitive and immersive approach to engaging with the artworks in the gallery and their local environment. This was substantiated in the books and essays they produced for the critical writing competition.

In the gallery, Anderson's work was described by many as being 'like a graveyard' or 'a city in another world' and often referred to as 'like something from the olden days'. Mackey's work was 'like a potion'. Parrott's work was 'like a ladder to a cloud castle'. Buick's work was 'like a pot for fairy dust'. The children made great attempts to recall place-based terminology and experiences they had learned in the classroom in order to apply them to their understanding of the exhibition.

There was a pace to the language in many of the critical writing competition entries that was engaging and dynamic: 'the floor is fire' and 'makes me feel scared and dream about a tower' being two examples. The first prize in the writing competition went to a seven-year-old child who had created a very detailed account of the show. All judges felt her essay articulated her embodied experience of the artworks and identified a sense of place:

I was inspired by all the colours and everything else. The black totem poles made me feel like they were enchanted fairy house. The test tubes looked like they were potions. The ladder made me feel like I wanted to climb up and go to a different dimension. The pot reminded me of my fairy pot. The colours in the test tubes reminded me of my garden. The painting of a pot it felt like I was there. I could feel the breeze. The totems made me feel like I was in a graveyard.

Even though many of the other essays were less detailed, they often included poetic and original imagery which demonstrated an enriched literary engagement with the locale in which the children lived.

'It made me think of the funny weird things I see. Bones and stones and pots and thoughts. Isn't that what is meant to be'.
  'Objects make you think! The gallery makes you dream and live'.
  'The gallery is a story'.

For the education officer at Oriel Myrddin, the Thin Place exhibition emphasised the interdisciplinary nature of artistic practice. Because it had many different entry points or themes through which to view the work, it made her more aware that '*everything* is an artist's business' (Stacey 2015).

Reggio Emilia schools give the artist, which they call the 'Atelierista', a particularly important role and see them as a crucial part of their pedagogical approach. The Atelierista is given a particular position, not of power, but of honour. This means they have a responsibility to foster a multiplicity of views and to engage with different forms of perception and expression, particularly through materials. This is seen to be an enormous benefit to the teachers in the schools, the children, their learning environment and the creative practices of the artists themselves. In many ways the Atelierista could be described as employing a thin approach to educational practice as they act as a mediator of worlds when working in the Reggio Emilia classroom.

According to the Reggio Emilia philosophy, the purpose of education is to provide an environment that fosters meaningful interpretative experiences that allow students to make their own meanings through material investigation and to understand how meanings and interpretations of reality are made. As such, responses to a place and its inhabitants, both human and ancestral, are led by wonder. This gives everyone involved in the Reggio Emilia schools the space and the potential to see the world in a relational light.

The learning activities I developed for the Thin Place exhibition were, therefore, like the Reggio Emilia approach: concerned with helping those involved in the education programme I devised, to move beyond individuated concerns and to see conceptual structures relationally. In Reggio Emilia education, as in thin curation, this thought–action–assessment axis endorses experimentation, failure, error and uncertainty as an inevitable part of unfolding growth.

5.5.2 Thin Place Publication

The Thin Place publication, which included specially commissioned texts from professionals who work in different disciplines outside of the art world, also endorsed Thin Curating. The texts were presented in both Welsh and English and were a critically important part of the project because they offered different lenses through which to read and engage with the artworks from the exhibition. They also provided a platform for the contributing writers to engage with new audiences. This was, in part, the point of the project, because these broader contextual interpretations offered multiple insights into how the artwork could be understood.

The publication also acted as a physical record and legacy, extending the 'duration' of the exhibition. Images of the artworks in the exhibition were placed between each textual contribution, creating an opportunity to interpret their meaning through multiple lenses. These multi-constituency dialogues produced a thriving – if sometimes precarious – cultural presence of their own. The intention of Thin Place was therefore to ensure that they were given a sense of critical solicitude and informed affirmation in the publication.

Cherry Smyth's poem 'Jumpcut' (2012), for example, described how the physical experience of being in a particular place can invoke a memory of inhabiting that place in the past. The poet is shocked that this past continues to dazzle her with its 'insouciant beauty' (2012:32).

Mark Jones, a psychotherapist, hypnotherapist and soul-centred astrologer, contributed an essay that describes his understanding of time and memory, in particular the journey of the soul through different lives. Reflecting on poetry by Rilke, he focused on the direction of the soul in this life into the next.

Dr Haley Gomez wrote about the formation and evolution of cosmic dust, which blocks out optical light and affects our view of the universe. Her research and discoveries, in the fields of astronomy and quantum physics, are currently destabilising and disrupting reductive scientism from within the field of science itself by positing that matter is, in fact, sentient and participatory.

Brother Joseph MacMahon, a theologian and member of the Franciscan Order in Ireland, wrote an essay reflecting on the Franciscan way of life as a way of acknowledging its legacy in Carmarthen today – even though the Friary remains buried under a large superstore.

The publication rubbed these very different interpretations and reflections on time, memory, place and its relationship to the numinous together with the artworks to see what would happen. I wanted to invite participants to think about the convergence of energy that this project represents, the different, yet relational, mind-sets it induces. I wanted participants and contributors to think about who they are when they are engaging with the work and to decide what they can and cannot digest.

5.5.3 Symposium

As well as presenting audience members with a multiplicity of worldviews from different disciplinary fields in the publication, I felt it important to include bodies of knowledge that were also situated outside of academia. This caused some challenges for audience members. Journalist Kirsten Hinks (2015), in her exhibition review, explained why:

Some found particular combinations of speakers at the symposium to be problematic – having academics or scientists speaking next to artists and even spiritual or religious speakers caused some consternation. There was a worry that one speaker may lend gravitas or authority to the next, who may be less empirical.

It was interesting to discover that quite a few members of the audience found it difficult to consider the less empirical presentations relationally with the more academic presentations. And yet, as Hinks (2015) reflected following conversations with those who attended the event:

It is possibly a more realistic depiction of how the mind really works. Holding different ideas and perspectives, even conflicting ones, at the same time ... Place is a particularly good example of such a state where it is possible to understand a particular place as holding religious significance, scientific interest and personal connections all at once despite that in a strictly factual sense one of these things may render another impossible. Yet in an educational, academic setting we are taught to pit one notion against another in order to attain the truth of the matter. Perhaps 'Thin Place' is an attempt to demonstrate a different way to think and a different way to learn.

The fact that these speakers compelled some members of the audience to confront their own dualisms was a deliberate choice for this symposium. Many came away energised by the slightly fractious, uncertain sense of community that came into being during the symposium. I felt that a fractured sense of community seems to be the kind of community we need to recognise if we are to effectively address climate change. The rituals that dominate our society are the rituals that increasingly suppress the kind of exchange that any genuine democracy depends on. While on the surface Thin Place might have seemed a long way from politics, for many, and for me, this project was an intensely political event because it was concerned with the friction of rubbing different ideas together, with new possibilities and ways of thinking that probably would not come together in any other context.

Everyone in some way took a risk by being involved in this project because everyone was confronted with the fact that they were more and less than the categories they believed defined them.

5.6 Conclusion

The information I gathered from questionnaires issued to visitors to the exhibition endorsed the key claims of my argument, but they did not offer sufficient material for analysis. This was because, on reflection, the questions I asked had a number of limitations: namely, that I did not ask specifically about the presence of animism, the archaic avant-garde, pragmatism and the holistic milieu in the exhibition. It was therefore difficult to say empirically that my thin curatorial approach had engendered an expanded sense of place and porosity between disciplines. There was however one key testimonial that endorsed the aims of my research:

All my life I have experienced slippage into (and out of) other spaces. I have consistently kept these experiences to myself because I was unsure of other people's reactions to this ability/experience. And now at last I feel *found*, and that I am part of a larger community. It is truly wonderful. Thank you.

Although Thin Place did propose concepts that were problematic and contentious for some, overall I felt that it offered an opportunity to navigate difference, opening audiences up to the possibility that 'we are not alone in the world' (Stengers 2012:183–192).

I recognise that the findings I uncovered during this research were not neutral because my methodology was informed by my own epistemology. It was also informed by my ontology by means of the perspective I was looking to find.

In terms of epistemology, I argued that the secularisation of society has not diminished the longing for spirituality and moral authenticity. I felt that the legacy of Enlightenment-based thinking has overlooked how many aspects of faith remain relevant even after their central tenets have been dismissed. I suggested that an enriched materialism could be encountered in a gallery space if multiplicities of worldviews are given a voice. Most importantly, I considered that this might foster an awareness of personhood beyond humans.

I found that when the ideas of the holistic milieu were embedded into curatorial practice, they expressed a conceptual imperative that perceives conjecture (based on thousands of years of storytelling and ritual) to be as valid a perception of reality as rationalism.

Through my ontological stance, I questioned the assumption that there is only one valid worldview – in particular, that the logical-positivist worldview, as a scientifically proven 'truth', should take precedence over other worldviews.

While Thin Place critiqued the conventional metanarratives derived from Enlightenment-based forms of thinking, it did not, however, make any grand claims about the superiority of an experiential approach over an objective scientific approach. It was more interested in creating a confluence between approaches.

All contributors to Thin Place exhibition drew connections between the material world and the numinous, prompting those who engaged with the exhibition, publication, symposium and education programme to see complexly and multiply. The thin places that the artworks in the exhibition described were effectively portals or thresholds between worlds and allowed for a certain kind of seepage from one to the other.

Admittedly, many of the participants, particularly those who came from more traditional academic and rationalist disciplines (such as astrophysics and archaeology), were already challenging the boundaries of reductive thinking within their own fields. Nevertheless, their positive responses to the exhibition and enthusiastic involvement in the project as a whole meant that their experiences could be relayed to their fellow practitioners, who, in turn, might become more open to regarding thin curating as a means of stepping outside of the security of their known disciplinary universe, to inhabit a polyverse, a *thin place*.

Thin Place offered audiences multiple invitations to go to different places. It offered invitations to go into the deep past and to think about what happened in the deep past in terms of the politics of the present. It offered invitations to go into molecular spaces, outer spaces, other lives, past lives. One of the most intriguing things about this project for most people was the difficulty of defining it. The symposium was a series of presentations, but it was not a series of presentations that made sense in any formal or conventional way.

The representation of different artistic, academic, scientific and theological viewpoints in the symposium was not an attempt to reveal an existing path but rather an attempt to invite people to enter what journalist Kirsten Hinks, in her reflections on the exhibition, called 'a woodland' – meaning a seeping, porous layering of diverse thoughts and ideas.

The journey of this project, while feeling distinctive and exciting, is only different in terms of location. It can be applied as an approach anywhere. Because if there are thin places, then I believe there are thin practices as well. So if we believe that thin places allow us to marvel, to see all and through the layers of here and now and then, without losing a sense of multiplicity and simultaneity of time and place, we enter the vibrancy of potentials.

I use the word 'potentials' guardedly, because this is more than a leaning towards futurity, only promising something in some possible future. Instead, it is an invitation to dwell attentively, with, and in, the moment. It is, as Haraway puts it, a 'staying with', a lingering on the threshold of this ecologically troubled world and its more numinous counterpart.

Acknowledgements

The author is grateful to all Staff at Oriel Myrddin Gallery, Carmarthen, and all artists, writers and other contributors to the Thin Place exhibition, symposium and education programme 2015. Special thanks are extended to Dr Iain Biggs and Sarah Bodman.

References

Adams, C. and Green, L. 2014. *Eco Feminism: Feminist Interactions with Other Animals and the Earth*. New York: Bloomsbury Academic.
Bennett, T. 1995. *The Birth of the Museum: History, Theory, Politics*. London: Routledge.
Biggs, I. 2013. *Grounding Ecosophy – reviewing Guattari and Ingold's neo-animism through the uncanny lens of the 'supernatural' Border ballads and the visions of Isobel Gowdie*. Paper presented at Invisible Scotland Conference, University of Dundee, 2013.
Bennett, J. 2010. *Vibrant Matter: A Political Ecology of Things*. Durham: Duke University Press.
Clifford, J. 1991. Four northwest coast museums: Travel reflections. In: Karp, I. and Lavine, S. (eds.), *Exhibiting Cultures: The Poetics and Politics of Museum Display*. Washington, DC: Smithsonian Institution Press, pp. 212–254.
O'Doherty, B. 2000. *Inside the White Cube: The Ideology of the Gallery Space*. Berkley: University of California Press.
Edwards, C., Gandini, L. and Forman, G., eds. 1998. *The Hundred Languages of Children: The Reggio Emilia Approach – Advanced Reflections*, 2nd edn. Westport, CT: Ablex Publishing.
Freire, P. 1993. *Pedagogy of the Oppressed*. London: Penguin.
Geoghegan, H. 2010. Museum geography: Exploring museums, collections and museum practice in the UK. *Geography Compass*, 4(10), 1462–1476.

Geoghegan, H., and Hess, A. 2015. Object-love at the Science Museum: Cultural geographies of museum storerooms. *Cultural Geographies*, 22(3), 445–465.
Guattari, F. 2008. *The Three Ecologies*. London: Continuum.
Haraway, D. 2016. *Staying with the Trouble: Making Kin in the Chthulucene*. Durham: Duke University Press.
Hinks, K. 2015. Thin Place- A Response, Oriel Myrddin website 17th March 2015. http://orielmyrddingallery.co.uk/cy/2015/03/thin-place-a-response-2/
Heelas, P. and Woodhead, L. 2005. *The Spiritual Revolution: Why Religion is Giving Way to Spirituality*. Malden, MA: Blackwell.
James, W. 1985. *The Varieties of Religious Experience*. Cambridge, MA: Harvard University Press.
James, W. 2000 [1907]. *Pragmatism*. London: Penguin Classics.
Jeffrey, B. and Craft, A. 2004. Teaching creatively and teaching for creativity: distinctions and relationships. *Educational Studies*. 30 (1), 77–87.
Jones, O. 2008. Stepping from the wreckage: Geography, pragmatism and anti- representational theory. *Geoforum*, 39, 1600–1612.
Kiberd, D. 2004. The Celtic revival – an Irish renaissance. National Museum of Ireland lecture series in conjunction with the Neo-Celtic Art Exhibition, 26 October.
Lear, R. 2015. Thin Place. *This Is Tomorrow*. 12 February. Available at: http://thisistomorrow.info/articles/thin-place
Melitopoulos, A. and Lazzarato, M. 2012. Assemblages: Félix Guattari and machinic animism [online]. *e-flux*. 36. Available at: www.e-flux.com/journal/assemblages-felix-guattari-and-machinic-animism/ [Accessed 29 April 2015].
Mulk, I. and Bayliss-Smith, T. 2007. Liminality, rock art and the Sami sacred landscape. *Journal of Northern Studies*, 1 (1–2), 95–122.
National Assembly for Wales Commission 2008. *Key Statistics for Carmarthenshire* [online]. Available at: www.assemblywales.org/carmarthenshire.pdf [Accessed 1 November 2013].
Obrist, Hans U. 2004. Curating Now: An International Symposium on Curating Contemporary Art in Public Museums and Galleries, IMMA, Dublin, 10–12 November.
Office for National Statistics 2013. *Personal Well-Being across the UK, 2012/13* [online]. Available at: www.ons.gov.uk/ons/dcp171778_328486.pdf [Accessed 1 November 2013].
Robb, G. 2013. *The Ancient Paths: Discovering the Lost Map of Celtic Europe*. London: Picador.
Robinson, K. 2006. *How Schools Kill Creativity* [video]. Available at: www.ted.com/talks/ken_robinson_says_schools_kill_creativity?language=en [Accessed 10 May 2012].
Serota, N. 2000. *Experience or Interpretation*. London: Thames and Hudson.
Smyth, C. 2012. *Test, Orange*. London: Pindrop Press.
Stacey, S., Feedback on Thin Place 31/3/2015.
Stengers, I. 2012. Reclaiming animism. In: Folie, S. and Franke, A. (eds.), *Animism Modernity Through the Looking Glass*. Koln: Walther Konig, pp. 183–192.
Thayer-Bacon, B. J. 2002. Using the 'R' word again: Pragmatism as qualified relativism. *Philosophical Studies in Education*. 33, 93–103.
Turner, V. 1986. *The Anthropology of Experience*. Chicago, IL: University of Illinois Press.
van Gennep, A. 1960. *The Rites of Passage*. Chicago, IL: University of Chicago Press.
Wilby, E. 2010. *Visions of Isobel Gowdie: Magic, Shamanism and Witchcraft in Seventeenth-Century Scotland*. Brighton: Sussex Academic Press.

Part II

Being in a Climate Change World

6

Multi-temporal Adaptations to Change in the Central Andes

JULIO C. POSTIGO

6.1 Introduction

Adaptation is a crucial factor in the survival of civilisations, populations and social-ecological systems (SES)[1] over time. It becomes particularly relevant when civilisations or states face compounding socio-environmental threats to their ways of life (Butzer 2012). Adaptation entails modifications in SES as a response to actual, or expected, impacts of climate change and interacting non-climatic processes (Moser and Ekstrom 2010). It is the dynamic between change and adaptation that shapes and re-shapes SES, and this relationship is crucial to a civilisation's ability to cope with disturbances that may exceed their collective resilience over the short or long term (Butzer 2012). Additionally, such disturbances may trigger positive and negative feedback with unpredictable consequences (Holling and Gunderson 2002; Bauch *et al.* 2016).

More specifically, adaptation at multiple levels, from civilisations to local communities, requires social structures and institutional arrangements, political and economic organisations, ideological continuity and leadership, resilient ecosystems, technology, access and control over the labour force and broad knowledge of resources, capabilities, threats and opportunities (Acemoglu and Robinson 2012; Butzer 2012). These elements change over time depending on internal dynamics, on interactions with external pressures (e.g. disease, war, markets and climatic events) and on normative views for desired SES (Zhang *et al.* 2011; von Uexkull *et al.* 2016). Moreover, civilisations hold and utilise knowledge on the characteristics of the landscapes on which they live, including knowledge on vegetation, climate, soils, topography and elevation. This provides information

---

[1] SES are integrated systems in which society interacts with the environment (Liu et al 2007, Ostrom 2009).

about which biophysical elements may require modification to create suitable conditions for living in general and for producing food in particular (deMenocal 2001; Turner II and Sabloff 2012).

Adaptation in the Andes is no different; it results from the interactions between civilisations and environmental change and perturbations over centuries. It expresses communities' abilities to respond to old and new, and known and unexpected, processes caused either by local or external events and actors. Civilisations in the Central Andes, including Peru, northern Chile and Argentina and the Bolivian Altiplano, have been adapting their landscapes to address social, biophysical, climatic and environmental conditions over millennia (Dillehay and Kolata 2004; Sietz and Feola 2016). Some adaptations, though ancient, are still in use, such as potato breeding, while others show different degrees of abandonment, for example terraces, sectoral fallow or modification, such as mobility of crops and livestock (Postigo 2013; Zimmerer and Rojas Vaca 2016). These responses have been implemented despite hundreds of years of pressure which started with the Colonial period and were followed by capitalist development and expansion. However, the continuous perturbations have left Andean populations with a weak institutional capacity, deteriorated environmental conditions, limited control over their resources and minimal government support (Bridge 2004; Postigo *et al.* 2013).

Despite the long history of perturbations and adaptive responses, it is recent, anthropogenically caused climate change that calls attention to the concept of adaptation (Noble *et al.* 2014; Steffen *et al.* 2015). Indeed, understanding Andean adaptation to changing climatic conditions is a deeply important question because Andean landscapes, rendered vulnerable by climate change, are crucial to the region's adaptive capacity to climatic and non-climatic challenges (Valdivia *et al.* 2010; Young *et al.* 2011; Postigo *et al.* 2012). It is, however, equally important to realise that focusing too narrowly on the connection between adaptation and present-day climate change may cause at least three issues. First, a narrow focus on climate change as a driver of adaptation would simplify the Andean civilisations' understanding that climate was but one factor among many that shaped their landscapes and which led to adaptations over centuries to the climatic conditions and modifications of the biophysical and environmental characteristics (Herrera and Lane 2006; Young 2009).

Second, a narrow focus may cause us to neglect the interconnections between climatic and non-climatic perturbations (Räsänen *et al.* 2016; Feola 2017). For example, access to resources, which is crucial for adaptation to climate change and other perturbations, varies along dimensions such as class, gender and ethnicity (Noble *et al.* 2014; Olsson *et al.* 2014). In other words, differentiated access is chiefly due to long-term social processes occurring at local to international levels

(Ribot 2010; Carmin *et al.* 2015). This variation in access to, and distribution of, resources generates different levels of adaptive capacity and exposure, or an uneven degree of vulnerability, to perturbations. Ignoring the social underpinnings of unequal vulnerabilities undermines our understanding of how and why adaptation to change may or may not happen (O'Brien *et al.* 2007; Crane *et al.* 2017). Moreover, research shows quite clearly that the drivers of climate change[2] are also weakening cultural elements, such as traditional knowledge and collective property, which previously enabled adaptations in local populations (Walsh-Dilley 2016; Feola 2017).

Third, a narrow focus may overlook the interplay among responses to different perturbations. Adaptations to one perturbation, for example, are known, at times, to use resources that ideally would be reserved for use to adapt to a later challenge (McDowell and Hess 2012). The relevance of the aforementioned issues, particularly the effects of interacting social, environmental, climatic and economic processes, is underscored by the multi-stressor (Wilbanks and Kates 2010; Feola *et al.* 2015; Räsänen *et al.* 2016) and double-exposure frameworks (Leichenko and O'Brien 2008).

The premise of this chapter is that climate change is not the only perturbation for Andean populations; it may even have less affect than pressures caused by mining, government neglect and marginalisation. The chapter argues that it is necessary, therefore, to broaden our understanding of the relationship between climate change and adaptation by showing how Andean adaptation to climate change interacts with non-climatic perturbations driven by capitalism. It begins by defining the concept of adaptation and illustrating its relationship to Andean civilisations' continuity over time. The chapter then presents two ancient Andean adaptations to biophysical and environmental characteristics, which, though originating thousands of years ago, have been modified over time. These cases demonstrate that adaptation is a normal part of people's livelihood strategies that rely on accessing natural and social resources (Valdivia *et al.* 2010; McDowell and Hess 2012). Following these cases, the chapter presents some important effects of climate change on Andean water resources, people's responses to such effects and then briefly illustrates how interacting multiple stressors are currently impacting on Andean households' capacity to adapt. Finally, the chapter explains how local knowledge and institutions enable crucial Andean responses to climatic and non-climatic perturbations. Ultimately, this chapter reveals that adaptation to climate change in the Andes and beyond should be understood through the dynamic interactions of changing environmental, climatic and social conditions over time.

---

[2] Understood as the societal characteristics that have most influence on climate dynamics (Rosa *et al.* 2015).

6.2 Adaptation of Andean Landscapes

Andean civilisations continually modify characteristics of their natural landscape, including local climate, slope and aspect. The landscape contains both the result of previous human–environment interactions and the potential to adjust to changing conditions (Herrera and Lane 2006). For instance, the changing conditions wrought by the Spanish invasion and colonial enterprise were major stressors on the Inka state and the landscapes on which it lived (Gade 1992). Local administrative and belief systems were destroyed, large portions of the population were killed, displaced and resettled (i.e. *reducciones*), forced labour in mines and textile workshops were expanded (i.e. the *mita* and *obraje*), new crops and animals were introduced, which displaced native cultivars to marginal lands and towns were located and designed according to Spanish planning and needs (Knapp 2007; Moore 2017). In such a context of drastically changing landscapes, some adaptations disappeared, or became outdated, whereas others remained and are still current or have been modified. Next, the chapter turns to reviewing two particularly relevant Andean adaptations: the domestication of plants and animals; and the access to, and control of, diverse ecological zones. These two fundamental adaptations are present across multiple landscapes; they survived both the Colonial period and the expansion of capitalism, although they have also been modified over time.

6.2.1 Domestication of Plants and Animals

Domestication is a dynamic adaptation between humans and the environment which enabled the emergence of farming systems and Andean civilisations (Piperno and Dillehay 2008; Young 2008). Domestication requires sophisticated technologies, social institutions and an understanding of the trends and patterns of precipitation and temperature, as well as the limitations and opportunities wrought by biophysical characteristics like soil quality, altitudinal gradients, topography, and animal and plant diversity.

Domestication of potatoes occurred around 6000 years ago (Brush *et al.* 1995). Domestication and selection of potato varieties provide natural support for risk-reduction strategies when facing climatic variability and extreme weather events by increasing the diversity of crops and planting locations (Nakashima *et al.* 2012). The dynamics of potato seeds thus reveal strategies adaptive to local environmental conditions and harsh climatic conditions (Arce *et al.* 2018). For example, in the Peruvian Central Andes (Cuzco, Peru) the patterns of exchange of native cultivars are characterised by households obtaining seeds from neighbours or, in the case of a new household, from the couple's parents (Zimmerer 1991). If there is a crop failure in a community, rendering local networks of exchange useless, Andean farmers obtain potato seeds from the nearest unaffected area. Additionally, farmers have guidelines for what climate and elevation the area where they are obtaining their seeds should have, based on the kind of crop failure they experienced (Zimmerer 1991). Notwithstanding these adaptations, native potatoes are threatened by a variety of factors, including the expansion of less diverse commercial varieties favoured by the market, seed certification schemes that overlook local mechanisms of seed exchange and the reduction of suitable land for wild potatoes because of contemporary climate change (Brush 1995; Lutaladio and Castaldi 2009). It is important to note that climate change is, therefore, just one among several factors that threaten domestication.

The transformation of wild animals into domestic ones is a socially driven process involving physiological and genetic changes to the species (Bettinger et al. 2009; Pearsall 2009). Domesticated animals are also an important adaptation in Andean farming systems, particularly at altitudes above the limits of agriculture. Humans living at 4000 m or higher adapted to their landscapes by transforming and using animals as efficiently as possible, given the specific natural resources available at such high elevations (Browman 1983). The domestication of camelids is one of the most important transformations carried out by pastoralist cultures in their long-term process of adaptation to the highlands (Izeta 2008). In the past 5000 years, most of the animal protein consumed by high-elevation Andean populations has been from domesticated camelids (Shimada et al. 1988), except those high-altitude communities with cervid populations within range of mobile hunters (Rick 1980; Aldenderfer 2001). The animal proteins complement calorific intake chiefly obtained from tuberous crops (Moore et al. 1988).

The two domesticated South American camelids are the alpaca (*Vicugna pacos*) and the llama (*Lama glama*) (Kadwell et al. 2001). The alpaca provides fibre, meat and hide, whereas the llama is a service provider as a beast of burden. Additionally, llamas provide wool for rope and meat for charqui, a form of jerky made from dried, salted meat. Moreover, in times of extreme scarcity, llamas are slaughtered for their meat. The Colonial period marked the onset of the displacement of alpacas and llamas to marginal arid lands in the high Andes (Baied and Wheeler 1993), and present-day pastoral SES are pressured by a variety of threats, including modifications to the extent of wetlands and the location of springs caused by habitat conversion and climate change, mining appropriation and pollution of land and water, government neglect and sustained poverty (Postigo 2013; Verzijl and Quispe 2013; Dangles et al. 2017).

6.2.2 Access to, and Control of, Diverse Ecological Zones

One of the best-known adaptations of Andean civilisations is the capacity to access the maximum number of ecological zones at different altitudes, also known as the model of verticality (Murra 1967). The notion of ecological zones draws on the concept of natural life zones, which are units defined by the following climatic variables: (i) mean annual biotemperature; (ii) mean annual precipitation and (iii) potential evapotranspiration ratio. Each combination of these three variables supports a group of possible plant communities and can be designated as a 'natural life zone' (Tosi Jr and Voertman 1964; Holdridge 1967). Ethnohistorical work shows that the verticality model is the ideal form of agrarian organisation in the Andes. Within this model each ethnic group aims to achieve self-sufficiency by simultaneously accessing diverse resources across different ecological zones, which form a kind of 'vertical archipelago' (Brush 1976). Ethnic territories largely disappeared with the Inka while archipelagos were severely disrupted by Spanish occupations. However, nowadays, there are both compact and extended variations of the ideal model still in place. The compact version applies to very steep valleys where the different ecological zones are next to each other along the altitudinal gradient (Fonseca and Mayer 1976). The extended version applies to large inter-Andean valleys where resources flow through long networks of exchange and trade, replacing direct control with circulation of goods (e.g. Custred 1974; Browman 1975). Control over ecological zones is also extended to zones for extracting non-farming resources, such as wood and salt (Núñez and Dillehay 1998; Valdiva and Ricard 2010).

A variation of the extended model was developed in the Central Andes in Bolivia, northern Chile and Argentina, particularly in the Altiplano (a large flat area), where controlling ecological zones hundreds of kilometres away from the village was impractical (Browman 1981). This variation of the model is characterised by caravans accessing distant zones and products through trade and exchange in settlements established by the caravans (Núñez and Dillehay 1998). In so doing, access to resources is achieved through economic means, rather than the exertion of political control (Janusek 2004). Communities from the Altiplano also provide an early example (1350–1850 BC) of regional trade of ceramics, sumptuary commodities, such as metals, stones and seashells and food, such as maize, coca leaves, potatoes and quinoa, in the area from the Nudo de Vilcanota to Lake Poopo in Oruro, Bolivia (Browman 1981). The Classic Tiwanaku civilisation (375–750 AD) perfected the Altiplano mode of accessing resources from distant ecological zones by adding specialised production by communities (e.g. textiles, pottery) and establishing networks of trade and periodic markets (Browman 1981; Nielsen 2009). The Tiwanaku's use of alpacas and llamas as beasts of burden was crucial for trading goods produced by populations located far away, for the production of meat and wool and for ceremonial purposes (Kolata 1993).

Interestingly, pastoralists ultimately modified the early Altiplano model as an adaptation to changing socio-political and environmental contexts, such as warfare, and the integration into the Inka state (Nielsen 2009). In doing so, three new types of exchange networks became identifiable: (i) long-distance caravans; (ii) open networks of neighbouring communities linked by sequences of exchange and (iii) exchange points where groups from different regions meet seasonally (Núñez and Dillehay 1998). Llamas are still currently used in caravans, although their primacy is challenged by the expansion of the road network (Tripcevich 2010). For instance, llamas are still used in caravans between Qochuama (Canchis province, Peru) and Marcapata (Cuzco) for exchanging livestock products for crops (Sendón 2009); Ccalaccapcha (Páucar del Sara Sara, Ayacucho, Peru) and Huaquirca (Antabamba, Apurímac, Peru) for transporting salt (Valdiva and Ricard 2010) and Huarhua (Cotahuasi, Arequipa, Peru) and southern Apurimac, Peru, also for carrying salt, which is then traded for maize and tubers (Tripcevich 2008). Llama caravans also trade different products, for example salt, meat, maize and potatoes, along their journeys (Tripcevich 2008; Tripcevich 2010). In doing so, caravans link people from different places and provide them with access to commodities produced far away and under different ecological conditions.

Alliances, in the form of marriage or symbolic kinship, are another way for pastoralists to access diverse ecological zones and exert some level of indirect control over them (Valdiva and Ricard 2010). Marriage is particularly important because it grants access to the in-laws' resources, such as land and labour force. For example, an analysis of the genealogy of farmers of the Lacco sector (Marcapata, Cusco) shows two trends in their marriage practices (Sendón 2016). First, they tend to marry herders from Phinaya (Pitumarca, Cusco) to access extensive pastures, while the Phinaya people gain access to farmland in Lacco. Second, Lacco farmers often marry farmers from Sahuancay (Marcapata, Cusco) in order to access lands for maize cultivation. In exchange, the Sahuancay families access lands for potato cultivation, land that is called Sahuancaypampa ('the land for Sahuancay people') and thus is named after them (Sendón 2016). Symbolic kinship secures a continued exchange of goods, such as meat for crops, in an arrangement in which the terms of trade, including commodity prices, are stable (Valdiva and Ricard 2010).

Currently, agro-pastoral households and extended families access different zones by moving their livestock around grazing areas, which allows for pasture rotation and decreases the threat of overgrazing. Grazing areas are used following a pattern of seasonal migration, whereby areas that are wetter or closer to water sources are used in the dry season (April–October) and areas that tend to be drier are used in the rainy season (November–March) (Inamura 1986; Postigo 2013). Having access to multiple grazing areas can be seen as a form of control of different ecological zones. There are also cases in which Andean pastoralists use grazing areas at different elevations, whereby animals graze at lower elevation areas in the rainy season and at higher altitude pastures during the dry season (Postigo et al. 2008; López-i-Gelats et al. 2015).

The forms of mobility outlined above depend on community-based institutional arrangements (López-i-Gelats et al. 2015). Households and extended families gaining and maintaining access to seasonal grazing areas draw on community institutions governing land allocation (Postigo 2013). The herding and maintenance tasks needed in grazing areas require household institutions for the division of labour by gender and generation (López-i-Gelats et al. 2015; Struelens et al. 2017). Community-level decisions and the relations between community and households are managed through explicit norms (Postigo et al. 2008; Verzijl and Quispe 2013).

6.3 Andean Responses to Climate Change and Their Interplay with Non-climatic Perturbations

So far, this chapter has discussed some of the major human responses in the Andes to dynamic environmental and biophysical characteristics. The following section presents, firstly, some of the major effects that climate change is anticipated to have on Andean water resources and, secondly, people's responses to such effects and how they are affected by capitalism.

6.3.1 Effects of Climate Change on Andean Water Resources

Climate change has multiple and interconnected impacts on mountains (Beniston 2003; Körner et al. 2005). Modified precipitation, temperature and atmospheric humidity are melting glaciers, changing soil composition and shifting altitudinal ranges of vegetation (Nogués-Bravo et al. 2007; Rosenzweig et al. 2007; López et al. 2017b), thereby changing, and sometimes degrading, mountain ecosystem functions and services (Celleri 2010; Anderson et al. 2011; Buytaert et al. 2011).

People whose livelihoods are dependent on mountain ecosystems in general, and on the Andes in particular, are vulnerable to the changes wrought by climate change (Valdivia et al. 2010). Colder nights and hotter days are negatively impacting agropastoral systems, as well as biodiversity in the Andes (Young 2009; Postigo 2014). Andean herders, agro-pastoralists, farmers and urban dwellers are vulnerable to nightly frosts, which inhibit crop growth, while intense heat during the day burns crops and dries pastures (FAO 2008; Seth et al. 2010).

The impact of climate change on the amount and availability of water is very important for the Andean region (Bradley et al. 2006; De Bièvre et al. 2012). The impacts on water resources are widely felt, as they involve urban and rural populations, and affect multiple uses, including pastoralism, mining and industry

(Bury *et al.* 2013; Buytaert *et al.* 2017). A lower availability of water will negatively impact irrigation, and the increasing variability of precipitation will compromise rain-fed agriculture and power generation, particularly in the dry season (Bury *et al.* 2013; Buytaert *et al.* 2017). Thus, the impact of climate change compromises peasants' and farmers' food security and subsistence activities since their productive capacity is hindered and their livelihoods are jeopardised (Postigo 2013; Montaña *et al.* 2016).

Pastoralists are also affected by climate change through glacial retreat (López-i-Gelats *et al.* 2016). This retreat may increase the amount of runoff in the short term and, in turn, may also increase the extent of pasturelands. However, in the long term, after passing the peak of the runoff (see Baraer *et al.* 2011), the effects may reverse the expanding of the wetlands and pastures, leading to negative impacts on herders' livelihoods (Dangles *et al.* 2017; Polk *et al.* 2017).

In addition to these effects, Andean livelihoods are exposed to non-climatic perturbations that make people vulnerable and may limit their capacity to adapt to climatic changes (Feola 2017; O'Brien *et al.* 2007; O'Brien *et al.* 2004). Further, the expansion of capitalism in the Andes has not only been driving and exacerbating most of the non-climatic perturbations but has also been destroying Andean adaptations (Postigo 2011; Walsh-Dilley 2016; Moore 2017). For instance, Peru has issued 1810 water rights to 331 mining companies, 248 of which are in perpetuity, with 31 located in 12 zones that are of high risk of drought (Salazar Vega 2018). Thus, increased vulnerability is generated by mining operations using water in drought-prone areas (Salazar Vega 2018), similar to situations in which agribusinesses over-extract from a desert aquifer and usurp control of water from herders in the highlands (Postigo *et al.* 2013; Marshall 2014). Even more extreme, the case of lead-poisoned children and polluted water and land due to mining operations in Cerro de Pasco (Peru) illustrates how external disturbances can modify the life expectancy of populations and destroy key livelihoods (Smuda *et al.* 2007; Dajer 2015; Graeter 2017).

6.3.2 Responses to the Effects of Climate Change and Their Interactions with Non-climatic Perturbations

Responses to the impacts of climate change on water resources include the introduction of polyculture growing techniques and switching to cash crops with shorter growing periods and higher tolerance to water stress (Montaña *et al.* 2016). In the Peruvian southern Andes, wheat and fava bean are replacing maize because the former are more resistant to cold spells. In addition, farmers are cultivating in furrows, which improves water management in the face of water scarcity (Postigo 2014). However, many communities want more permanent solutions, such as infrastructure to store water, and modern irrigation systems to water pastures and crops. A potential downside of these solutions is that converting pastures into alfalfa fields to feed dairy livestock will displace grazing alpacas to areas located at higher elevations (Postigo 2014). This would have the effect of pushing herders to increasingly marginal lands, thus potentially putting them in closer proximity to mining companies and rendering pastoralists' households with fewer resources and capacity for adaptation.

In the high Andes, pastoralist households respond to the drying up of grazing areas, which is a result of climate change, by increasing livestock mobility within their pastures, creating and expanding wetlands through irrigation, limiting the allocation of wetlands to new households and cultivating grasses (Postigo 2013; Verzijl and Quispe 2013). These responses, however, are severely compromised by socio-economic pressures on natural resources, especially water and land, and the lack of capacity to adapt. For instance, mining concessions located on the headwaters of watersheds limit control of and access to water and threaten water quality. Of all the mining concessions in Peru, more than 30 per cent are over 4000 metres above sea level (masl) and more than 55 per cent are above 3000 masl (Bebbington and Bury 2009).

Access to land, a crucial resource for household livelihoods, is transferred from communities to mining companies in exchange for money, employment in the mines and development projects, thus leaving many households in livelihood insecurity (Bebbington and Bury 2009). Another pressure on pastoralists' livelihoods is the increase in extraction of water from the highlands to irrigate coastal crops for export (Oré et al. 2012; Postigo et al. 2013). This limits pastoralists' control of and access to water for irrigation of pasture and crops. Moreover, on the coast, land scarcity drives both expansion of the agrarian frontier into the desert and agricultural intensification leading to soil degradation, which jeopardises the sustainability of this critical resource (Marshall 2014).

6.4 Enablers of Andean Responses to Multiple Perturbations

This section shows that local knowledge and institutions enable and support fundamental Andean responses to climatic and non-climatic perturbations.

6.4.1 Local Knowledge

It is well established that knowledge is an important element in local responses to climatic and other changes (Berkes et al. 2000). One of the most noticeable elements of local knowledge is the understanding of atmospheric and biological indicators for weather forecasting and farming decision-making (Gilles and

Valdivia 2009; López et al. 2017a). Andean farmers, from central Peru to the Bolivian Andes, observe the Pleiades, a star cluster in the Taurus constellation, immediately after the winter solstice to forecast the rainy season. During this period, the brightness of the constellation is observed with the goal of determining when the rainy season will start and how much rain can be expected (Orlove et al. 2000; Orlove et al. 2002). These assessments allow farmers to adjust their farming practices, such as the date on which to begin planting, in order to accommodate inter-annual precipitation variability (Orlove et al. 2000). In Apurímac and Cusco, farmers forecast whether it will be a good year for farming potatoes and maize based on whether or not the *huaraco* (*Austrocylindropuntia, Cactaceae*) has flowers (Servicio Nacional de Meteorología e Hidrología del Perú (SENAMHI) 2017). In the Lake Titicaca region, communities use the location of birds' nests as an indicator of the amount of precipitation to expect; the farther (or higher) the nests are, the higher the lake will rise due to heavy precipitation (Claverías 2000).

Local knowledge about the landscape is crucial if farmers are to take advantage of potential opportunities wrought by climate change, such as improved land suitability due to warmer conditions. In this context, there is a need for knowledge about soil characteristics for cultivating crops at higher elevations where previous climatic conditions have prevented agriculture (Vos and del Callejo 2010). Furthermore, the upwards expansion of farming may be seen as an illustration of accessing a new suitable ecological zone for crops usually cultivated at lower elevations. This expansion and its caveats are further explored below at the household and macro levels.

Households, for instance, have responded to a changing climate by cultivating new crops, like tomatoes, since 2003 in the isle of Anapia (Lake Titicaca, Peru) (Flores 2009). Elsewhere, peasants continue to improve their productive systems with crops such as maize, broad beans, green peas, amaranth (*Amaranthus*) and sweet granadilla (*Passiflore ligularis*), the altitudinal limits of which have shifted upwards (Halloy et al. 2006; Araujo 2008). In addition to improved climatic conditions and local knowledge, these adaptations might be supported by institutions that can address key questions such as: How are new farming lands allocated? To what extent does land use change compromise or enhance the resources that might eventually be needed to respond to other challenges? Furthermore, it is worth considering the possible effects of introducing new crops into household diets, the economy of the community and the farming system, such as the incidence of pests, costs and the types of agricultural inputs required.

Local knowledge, however, is challenged by the increasing dominance of market-oriented crops and livestock, the lack of interest shown by younger generations but whose subsequent migration out of the Andes compromises intergenerational knowledge transmission, the speed and intensity of climate change and knowledge marginality (Gilles *et al.* 2013). The education system hardly recognises local knowledge and, if it is mentioned at all, it is merely as devalued folktales that belong to old indigenous peasants (Gilles and Valdivia 2009; Valdivia *et al.* 2010). Moreover, there is little effort to combine scientific and local knowledge for the generation of more robust weather forecasting with higher levels of acceptance and use (Gilles and Valdivia 2009; López *et al.* 2017a).

6.4.2 Institutions

Changes in Andean farming practices are implemented and supported by institutional changes. In Bolivia, maize and peaches are rearranged to form clusters of plots in order to improve labour productivity of caretakers[3] who are farming due to high emigration from the area (Zimmerer and Rojas Vaca 2016). Like sectoral fallow, this clustering diminishes walking time among fields, enables coordination-based activities, such as renting tractors and irrigation, and improves protection of crops from intruders. Moreover, the livestock may graze on the maize field after harvest, which requires coordination of planting and harvesting time to prevent crop damage by the livestock. Though the fields do not lie fallow as they used to in the old system, the traditional Andean practice of clustering and coordination is used as a response to the scarcity of a labour force (Zimmerer and Rojas Vaca 2016). The modification of the farming calendar is an adjustment of agricultural tasks to changes in rainfall regimes and temperatures (Gilles and Valdivia 2009). The downside of this practice is twofold. First, the sowing season is concentrated into a short period of time, which generates a great demand for water that is not always available. Second, a shorter sowing season limits the seed growth period, yielding a smaller harvest.

The pressure and demands on institutions for water use are inversely related to water availability. Decreasing volumes of the resource increases the tensions among users and the pressures that they are willing to put on higher levels of government (Boelens *et al.* 2002; Meinzen-Dick 2007). The increasing water demanded by high-productivity agriculture spurs tensions among farmers (Oré *et al.* 2012; Montaña *et al.* 2016). In turn, high-productivity agriculture is brought about by the need to generate profits that will enable farmers to pay off their debts to the financial system (Postigo *et al.* 2013). Organisations addressing these tensions require water governance institutions, such as Watershed Councils, strengthened with conflict resolution mechanisms to help prevent serious conflict (Boelens *et al.* 2002).

---

[3] People left in charge by the emigrant farmer to cultivate the land.

Farmers expect that institutional improvements for better water management adapt to socio-economic pressures and the effects of climate change. However, this is not occurring at present. For example, in Arequipa (Peru), the prevalence of agriculture over herding means that the water board grants priority to agricultural irrigation over pastoral irrigation. As a result, pastoralists are not allowed to use water from rivers and streams that is being used to irrigate crop fields in the valleys. Access restrictions and water scarcity fuel conflicts over the control of water resources between upstream and downstream users (Prado 2011; Red Muqui 2011), farmers and peasants, agriculturalists and pastoralists and between subsistence and market economies (Boelens et al. 2002; Alerta Perú 2011).

Institutional improvements include increasing the capacity of water authorities to enforce norms and standards in a fair and rigorous manner. However, an institutional change from within seems hard to make because, in the face of water scarcity, trust between farmers and local authorities has diminished and labour exchanges for irrigation have faded. Not surprisingly, irrigation rules frequently fail and norms of enforcement are weak (Oré and Geng 2014). Thus, despite Andean knowledge, institutional adaptive governance of natural resources in response to increasing socio-economic and climatic pressures needs to (re)connect individuals and organisations at multiple levels, both outside and within the government (Montaña et al. 2016; Feola 2017). Furthermore, institutional arrangements should provide peasants with a voice equal to other, more powerful, stakeholders (Doughty 2016).

6.5 Conclusion

Andean capacities and strategies for adaptation are forged by agrarian civilisations through their longstanding interactive and mutually transformative relations with their landscape (Dillehay and Kolata 2004; Herrera et al. 2006). However, as this chapter has argued, the intensity and rapid pace of current climatic change and the loss of some productive strategies due to social and economic conditions (Valdivia et al. 2010; Sietz and Feola 2016) are likely to overcome peasants' adaptive capacities and resilient productive strategies, thus increasing their vulnerability (Sperling et al. 2008; Stadel 2008; McDowell and Hess 2012).

Archeological research shows that pre-Columbian Andean civilisations exhibited resilience and adaptive capacity because they were part of the Inka state, which had a landscape perspective towards identifying and responding to socio-environmental change. Many of these strategies, although modified, have become part of the Andean culture as farming practices. For instance, households manage plots in different ecological zones, although sectoral fallow in the highlands is communally managed. The diversity of practices and strategies exhibited by farmers is supported by dynamic institutional arrangements that define norms concerning access, control and the allocation of water and land. The latter resource, land, has become particularly relevant as high numbers of young people leave their communities as a part of widespread emigration. Even more dramatic has been community resistance to mining operations that aim to use water from lakes considered sacred and which threaten the headwaters of multiple basins.

Some adaptations may compromise the natural and social resources needed for adaptation to current and future perturbations. Examples include emigration from the Andes, which limits the labour force available for farming and adaptive activities, and the 'quinoa boom', which has caused soil degradation, diminished crop variety and weakened communal land tenure institutions (Jacobsen 2011; Walsh-Dilley 2016). Moreover, expansion of urban areas, agrarian activities and mining increases the demand for those land and water resources that are desperately needed for Andean agricultural adaptation (Polk *et al.* 2005; Bury *et al.* 2013; Salazar Vega 2018). Furthermore, poverty and marginalisation undermine resources for adaptation while increasing the vulnerability of the Andean population (López *et al.* 2017a). Thus, research is needed to understand the complex interconnections between access to resources, adaptation and perturbations from multiple stressors, including climate change, but not restricted to it (Agrawal and Lemos 2015; Nelson *et al.* 2016).

In the transformative interaction between nature and society, the farming of crops and animals is both the link between the interacting parts and a consequence of such interaction. This transformation of the landscape involves domestication of animals, agricultural terraces, irrigation and the introduction of non-native species (Moseley 2001; Knapp 2007; Postigo *et al.* 2008). In bringing about these transformations, the landscape becomes the livelihood of Andean populations. This landscape, however, has also provided resources for capitalism over centuries, and this profit-driven way of organising people and places has deepened the uneven distribution of resources among people and their access to them (Thorp and Bertram 1978; Moore 2017). This specific form of organising nature–society interactions limits both livelihood strategies and adaptive capacity and enhances the vulnerability of marginal populations (Postigo and Young 2016; Sietz and Feola 2016).

At the macro level, there is considerable potential for expanding the agrarian frontier to higher altitudes and increasing farmers' crop portfolios as a result of climate change in Andean countries. However, beyond those individuals or communities who possess the resources and capacities to innovate, this potential benefit will only come about with the introduction of state policies and resources for a major transformation in small-scale farming. The required policies include modernisation, expanding sustainable agriculture, strengthening small-scale farming, improving infrastructure, technology, rural services and financial programs, building research on potentially adaptive varieties of crops and developing new energy sources (Hoffmann and Requena 2012; Postigo *et al.* 2012).

This chapter demonstrates the continuity of cultural elements since the pre-Columbian era. It also underscores the urgency for the introduction of national and sub-national policies that not only address resilience in the face of climate change but also tackle the non-climatic dimensions of vulnerability, such as poverty, hunger, exclusion and marginality, while challenging current social structures and the dominant political-economic model of development (Postigo 2011; Olsson *et al.* 2014; Feola 2017). Broad policies that build a safety net for rural populations, including conditional cash transfers, have been effective in moving families out of poverty while also improving health and schooling outcomes (Rawlings and Rubio 2005; De Janvry *et al.* 2006). Other policies that might be effective in targeting farmers may include programs for improving rural health and housing, which will diminish households' vulnerabilities and enhance their resilience. Programs should also endeavour to support farmers by correcting market imperfections that lead to unfair competition with imports subsidised in their country of origin. Finally, a locally delivered public extension service and land security that protect family farming and limit large-scale land acquisitions from taking control of land, water and political power are also necessary. Without these policies, local responses are largely insufficient to counter the long-term, compounding socio-economic, political and climatic changes that are increasing the vulnerability of Andean households and communities.

Acknowledgements

I would like to thank my colleagues Ignacio Cancino, Santiago López, Pablo Palomino, Manuel Peralvo and the editors for their helpful comments on this chapter.

References

Acemoglu, D. and Robinson, J. A. 2012. *Why Nations Fail. The Origins of Power, Prosperity, and Poverty*, New York, NY: Crown Business.

Agrawal, A. and Lemos M. C. 2015. Adaptive development. *Nature Climate Change*, 5, 185.

Aldenderfer, M. 2001. Andean pastoral origins and evolution: The role of ethnoarchaeology. In: Kuznar, L. A. (ed.) *Ethnoarchaeology of Andean South America: Contributions to Archaeological Method and Theory*. Ann Arbor, MI: International Monographs in Prehistory.

Aldenderfer, M., Craig, N. M., Speakman, R. J., and Popelka-Filcoff, R. 2008. Four-thousand-year-old gold artifacts from the Lake Titicaca basin, southern Peru. *Proceedings of the National Academy of Sciences*, 105, 5002–5005.

Alerta, Perú. 2011. Majes Siguas no es un conflicto entre regiones. Available: http://alertaperu .pe/publicar/nacionales/1732-majes-siguas-no-es-un-conflicto-entre-regiones.html [Accessed 05/04/2011].

Anderson, E. P., Marengo, J., Villalba, R., Halloy, S., Young, B. E., Cordero, D., Gast, F., Jaimes, E., and Ruiz, D. 2011. Consequences of climate change for ecosystems and ecosystem services in the tropical Andes. In: Herzog, S. K., Martïnez, R., Jorgensen, P. M., and Tiessen, H. (eds.) *Climate Change and Biodiversity in the Tropical Andes*. Montevideo, Uruguay: Inter-American Institute for Global Change Research (IAI) and Scientific Committee on Problems of the Environment (SCOPE).

Araujo, H. G. 2008. Estrategias de las comunidades campesinas altoandinas frente al cambio climático. In: Araujo, H. G. (ed.) *Los Andes y las Poblaciones Altoandinas en la Agenda de la Regionalización y la Descentralización*. Lima: CONCYTEC.

Arce, A., De Haan, S., Burra, D. D., and Ccanto, R. 2018. Unearthing unevenness of potato deed networks in the high Andes: A comparison of distinct cultivar groups and farmer types following seasons with and without acute stress. Frontiers in Sustainable Food Systems, 2, 43.

Baied, C. A., and Wheeler, J. C. 1993. Evolution of high Andean puna ecosystems: Environment, climate, and culture change over the last 12,000 years in the Central Andes. *Mountain Research and Development*, 13, 145–156.

Baraer, M., Mark, B. G., and McKenzie, J. M. 2011. Past peak water in Peru's Cordillera Blanca: Diagnosing the demise of glacier influence on stream discharge. San Francisco, CA: AGU.

Bauch, C. T., Sigdel, R., Pharaon, J., and Anand, M. 2016. Early warning signals of regime shifts in coupled human–environment systems. Proceedings of the National Academy of Sciences, 113, 14560–14567.

Bebbington, A. J., and Bury, J. T. 2009. Institutional challenges for mining and sustainability in Peru. *Proceedings of the National Academy of Sciences*, 106, 17296–17301.

Beniston, M. 2003. Climatic change in mountain regions: A review of possible impacts. *Climatic Change*, 59, 5–31.

Berkes, F., Colding, J., and Folke, C. 2000. Rediscovery of traditional ecological knowledge as adaptive management. *Ecological Applications*, 10, 1251–1262.

Bettinger, R., Richerson, P., and Boyd, R. 2009. Constraints on the development of agriculture. *Current Anthropology*, 50, 627–631.

Boelens, R., Dourojeanni, A., Durán, A., and Hoogendam, P. 2002. Water rights and watersheds. Managing multiple water uses and strengthening stakeholder platforms. In: Boelens, R., and Hoogendam, P. (eds.) *Water Rights and Empowerment*. Assen, the Netherlands: Van Gorcum.

Bradley, R. S., Vuille, M., Diaz, H. F., and Vergara, W. 2006. Threats to water supplies in the tropical Andes. *Science*, 312, 1755–1756.

Bridge, G. 2004. Contested terrain: Mining and the environment. *Annual Review of Environment and Resources*, 29, 205.

Browman, D. L. 1975. Trade patterns in the central highlands of Peru in the first millennium B.C. *World Archaeology*, 6, 322–329.

Browman, D. L. 1981. New Light on Andean Tiwanaku: A detailed reconstruction of Tiwanaku's early commercial and religious empire illuminates the processes by which states evolve. *American Scientist*, 69, 408–419.

Browman, D. L. 1983. Andean arid land pastoralism and development. *Mountain Research and Development*, 3, 241–252.

Brush, S. B. 1976. Man's use of an Andean ecosystem. *Human Ecology*, 4, 147–166.

Brush, S. B. 1995. In situ conservation of landraces in centers of crop diversity. *Crop Science*, 35, 346–354.

Brush, S., Kesseli, R., Ortega, R., Cisneros, P., Zimmerer, K., and Quiros, C. 1995. Potato diversity in the Andean center of crop domestication. *Conservation Biology*, 9, 1189–1198.

Bury, J., Mark, B. G., Carey, M., Young, K. R., McKenzie, J. M., Baraer, M., French, A., and Polk, M. H. 2013. New geographies of water and climate change in Peru: Coupled natural and social transformations in the Santa River watershed. *Annals of the Association of American Geographers*, 103, 363–374.

Butzer, K. W. 2012. Collapse, environment, and society. *Proceedings of the National Academy of Sciences*, 109, 3632–3639.

Buytaert, W., Cuesta-Camacho, F., and Tobón, C. 2011. Potential impacts of climate change on the environmental services of humid tropical alpine regions. *Global Ecology and Biogeography*, 20, 19–33.

Buytaert, W., Mould, S., Acosta, L., De Biévre, B., Olmos, C., Villacis, M., Tovar, C., and Verbist, K. 2017. Glacier melt content of water use in the tropical Andes. *Environmental Research Letters*, 12, 114014.

Carmin, J., Tierney, K., Chu, E., Hunter, L. M., Roberts, J. T., and Shi, L. 2015. Adaptation to climate change. In: Dunlap, R. E., and Brulle, R. J. (eds.) *Climate Change and Society. Sociological Perspectives*. New York, NY: Oxford University Press.

Celleri, R. 2010. Estado del conocimiento técnico-científico sobre los servicios ambientales hidrológicos generados en los Andes. In: Quintero, M. (ed.) *Servicios ambientales hidrológicos en la región Andina. Estado del conocimiento, la acción y la política para asegurar su provisión mediante esquemas de pago por servicios ambientales.* Lima: IEP/ CONDESAN.

Claverías, R. 2000. Conocimientos de los campesinos andinos sobre los predictores climáticos: elementos para su verificación. *Seminario-Taller organizado por Proyecto NOAA (Missouri)*. Chucuito-Puno.

Crane, T. A., Delaney, A., Tamás, P. A., Chesterman, S., and Ericksen, P. 2017. A systematic review of local vulnerability to climate change in developing country agriculture. *Wiley Interdisciplinary Reviews: Climate Change*, 8, e464.

Custred, G. 1974. Llameros y comercio interregional. In: Alberti, G., and Mayer, E. (eds.) *Reciprocidad e intercambio en los andes peruanos*. Lima: Instituto de Estudios Peruanos.

Dajer, T. 2015. High in the Andes, a mine eats a 400-year-old city. *National Geographic* [Online]. Available: https://news.nationalgeographic.com/2015/12/151202-Cerro-de-Pasco-Peru-Volcan-mine-eats-city-environment/ [Accessed 10/30/2017].

Dangles, O., Rabatel, A., Kraemer, M., Zeballos, G., Soruco, A., Jacobsen, D., and Anthelme, F. 2017. Ecosystem sentinels for climate change? Evidence of wetland cover changes over the last 30 years in the tropical Andes. *PLOS ONE*, 12, e0175814.

De Biévre, B., Bustamante, M., Buytaert, W., Murtinho, F., and Armijos, M. T. 2012. Síntesis de los impactos de los efectos del cambio climático en los recursos hídricos en los Andes Tropicales y las estraegias de adaptación desarolladas por los pobladores. In: Cuesta, F., Bustamante, M., Becerra, M. T., Postigo, J., and Peralvo, M. (eds.) *Panorama andino de cambio climático: Vulnerabilidad y adaptación en los Andes Tropicales*. Lima: CONDESAN, SGCAN.

De Janvry, A., Finan, F., Sadoulet, E., and Vakis, R. 2006. Can conditional cash transfer programs serve as safety nets in keeping children at school and from working when exposed to shocks? *Journal of Development Economics*, 79, 349–373.

Demenocal, P. B. 2001. Cultural responses to climate change during the late Holocene. *Science*, 292, 667–673.

Dillehay, T. D., and Kolata, A. L. 2004. Long-term human response to uncertain environmental conditions in the Andes. *Proceedings of the National Academy of Sciences*, 101, 4325–4330.

Doughty, C. 2016. Building climate change resilience through local cooperation: A Peruvian Andes case study. *Regional Environmental Change*, 16, 2187–2197.

FAO 2008. *Análisis del impacto de los eventos fríos (friaje) del 2008 en la agricultura y ganadería alto andina en el Perú*. Rome, Italy: FAO.

Feola, G. 2017. Adaptive institutions? Peasant institutions and natural models facing climatic and economic changes in the Colombian Andes. *Journal of Rural Studies*, 49, 117–127.

Feola, G., Agudelo Vanegas, L. A., and Contesse Bamón, B. P. 2015. Colombian agriculture under multiple exposures: A review and research agenda. *Climate and Development*, 7, 278–292.

Flores, H. S. 2009. Cambio climático en la isla Anapia del Lago Titicaca. Puno. In: PRATEC (ed.) Cambio climático y sabiduría Andino Amazónica – Perú: Prácticas, percepciones y adaptaciones indígenas. Lima: PRATEC.

Fonseca, C., Mayer, E. 1976. Sistemas Agrarios y Ecología en la Cuenca del Rio Cañete. Lima: Departamento de Ciencias Sociales, Pontificia Universidad Católica del Perú.

Gade, D. W. 1992. Landscape, system, and identity in the post-conquest Andes. *Annals of the Association of American Geographers*, 82, 460–477.

Gilles, J. L., and Valdivia, C. 2009. Local forecast communication in the Altiplano. *Bulletin of the American Meteorological Society*, 90, 85–91.

Gilles, J. L., Yucra, E., García, M., Quispe, R., Yana, G., and Fernández, H. 2013. Factores de pérdida de los conocimientos sobre el uso de los indicadores locales en comunidades del Altiplano norte y central. In: Jimenez Zamora, E. (ed.) *Cambio climático y adaptación en el Altiplano boliviano*. La Paz, Bolivia: CIDES-UMSA.

Graeter, S. 2017. To revive an abundant life: Catholic science and neoextractivist politics in Peru's Mantaro Valley. *Cultural Anthropology*, 32, 117–148.

Halloy, S., Seimon, A., Yager, K., and Tupayachi, A. 2006. Multidimensional (Climatic, biodiversity, socioeconomic) changes in land use in the Vilcanota watershed, Peru. In: Spehn, E., Liberman, M., and Körner, C. (eds.) *Land Use Change and Mountain Biodiversity*. Boca Raton, FL: CRC Press.

Herrera, A., and Lane, K. 2006. La complejidad social en la arqueología de la Sierra de Ancash. In: Herrera, A., Orsini, C. and Lane, K. (eds.) *La complejidad social en la Sierra de Ancash: ensayos sobre paisaje, economía y continuidades culturales*. Milan/ Lima: Civiche Raccolte d'Arte Aplicada del Castello Sforzesco; PUNKU Centro de Investigación Andina.

Herrera, A., Orsini, C., and Lane, K. (eds.) 2006. *La complejidad social en la Sierra de Ancash: ensayos sobre paisaje, economía y continuidades culturales*, Milan/ Lima: Civiche Raccolte d'Arte Aplicada del Castello Sforzesco; PUNKU Centro de Investigación Andina.

Hoffmann, D., and Requena, C. 2012. *Bolivia en un mundo 4 grados más caliente. Escenarios sociopolíticos ante el cambio climático para los años 2030 y 2060 en el altiplano norte*, La Paz: Instituto Boliviano de la Montaña / Fundación PIEB.

Holdridge, L. R. 1967. *Life Zone Ecology*, San Jose, CR: Tropical Science Center.

Holling, C., and Gunderson, L. 2002. Resilience and adaptive cycles. In: Gunderson, L., and Holling, C. S. (eds.) *Panarchy: Understanding Transformations in Human and Natural Systems*. Washington, DC: Island Press.

Inamura, T. 1986. Relaciones estructurales entre pastores y agricultores de un distrito altoandino en el sur del Perú. In: Masuda, S. (ed.) *Etnografía e historia del mundo andino: Continuidad y cambio*. Tokio: Universidad de Tokio.

Izeta, A. D. 2008. Late Holocene camelid use tendencies in two different ecological zones of Northwestern Argentina. *Quaternary International*, 180, 135–144.

Jacobsen, S. E. 2011. The situation for quinoa and its production in southern Bolivia: From economic success to environmental disaster. *Journal of Agronomy and Crop Science*, 197, 390–399.

Janusek, J. W. 2004. Tiwanaku and its precursors: Recent research and emerging perspectives. *Journal of Archaeological Research*, 12, 121–183.

Kadwell, M., Fernandez, M., Stanley, H. F., Baldi, R., Wheeler, J. C., Rosadio, R., and Bruford, M. W. 2001. Genetic analysis reveals the wild ancestors of the llama and the alpaca. *Proceedings of the Royal Society of London Series B-Biological Sciences*, 268, 2575–2584.

Knapp, G. 2007. The legacy of European colonialism. In: Veblen, T. T., Young, K. R., and Orme, A. R. (eds.) *The Physical Geography of South America*. Oxford, UK: Oxford University Press.

Kolata, A. L. 1993. *The Tiwanaku: Portrait of an Andean Civilization*, Cambridge, MA: Blackwell.

Körner, C., Ohsawa, M., Spehn, E., Berge, E., Bugmann, H., Groombridge, B., Hamilton, L., Hofer, T., Ives, J., Jodha, N., Messerli, B., Paratt, J., Prince, M., Reasoner, M., Rodgers, A., Thonell, J., Yoshino, M., Baron, J., Barry, R., Blais, J., Bradley, R., Hofstede, R., Kapos, V., Leavitt, P., Monson, R., Nagy, L., Schindler, D., Vinebrooke, R., and Watanabe, T. 2005. Mountain Systems. In: Hassan, R., Scholes, R., and Ash, N. (eds.) *Ecosystems and Human Well-being: Current State and Trends, Volume 1*. Washington, Covelo, London: Island Press.

Leichenko, R. M., and O'Brien, K. L. 2008. *Environmental Change and Globalization: Double Exposures*, New York, NY: Oxford University Press.

Liu, J., Dietz, T., Carpenter, S. R., Alberti, M., Folke, C., Moran, E., Pell, A. N., Deadman, P., Kratz, T., Lubchenko, J., Ostrom, E., Ouyang, Z., Provencher, W., Redman, C. L., Schneider, S. H., and Taylor, W. W. 2007. Complexity of coupled human and natural systems. *Science*, 317, 1513–1516.

López, S., Jung, J.-K., and López, M. F. 2017a. A hybrid-epistemological approach to climate change research: Linking scientific and smallholder knowledge systems in the Ecuadorian Andes. *Anthropocene*, 17, 30–45.

López, S., Wright, C., and Costanza, P. 2017b. Environmental change in the equatorial Andes: Linking climate, land use, and land cover transformations. *Remote Sensing Applications: Society and Environment*, 8, 291–303.

López-i-Gelats, F., Contreras Paco, J. L., Huilcas Huayra, R., Siguas Robles, O. D., Quispe Peña, E. C., and Bartolomé Filella, J. 2015. Adaptation strategies of Andean pastoralist households to both climate and non-climate changes. *Human Ecology*, 43, 267–282.

López-i-Gelats, F., Fraser, E. D. G., Morton, J. F., and Rivera-Ferre, M. G. 2016. What drives the vulnerability of pastoralists to global environmental change? A qualitative meta-analysis. *Global Environmental Change*, 39, 258–274.

Lutaladio, N., and Castaldi, L. 2009. Potato: The hidden treasure. *Journal of Food Composition and Analysis*, 22, 491–493.

Marshall, A. 2014. *Apropiarse del desierto. Agricultura globalizada y dinámicas socio-ambientales en la costa peruana. El caso de los oasis de Virú e Ica-Villacuri*. Lima: IFEA/ird.

McDowell, J. Z., and Hess, J. J. 2012. Accessing adaptation: Multiple stressors on livelihoods in the Bolivian highlands under a changing climate. *Global Environmental Change*, 22, 342–352.

Meinzen-Dick, R. 2007. Beyond panaceas in water institutions. *Proceedings of the National Academy of Sciences*, 104, 15200–15205.

Montaña, E., Diaz, H., and Hurlbert, M. 2016. Development, local livelihoods, and vulnerabilities to global environmental change in the South American Dry Andes. *Regional Environmental Change*, 16, 2215–2228.

Moore, J. W. 2017. The Capitalocene, Part I: On the nature and origins of our ecological crisis. *The Journal of Peasant Studies*, 44, 594–630.

Moore, K. M., Wing, E. S., and Wheeler, J. C. 1988. Hunting and herding economies on the Junin Puna. *Economic Prehistory of the Central Andes*. Oxford: BAR.

Moseley, M. E. 2001. *The Incas and Their Ancestors. The Archaeology of Peru*, London: Thames and Hudson.

Moser, S. C., and Ekstrom, J. A. 2010. A framework to diagnose barriers to climate change adaptation. *Proceedings of the National Academy of Sciences of the United States of America*, 107, 22026–22031.

Murra, J. V. 1967. El 'control vertical' de un máximo de pisos ecológicos en la economía de las sociedades Andinas. In: Ortiz de Zuñiga, I., and Murra, J. V. (eds.) *Visita de la provincia de León de Huánuco en 1562*. Huánuco, Perú: Universidad Nacional Hermilio Valdizán, Facultad de Letras y Educación.

Nakashima, D., McLean, K. G., Thulstrup, H., Ramos Castillo, A., and Rubis, J. 2012. *Weathering Uncertainty: Traditional Knowledge for Climate Change Assessment and Adaptation*, Paris: UNESCO.

Nelson, D. R., Lemos, M. C., Eakin, H., and Lo, Y.-J. 2016. The limits of poverty reduction in support of climate change adaptation. *Environmental Research Letters*, 11, 094011.

Nielsen, A. E. 2009. Pastoralism and the non-pastoral world in the late pre-Columbian history of the southern Andes (1000–1535). *Nomadic Peoples*, 13, 17–35.

Noble, I. R., Huq, S., Anokhin, Y. A., Carmin, J., Goudou, D., Lansigan, F. P., Osman-Elasha, B., and Villamizar, A. 2014. Adaptation needs and options. In: Field, C. B., Barros, V. R., Dokken, D. J., Mach, K. J., Mastrandrea, M. D., Bilir, T. E., Chatterjee, M., Ebi, K. L., Estrada, Y. O., Genova, R. C., Girma, B., Kissel, E. S., Levy, A. N., MacCracken, S., Mastrandrea, P. R., and White, L. L. (eds.) *Climate Change 2014: Impacts, Adaptation, and Vulnerability. Part A: Global and Sectoral Aspects. Contribution of Working Group II to the Fifth Assessment Report of the Intergovernmental Panel of Climate Change*. Cambridge, UK and New York, NY: Cambridge University Press.

Nogués-Bravo, D., Araújo, M. B., Errea, M. P., and Martínez-Rica, J. P. 2007. Exposure of global mountain systems to climate warming during the 21st Century. *Global Environmental Change*, 17, 420–428.

Núñez, L., and Dillehay, T. 1998. *Movilidad Giratoria, Armonía Social y Desarrollo en los Andes Meridionales: Patrones de Tráfico e Interacción Ecconómica*, Antofagasta: Universidad Católica del Norte.

O'Brien, K., Eriksen, S. E. H., Schjolden, A., and Nygaard, L. P. 2004. What's in a word? Conflicting interpretations of vulnerability in climate change research. *CICERO Working Paper*.

O'Brien, K., Eriksen, S., Nygaard, L. P., and Schjolden, A. 2007. Why different interpretations of vulnerability matter in climate change discourses. *Climate Policy*, 7, 73–88.

Olsson, L., Opondo, M., Tschakert, P., Agrawal, A., Eriksen, S. H., Ma, S., Perch, L. N., and Zakieldeen, S. A. 2014. Livelihoods and poverty. In: Field, C. B., Barros, V. R.,

Dokken, D. J., Mach, K. J., Mastrandrea, M. D., Bilir, T. E., Chatterjee, M., Ebi, K. L., Estrada, Y. O., Genova, R. C., Girma, B., Kissel, E. S., Levy, A. N., MacCracken, S., Mastrandrea, P. R., and White, L. L. (eds.) *Climate Change 2014: Impacts, Adaptation, and Vulnerability. Part A: Global and Sectoral Aspects. Contribution of Working Group II to the Fifth Assessment Report of the Intergovernmental Panel of Climate Change.* Cambridge, UK and New York, NY: Cambridge University Press.

Oré, M. T., and Geng, D. 2014. Políticas públicas del agua en las regiones: las vicisitudes para la creación del Consejo de Recursos Hídricos de la cuenca Ica-Huancavelica. In: Oré, M. T., and Damonte, G. (eds.) *¿Escasez de agua? Retos para la gestión de la cuenca del río Ica.* Lima: PUCP, pp. 269–332.

Oré, M. T., Bayer, D., Chiong, J., and Rendón, E. 2012. La 'guerra' por el agua en Ica, Perú: el colapso del agua subterránea. In: Isch L., E., Boelens, R., and Peña, F. (eds.) *Agua, injusticia y conflictos.* Lima: Justicia Hídrica/CBC/Fondo Editorial PUCP/IPE.

Orlove, B. S., Chiang, J. C. H., and Cane, M. A. 2000. Forecasting Andean rainfall and crop yield from the influence of El Nino on Pleiades visibility. *Nature*, 403, 68–71.

Orlove, B. S., Chiang, J. C. H., and Cane, M. A. 2002. Ethnoclimatology in the Andes. *American Scientist*, 90, 428–435.

Ostrom, E. 2009. A general framework for analyzing sustainability of social-ecological systems. *Science*, 325, 419–422.

Pearsall, D. M. 2009. Investigating the transition to agriculture. *Current Anthropology*, 50, 609–613.

Piperno, D. R., and Dillehay, T. D. 2008. Starch grains on human teeth reveal early broad crop diet in northern Peru. *Proceedings of the National Academy of Sciences*, 105, 19622–19627.

Polk, M. H., Young, K. R., and Crews-Meyer, K. A. 2005. Biodiversity conservation implications of landscaped change in an urbanizing desert of Southwestern Peru. *Urban Ecosystems*, 8, 313–332.

Polk, M. H., Young, K. R., Baraer, M., Mark, B. G., McKenzie, J. M., Bury, J., and Carey, M. 2017. Exploring hydrologic connections between tropical mountain wetlands and glacier recession in Peru's Cordillera Blanca. *Applied Geography*, 78, 94–103.

Postigo, J. C., and Young, K. R. (eds.) 2016. *Naturaleza y sociedad: Perspectivas socioecológicas sobre procesos globales en América Latina.* Lima: desco, IEP, INTE-PUCP.

Postigo, J. C. 2011. Capitalismo, cambio climático, y las trampas de las soluciones locales. Ruth Cuadernos de Pensamiento Crítico. Special Issue: Cambio climático: enfoques desde el sur.

Postigo, J. C. 2013. Adaptation of Andean herders to political and climatic changes. In: Lozny, L. R. (ed.) *Continuity and Change in Cultural Adaptation to Mountain Environments.* New York, NY: Springer.

Postigo, J. C. 2014. Perception and resilience of Andean populations facing climate change. *Journal of Ethnobiology*, 34, 383–400.

Postigo, J. C., Montoya, M., and Young, K. R. 2013. Natural resources in the subsoil and social conflicts on the surface: Perspectives on Peru's subsurface political ecology. In: Bebbington, A., and Bury, J. (eds.) *Subterranean Struggles. New Dynamics of Mining, Oil, and Gas in Latin America.* Austin, TX: University of Texas Press.

Postigo, J. C., Peralvo, M., López, S., Zapata-Caldas, E., Jarvis, A., Ramirez, J., and Lau, C. 2012. Adaptación y vulnerabilidad en los sistemas productivos andinos. In: Cuesta, F., Bustamante, M., Becerra, M. T., Postigo, J., and Peralvo, M. (eds.) *Panorama andino de cambio climático: Vulnerabilidad y adaptación en los Andes Tropicales.* Lima: CONDESAN, SGCAN.

Postigo, J. C., Young, K. R., and Crews, K. A. 2008. Change and continuity in a pastoralist community in the high Peruvian Andes. *Human Ecology*, 36, 535–551.

Prado, E. 2011. Quieren imponernos el proyecto Majes-Siguas II. *La República*, 26 February 2011, p. 12.

Räsänen, A., Juhola, S., Nygren, A., Käkönen, M., Kallio, M., Monge Monge, A., and Kanninen, M. 2016. Climate change, multiple stressors and human vulnerability: A systematic review. *Regional Environmental Change*, 16, 2291–2302.

Rawlings, L. B., and Rubio, G. M. 2005. Evaluating the impact of conditional cash transfer programs. *The World Bank Research Observer*, 20, 29–55.

Red Muqui. 2011. Sigue latente conflicto del proyecto Majes Siguas II después de su paralización. 2011. Available: www.conflictosmineros.net/contenidos/19-peru/7113-sigue-latente-conflicto-del-proyecto-majes-siguas-ii-despues-de-su-paralizacion [Accessed 05/04/2011].

Ribot, J. 2010. Vulnerability does not fall from the sky: Toward multiscale, pro-poor climate policy. In: Mearns, R., and Norton, A. (eds.) *Social Dimensions of Climate Change. Equity and Vulnerability in a Warming World*. Washington, DC: The World Bank.

Valdivia, G., and Ricard, X 2009. *Tejedores de espacio en los Andes. Itinerarios agropastoriles e integración regional en el sur peruano*, Cuzco, Centro de Estudios Regionales Andino Bartolomé de las Casas / Grupo Voluntariado Civil de Italia.

Rick, J. W. 1980. *Prehistoric Hunters of the High Andes*, New York, NY: Academic Press.

Rosenzweig, C., Casassa, G., Karoly, D., Imeson, A., Liu, C., Menzel, A., Rawlins, S., Root, T. L., Seguin, B., and Tryjanowsky, P. 2007. Assessment of observed changes and responses in natural and managed systems. In: Parry, M. L., Canziani, O. F., Palutifok, J. P., Van Der Linden, P. J., and Hanson, C. E. (eds.) *Climate Change 2007: Impacts, Adaptation and Vulnerability. Contribution of Working Group II to the Fourth Assessment Report of the Intergovernmental Panel on Climate Change*. Cambridge, UK: Cambridge University Press.

Salazar Vega, E. 2018. Mineras extraen agua de zonas en riesgo de sequía. *Ojo Público* [Online]. Available: https://duenosdelagua.ojo-publico.com/especiales/mapadelagua/ [Accessed 02/03/2018].

Sendón, P. F. 2009. Mountain pastoralism and spatial mobility in the South-Peruvian Andes in the age of state formation (1880–1969 and beyond). *Nomadic Peoples*, 13, 51–64.

Sendón, P. F. 2016. *Ayllus del Ausangate: parentesco y organización social en los andes del sur peruano*, Cusco/Lima, Centro de Estudios Regionales Andinos Bartolomé de Las Casas, Instituto de Estudios Peruanos, Pontificia Universidad Católica del Perú.

Servicio Nacional de Meteorología e Hidrología del Perú (SENAMHI) 2017. *Willay. Midiendo el tiempo sin instrumentos*, Lima: SENAMHI.

Seth, A., Thibeault, J., Garia, M., and Valdivia, C. 2010. Making sense of twenty-first-century climate change in the Altiplano: Observed trends and CMIP3 projections. *Annals of the Association of American Geographers*, 100, 835–847.

Shimada, M., Wing, E. S., and Wheeler, J. C. 1988. Prehistoric subsistence in the North Highlands of Peru: Early horizon to late intermediate. *Economic Prehistory of the Central Andes*. Oxford: BAR.

Sietz, D., and Feola, G. 2016. Resilience in the rural Andes: Critical dynamics, constraints and emerging opportunities. *Regional Environmental Change*, 16(8), 2163–2169.

Smuda, J., Dold, B., Friese, K., Morgenstern, P., and Glaesser, W. 2007. Mineralogical and geochemical study of element mobility at the sulfide-rich Excelsior waste rock dump from the polymetallic Zn–Pb–(Ag–Bi–Cu) deposit, Cerro de Pasco, Peru. *Journal of Geochemical Exploration*, 92, 97–110.

Sperling, F., Valdivia, C., Quiroz, R., Valdivia, R., Angulo, L., Seimon, A., and Noble, I. 2008. *Transitioning to Climate Resilient Development. Perspectives from Communities of Peru. Climate Change Series No. 115.* Washington, DC: The World Bank Enviromental Department Papers.

Stadel, C. 2008. Vulnerability, resilience and adaptation: Rural development in the tropical Andes. *Pirineos*, 163, 15–36.

Steffen, W., Richardson, K., Rockström, J., Cornell, S. E., Fetzer, I., Bennett, E. M., Biggs, R., Carpenter, S. R., De Vries, W., De Wit, C. A., Folke, C., Gerten, D., Heinke, J., Mace, G. M., Persson, L. M., Ramanathan, V., Reyers, B., and Sörlin, S. 2015. Planetary boundaries: Guiding human development on a changing planet. *Science*, 347.

Struelens, Q., Pomar, K. G., Herrera, S. L., Huanca, G. N., Dangles, O., and Rebaudo, F. 2017. Market access and community size influence pastoral management of native and exotic livestock species: A case study in communities of the Cordillera Real in Bolivia's high Andean wetlands. *PLOS ONE*, 12, e0189409.

Thorp, R., and Bertram, G. 1978. *Peru, 1890–1977: Growth and Policy in an Open Economy.* London: MacMillan.

Tosi JR, J. A., and Voertman, R. F. 1964. Some environmental factors in the economic development of the tropics. *Economic Geography*, 40, 189–205.

Tripcevich, N. 2008. Llama caravan transport: A study of mobility with a contemporary Andean salt caravan. *73th Annual Meeting of the Society for American Archaeology.* Vancouver, B.C.

Tripcevich, N. 2010. Exotic goods, Chivay obsidian, and sociopolitical change in the South-Central Andes. In: Dillian, C. D., and White, C. L. (eds.) *Trade and Exchange.* New York, NY: Springer.

Turner II, B. L., and Sabloff, J. A. 2012. Classic Period collapse of the Central Maya Lowlands: Insights about human–environment relationships for sustainability. *Proceedings of the National Academy of Sciences*, 109, 13908–13914.

Valdivia, C., Seth, A., Gilles, J. L., Garcia, M., Jimenez, E., Cusicanqui, J., Navia, F., and Yucra, E. 2010. Adapting to climate change in Andean ecosystems: Landscapes, capitals, and perceptions shaping rural livelihood strategies and linking knowledge systems. *Annals of the Association of American Geographers*, 100, 818–834.

Verzijl, A., and Quispe, S. G. 2013. The system nobody sees: Irrigated wetland management and alpaca herding in the Peruvian Andes. *Mountain Research and Development*, 33, 280–293.

Von Uexkull, N., Croicu, M., Fjelde, H., and Buhaug, H. 2016. Civil conflict sensitivity to growing-season drought. *Proceedings of the National Academy of Sciences*, 113, 12391–12396.

Vos, J., and Del Callejo, I. 2010. El riego campesino, la seguridad hídrica y la seguridad alimentaria en los Andes. In: Vos, J. (ed.) *Riego campesino en los Andes: seguridad hídrica y seguridad alimentaria en Ecuador, Perú y Bolivia.* Lima: IEP/ Concertación.

Walsh-Dilley, M. 2016. Tensions of resilience: Collective property, individual gain and the emergent conflicts of the quinoa boom. *Resilience*, 4, 30–43.

Wilbanks, T. J. and Kates, R. W. 2010. Beyond adapting to climate change: Embedding adaptation in responses to multiple threats and stresses. *Annals of the Association of American Geographers*, 100, 719–728.

Young, B. E., Young, K. R., and Josse, C. 2011. Vulnerability of tropical andean ecosystems to climate change. In: Herzog, S. K., Martínez, R., Jorgensen, P. M., and Tiessen, H.

(eds.) *Climate Change and Biodiversity in the Tropical Andes*. Montevideo, Uruguay: Inter-American Institute for Global Change Research (IAI) and Scientific Committee on Problems of the Environment (SCOPE).

Young, K. R. 2008. Stasis and flux in long-inhabited locales: Change in rural Andean landscapes. In: Millington, A., and Jepson, W. (eds.) *Land-Change Science in the Tropics: Changing Agricultural Landscapes*. New York, NY: Springer.

Young, K. R. 2009. Andean land use and biodiversity: Humanized landscapes in a time of change. *Annals of the Missouri Botanical Garden*, 96, 492–507.

Zhang, D. D., Lee, H. F., Wang, C., Li, B., Pei, Q., Zhang, J., and An, Y. 2011. The causality analysis of climate change and large-scale human crisis. *Proceedings of the National Academy of Sciences*, 108, 17296–17301.

Zimmerer, K. S. 1991. The regional biogeography of native potato cultivars in Highland Peru. *Journal of Biogeography*, 18, 165–178.

Zimmerer, K. S., and Rojas Vaca, H. L. 2016. Fine-grain spatial patterning and dynamics of land use and agrobiodiversity amid global changes in the Bolivian Andes. *Regional Environmental Change*, 16, 2199–2214.

7

Not for the Faint of Heart
Tasks of Climate Change Communication in the Context of Societal Transformation

SUSANNE C. MOSER

7.1 Introduction

Nearly two decades into the twenty-first century, climate communication research and practice has been established as a solid field of work. Numerous anthologies, encyclopaedias, journals and practice-oriented clearinghouses are dedicated to building up the foundations of 'best practices' as well as expanding the scope and diversity of what climate change communication is all about (Corner and Clarke 2014; Priest 2014; Pearce *et al.* 2015; Moser 2016; Nisbet 2017). But it has not yet ventured into what communication of and for the transformative changes entailed in climate change might look like. This chapter aims to open that door.

As that field is maturing, climate scientists and policy-makers place before us increasingly stark realities with their own advances in understanding of the climate change challenge. It is increasingly clear that it is superbly ambitious to think that the world can reduce its collective greenhouse gas emissions to a level where global average temperature increases remain below or return to 2°C (or even less). Modern society would essentially require the energy equivalent of a global 'blood transfusion' in which fossil fuel energy sources are replaced with non-greenhouse-gas-emitting sources by about 80 per cent by mid-century and completely by the end of the twenty-first century (IPCC 2014). Such numbers imply a fundamental rethinking and restructuring of the globalised economy, profound changes in people's consumption thinking and behaviours and the support from highly functional institutions to govern such an energy transition. Simultaneously, communities the world over would need to implement comparatively moderate amounts of adaptation to the climate change impacts expected with a 2°C degrees average warming (Field *et al.* 2014).

Overshooting that 2°C goal appears equally challenging, as climate change-related changes and disruptions from even more extreme events cause greater and greater damage and consequently place greater demands on society with regard to adaptation. At the same time, one can assume that society would not merely stand by such a trend towards increasing losses from climatic disasters and insidious changes but would attempt to prevent worsening of the situation by implementing more radical emission reductions. Given the growing interest in geoengineering in recent years (Markusson *et al.* 2014; National Research Council 2015; Rabitz 2016; Pasztor *et al.* 2017), it is also conceivable that demands to implement geoengineering schemes to try to protect against the undesirable consequences of climate change would grow in urgency. Together, simultaneously stepped-up adaptation, mitigation and geoengineering would result in an extremely demanding, complex situation for global policy-makers and for publics, markets and communicators everywhere.

Finally, whether arrived at by failure of institutions, by reinforcing feedback loops in the Earth system or by other political, economic and cultural driving forces, the even more dangerous world in which global average temperatures are 3°C, 4°C, 5°C or even 6°C warmer than historically portends even more extensive climatic changes and widespread increases in the frequency and severity of climate extremes, resulting shifts in production sites, markets and trade patterns, and massive migration of people from areas too challenging to eke out a living (or outright uninhabitable) to safer locations. Everywhere, society would be compelled to adapt. Yet in the unstable and continuously changing social, ecological, economic and physical environment, communities would struggle to make such adaptive changes with much predictive certainty. Living with cascading impacts and the second- and third-order consequences of adaptation, mitigation and geoengineering would create extraordinarily more complex and utterly different life circumstances than modern society is familiar with at present.

Each of these three scenarios makes clear that society stands at the brink of a transformative imperative, one that will only grow as it moves along any one of the three possible pathways (Table 7.1).

What does climate change communication research have to tell us about how to communicate the transformative imperative before us, and how to do so while being in it?

This chapter asks the question of what tasks communication has if it aims to actively support and participate in the far-reaching societal transformations that can be expected from deep emission cuts and adaptation to climate change. An important underlying premise here is that these mitigation and adaptation-related changes constitute not merely a complex technological challenge but a cultural one in the sense that the necessary changes will entail profound political, social, economic, financial,

Table 7.1 *The transformation imperative: facing the scope of the challenge before us*

To reach the goals of the Paris Accord: 2°C (or closer to 1.5°C)	To miss the goals of the Paris Accord: >2°C	To miss the goals of the Paris Accord significantly: >3–6°C
• The coming 'blood transfusion' of modern society • Fundamental restructuring of the globalised economy • Profound changes in people's consumption thinking and behaviour • Extraordinary demands on functioning institutions • Adaptation to 'moderate' climate changes	• More extensive climate changes and catastrophes + emission reduction efforts + adaptation efforts + geoengineering • Unprecedented political, legal and military complexity	• Extensive, deep climate changes and catastrophes • Global shift in markets and production sites • Massive migration of people • Attempts to adapt • Unpredictable consequences of climate change and of adaptation efforts in all areas and sectors of society

psychological, legal, institutional and environmental changes – all of which have foundational links to cultural worldviews, norms, beliefs, practices and artefacts. The challenges of such a transformation are not yet clear to most, and to some extent – at least as far as concrete events and problems are concerned – are unpredictable. In this chapter, I will attempt to identify some of the associated challenges categorically, and thus circumscribe a set of tasks for communication. In principle, this is not only about motivating people to actively participate in the transformative changes underway but also to support them in the psychological, social and cultural processes involved in fundamental (systemic, societal and environmental) change.

In Section 7.2, I will begin this exploration by a look at the existing literature on climate change communication and transformation. This section contextualises the exploration that follows. Section 7.3 defines and characterises the transformative change before us. And Section 7.4 – the heart of the chapter – then lays out ten tasks that a transformative communication might take on. Finally, Section 7.5 summarises and describes the work of transformative communication as cultural work, in which a society searches, reckons with and redefines its course.

7.2 The Dearth of Scientific Understanding of Communication amidst Societal Transformations

Writing this chapter began with a literature search, using the Web of Science citation system. The search employed the simple search terms (TI=communicat*

AND transform* AND climat*), without date- or publication-type restrictions. It yielded a total of two results (Izdebski *et al.* 2016; Tàbara *et al.* 2017). Neither, however, addressed the question this chapter tries to address.

A less restrictive search (TI=communicat* AND transform*) yielded 851 citations, of which no more than a half dozen related to climate change-related issues, such as energy, the environment or sustainability. The exercise did, however, enable a number of observations about the intersection of research interests in communication and its role in contributing to societal transformations on the one hand and about research interests in transformation and its implications for communication on the other hand (Figure 7.1).

Among the papers that were most centrally focused on communication, key interests were in the roles and changes in traditional forms of communication, as well as in modern changes in communication technologies such as digitisation, the creation of virtual realities, powerful visualisations, internet networking and the mediatisation of the world. These papers then examined the transformative implications of these changes on society, including radical changes in the arts and culture, social relations, political processes (both positive and negative for democracy), spatial relationships and the experience of geographic distance. Others explored the implications for educational opportunities, processes and contents, economic relationships and processes and the prospects of epistemic justice (all shown as topical clusters on the right-hand side of Figure 7.1).

By contrast, papers primarily focused on transformation examined transformational processes or transformative pathways such as globalisation, capitalism, political and governance changes with far-reaching implications, social movements (including sometimes revolutionary movements), education and psychological emancipatory processes, the role of leadership and conflict resolution and mediation. These papers intersected with communication in that they either explored the implications for communication or focused on the demands on communication, with a wide range of consequences for how a transforming and transformed society communicates. For example, many papers addressed the implications of transformation on the homogeneity or diversity of society and how this affects the ability to communicate. Others examined changing knowledge ecologies and the changing landscape of 'voice' and 'rights'. Yet others focused on changing meta-narratives, changes in power relations, profound shifts in political and civic participation, people's changing understanding of time and space, as well as changes in conversational ethics and the very idea of intimacy.

Overall, while clearly not exhaustive of what communication and transformation research addresses, this cursory review of the intersection of the two is revealing.

Clusters of Scientific Interest at the Intersection of Communication and Transformation

Consequences
- Homogeneity and diversity
- Knowledge ecologies
- Voice/rights
- Meta-narratives
- Power distribution and hierarchies
- Political and civic participation
- Understanding of time and space
- Conversational ethics and intimacy

Communication

Processes
- Traditional forms of communication
- Digitisation
- Creation of virtual realities
- Visualisation
- Networking
- Mediatisation of the world

Transformation

Processes
- Globalisation and capitalism
- Politics and governance
- Social movements and revolutions
- Conflict resolution and mediation
- Education
- Psychological emancipation
- Leadership

Consequences
- Art and culture
- Social relations
- Political/democratic processes
- Changes in various sectors
- Spatial structure
- Education (contents, process)
- Economic activities and relationships
- Scientific/epistemic justice

Climate Change and The Search for Sustainability

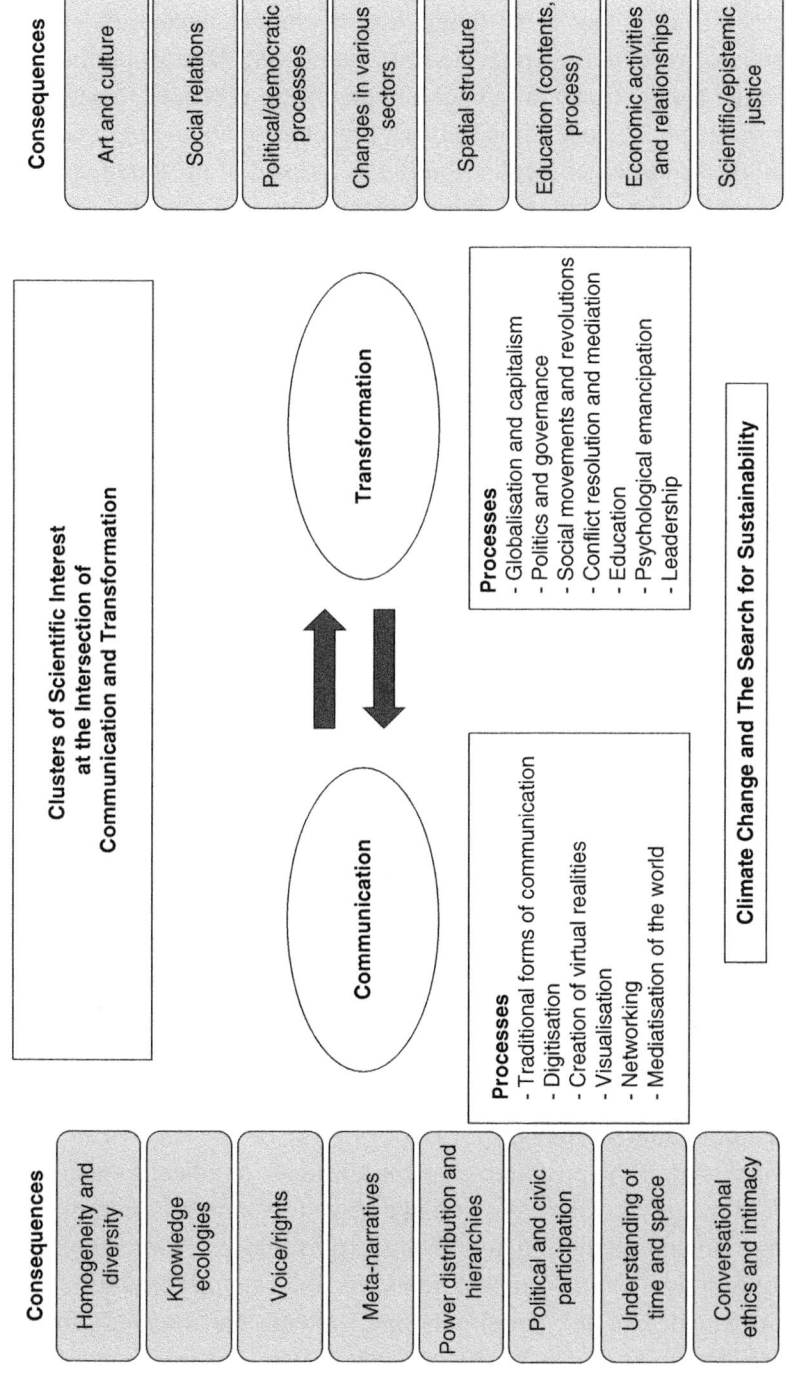

Figure 7.1 Clusters of scientific interest at the intersection of communication and transformation (based on a Web of Science search, details described in text).

The tenor of this broad body of work conveys a perpetual atmosphere of change, uncertainty and often insecurity, but also of possibilities of what might be or emerge and impossibilities – that is, of things lost, no longer tenable or viable. As such, the clusters of research interests do draw some very helpful outlines of what a climate change communication science and practice might need to embrace if it were to be of service to society going through profound change. For example, what are the possibilities of power (re)distribution as a result of profound climate change and associated policy changes? Which meta-narratives are empowering a liveable future? How can art help open up the imagination of the possible? What does epistemic justice in the midst of an unprecedented societal transformation look like? It is notable, however, that as much as the field of climate communication has matured against a backdrop of worsening news from the climate science community, it has not much to offer in the way of well-established knowledge on how to communicate in a rapidly transforming world.

7.3 The Depth of Change Afoot and Yet to Come

The rapidly expanding literature on societal transformation in the context of climate change and other sustainability challenges offers many definitions of 'transformation' that speak to the profundity of change involved, including changes in societal norms, narratives, structures, activities, identities, livelihoods and ways of being (O'Brien 2012; Feola 2015; McAlpine *et al.* 2015; O'Brien and Selboe 2015; Waddell *et al.* 2015; Few *et al.* 2017; Pichler *et al.* 2017). Few, however, lay bare the experience of going through a transformative change (Berzonsky and Moser 2017).

I turn instead here to Ulrich Beck, who – in his last (posthumously published) book *Metamorphosis of the World* (Beck 2016) – began to sketch some of the contours of living in a world of profound change. He in fact rejected the word 'transformation' to describe that world of change as he felt that (already overused) word did not sufficiently convey the radical and all-encompassing nature of the change that can be expected from climate change and other globalising changes. Instead, he proposed the term 'metamorphosis', a word with deep etymological roots and cultural connotations (Textbox 7.1).

Beck (2016) conceived of the profound changes ahead as 'a radical transformation in which the old securities fall away and make place for something completely new' (p.3), as 'an epochal change in worldview' (p.5) and as 'something that happens; not a program' (p.18). Importantly, he viewed the coming metamorphosis as the logical consequence of the paradigms and systems that created climate change in the first place or, as he put it, 'a global revolution of side effects [of the successes of modernity], that unfolds in the shadows of speechlessness' (p.29).

> **Textbox 7.1: Poetic meditation on the word 'transformation'**
>
> Transformation. Sea change. Seeing change. Shakespeare was the first to use the word 'sea change' – a term already needed apparently in 1610. 'The Tempest'. Sometimes it takes a storm ... Because the shape of something really has to change. Interestingly enough a story, some say, about the struggle between rationality and magic. We may hope rationality will take us across, but is it big enough? Sometimes it may take violent intervention, sometimes innate forces that demand evolution from one state to another. Like the larva in a cocoon becoming butterfly. Metamorphosis. Change of form. From Morpheus – name of the god of dreams. Dreaming the 'across, beyond' into being. To go beyond. To leave behind. To commit. Imagination. Endurance. Faith. Surrender. Loss. And novel gain.
> S. Moser, Reflections on 'transformation', written for the Transformations conference, Oslo (June 2013)

Without playing down the immense suffering that climate change could bring, he surmised that the metamorphising world would offer the possibility of emancipatory catastrophes (pp.115–118) by fundamentally challenging 'our way of being in the world, thinking about the world, and imagining and doing politics' (p.20). In both the most nightmarish and most beneficial sense imaginable, he characterised the experience of going through a metamorphosis – at inconceivable speed – as one in which 'what was utterly unthinkable yesterday, becomes possible and real today' (p.40).

It is in that profoundly transforming, metamorphising world that communication will be tasked to offer motivation and direction, consolation and support, orientation and guidance. In the remainder of this chapter, I turn to these tasks and explore the possibilities of what communication in support of and amidst a societal transformation may offer. To do so, I draw on the humanities and the thought leadership of public intellectuals – which, together, offer perceptive insights and touch a deeper note than mainstream social science.

7.4 The First Ten Tasks of (Climate) Communication amidst a Societal Transformation

The list of ten tasks of a (climate) communication in support of and accompanying a societal transformation offered here emerged inductively from the synthesis of the literature cited below. Generally speaking, the ten categories are informed by common tasks expected of communication (e.g. naming and framing; fostering constructive engagement), but also by a psychological and political reading of the transformation process itself (e.g. deconstructing certainties, empathy). Finally, it is informed by the emerging needs I increasingly notice in communicating with North American and European audiences already experiencing, or awaking to the need for, profound change (e.g. fostering hope, sense-making). While anecdotal evidence is emerging from across the world as to the profound challenges of transformative change at the frontlines of climate change, my own experience does not allow me to make any claims as to what those in lower-income, non-Westernised regions of the world might need or want from communication in their transformative contexts.

These caveats notwithstanding, the list of communicative tasks I offer does not aim to be complete, nor is it to be viewed as a sequential or prioritised catalogue of strategies. Rather, it is merely a conveniently round number of tasks to be fulfilled repeatedly in whatever sequence a particular situation demands (a quick overview is provided in Textbox 7.2). Each task is described in more detail below. For the purposes of this chapter, I call these tasks of a 'transformative communication'.

Textbox 7.2: Ten tasks of communication amidst a societal transformation – overview

1. Naming and Framing the Depth, Scale, Nature and Outline of (Necessary) Change
2. Fostering the Transformative Imagination
3. Mirroring Change Empathetically
4. Distinguishing (and Deconstructing) Valuable (Un)Certainties
5. Orienting and Course-Correcting Towards the Difficult
6. Helping People Resist the Habit of Acquiescing to Going Numb
7. Sense- and Meaning-Making of Difficult Change through Story (not Facts)
8. Fostering Authentic and Radical Hope
9. Fostering Generative Engagement in Building Dignified Futures for All
10. Promoting and Actively Living a Public Love

A word should also be said about who the communicators and audiences of such communication might be. What is envisioned and explored here is quite different from more traditional forms of science communication or, more specifically, climate change communication or journalistic forms of communication. As I will argue later, the kind of communication proposed here is transformative work and itself transformational when compared to the often more formal and traditional forms and practices of communication in which someone is tasked with 'communicating' and someone else is positioned as 'the audience'. In contrast to this, the communication envisioned here is often more dialogic, reciprocal and not primarily or necessarily educational or informative. Rather, it is a form of creating the future, opening up the space for a life worth having; it is social, psychological and cultural in nature – and more often than not counter-cultural – in that it questions, provokes, demands, gives, stops, listens, appreciates, reflects; it is self- and category-transcending and, as such, difficult work; it is actively embracing and grappling with deep change and, in doing so, courageous. While some may be gifted in this work, and many more inclined, it is not only carried out by a handful of specialised experts. It can and must be done by many. Some may become recognised as transformational leaders because of their skill in the work described below, but the radius of their influence may vary from a circle of friends to a company, from a township to a nation. To reach out, they may utilise any of the existing communication technologies, from the spoken and written word, to social and traditional media and to others yet to be invented or rediscovered. But to succeed hinges more on the authenticity and integrity of those involved, and their courage and willingness to acknowledge not-knowing, than on the resources and means to reach the masses.

Transformative communication, as sketched out in the following (first) ten tasks, is thus not a luxury amidst, distraction from or sideshow to, the material, practical work of transforming society to sustainability, but an essential part of that process. What might it entail?

7.4.1 Naming and Framing the Depth, Scale, Nature and Outline of (Necessary) Change

One critical early and probably never completed task of transformative communication is to name and frame the transformation before and around us. Something becomes more imaginable, more tangible, more doable if we can verbally 'put our arms around' this amorphous and, in many ways, unprecedented change. Many people from across a wide spectrum of societal arenas have begun to do so, and their reach needs to be expanded. There is no scientific evidence telling us which framings of a grand transformation are most resonant with most people. In fact, there is a risk, and some experiential evidence, that certain framings of transformative change evoke fear, such as widespread notions and popular interpretations of 'collapse' and 'apocalypse' (Diamond 2005; Foust and O'Shannon 2009; Swyngedouw 2010; Hoggett 2011; Pearson and Pearson 2012; Turner 2012; Scheffer 2016). They do so because they do not involve any desirable or enticing outcome or 'ending' of the transformative story.

There are alternative stories, however, that convey a narrative of necessary or inevitable decline followed by renaissance, or – as some would say – stories that follow the archetypal death-rebirth arc (Berzonsky and Moser 2017). Based largely on work done by depth psychologist Carol Berzonsky, we have a collection of various framings of that story (Berzonsky 2016). For example, the American Buddhist scholar and antinuclear activist Joanna Macy calls this transformation 'The Great Turning' (Macy 2009); Australian depth psychologist Sally Gillespie calls it a 'descent in the time of climate change' (Gillespie 2009); philosopher and cultural historian Rick Tarnas believes this time is a 'global rite of passage' (Tarnas 2001); similarly, and echoing Beck, social historian Barry Spector examines this 'mad' time of change through a mythological lens (Spector 2010) and the German climate scientist Hans-Joachim Schellnhuber, while not framing the entire transformation, names modern society's journey into unknown challenges, by pointing to the *terra quasi incognita* into which unmitigated climate change is hurling us (Schellnhuber 2009). The Danish anthropologist Bjørn Thomassen brings attention to that liminal time between the untenable present and the desired but unknown future (Thomassen 2014). And American philosopher Jonathan Lear walks us through a radical transformation to describe how means, ends and judgement and meaning-making about both transform in the course of fundamental change (Lear 2006).

These examples are not meant to be exhaustive. Rather, they are exemplary of the types of framings available that are capable of embracing the size and scope of the metamorphosis now underway. They also begin to outline the work of transformation, and as they are explored in greater depth, they provide a guard rail or mental map of that *terra quasi incognita*. Communicators can draw on them to help their audiences hold the immensity of what is unfolding without collapsing under its weight. Naming and framing it in this way is particularly crucial in the early stages of transformation when growing numbers of members of society experience the unease of profound change getting underway, but when many others still either deny or resist such change. As such, naming and framing the scope, scale and nature of transformative change is at once supportive and motivational, eye-opening and contributing to the changes underway.

7.4.2 Fostering the Transformative Imagination

One of the most difficult challenges for climate communicators has been to make the abstract and psychologically distant problem of climate change real and tangible in the here and now (Spence *et al.* 2012; McDonald *et al.* 2015; Jones *et al.* 2017). In my own work in communication and climate change adaptation with a wide variety of scientists and practitioners, I have found a persistent parallel difficulty, namely to imagine alternative futures, and in particular to dare to envision desirable futures, especially ones that break with current patterns of living, being, land use and so on. Where communities have been able to envision and articulate a meaningful, worthy goal to work towards, energy is usually freed up to undertake the necessary hard work of moving towards it. Without an imagined worthwhile goal, little movement occurs – at least not prior to the moment when change becomes imposed by circumstance and thus inevitable.

There is a critical task for communicators and anyone facilitating public engagement related to climate change work to help individuals expand their horizons of what they think is possible. A variety of tools and processes are available to do so, ranging from visioning exercises (Meadows 1994; Ames 2001; Burch et al. 2013; Wiek and Iwaniec 2014; Baker et al. 2017), to scenario planning (Peterson et al. 2003; Biggs et al. 2010; Tevis 2010; Marchais-Roubelat and Roubelat 2011; Amer et al. 2013; Ossewaarde 2017), to raising 'futures consciousness' (Adam and Groves 2011; Sharpe 2013; Wilenius 2014; Kunseler et al. 2015), to engaging all forms of art to free up imaginative powers and creativity (Shein et al. 2015; Nurmis 2016; Raven 2017) and more.

The work of the International Futures Forum may serve as a useful example of how the transformative imagination can be fostered. Sharpe (2013) suggests that the work of imagination spans three horizons: (1) the current way of doing things; (2) the future/new way of doing things and (3) the transition or transformation zone between them. Sharp and his colleagues offer facilitation approaches to engage each in depth, to explore, expand and define transformative horizons. Their work also helps train the eye of participants on the edges (not the centre) of society to identify and elevate innovative ideas.

Communicators can also help create dialogic forums in which those engaged discover not just each other's pre-existing ideas about what they think is possible (an important step, as different people have very different ideas about that), but they can be encouraged to push beyond those pre-existing ideas to 'the adjacent possible'. Both to motivate people to stretch beyond the familiar and comfortable and to anchor them through the difficulties involved in the often lengthy transformative process, it is furthermore important to ground people in what they most deeply care about. These motivations are deeper than monetary gains, even deeper than status gains. They involve the deepest sources of meaning and they vary among individuals (love, spirituality, belonging, soul and so on). Making space and giving voice to these deeper human dimensions (through silence, listening, music, poetry, rituals, community, presence and other practices) helps wake people up, motivate them and support them in their processing of the losses and gains involved in deep change.

7.4.3 Mirroring Change Empathetically

A transformative communication must mirror the change that is occurring, explain the responses to that change and enable people to participate in it constructively and effectively. What that means specifically cannot be detailed pro forma, given the multitude of situations, contexts and layers of transformation we may anticipate (although the framing and underlying understanding of transformation processes can provide critical guidance!). In any event, mirroring and explaining is not easy but will be aided if communication experts become allies and partners of transformation experts and actors. (Professionally, these worlds of expertise live rather side by side at present, leaving a gap of missed opportunities and mishaps between them!)

As a force of motivation and support through difficult times, transformative communication must gain far greater comfort with and fluency in the emotional experience of going through a transformative shift (Moser 2012, 2013, 2014; Berzonsky and Moser 2017). After all, deep change is – first and foremost – experienced and processed emotionally. There is fear, grief, anger, hope, resentment, regret, guilt, frustration and any number of other emotional responses as lives are as profoundly upended as Beck describes. Communication can help reflect back what these experiences are and, as such, support individuals' reflexivity, processing and learning while countering tendencies of unguarded reactivity or a sense of overwhelmedness, hopelessness and apathy.

Importantly, communicators must expect traumatic experiences, especially among the poorer and marginalised groups of society. This includes those who are most directly exposed to climate change threats, have the least resources to protect themselves against them or who cannot resiliently return to a prior status or even advance to a better situation (Bonanno 2004; Dominey-Howes 2015; Eriksen and Ditrich 2015; White 2015; Davidson 2016). This includes those who will lose their homes, their professional identities or even their livelihoods in industries that will transform due to extensive mitigation and adaptation efforts.

Practically, this means that communicators must build their own capacities, and support others in building theirs, for dealing with strong emotions. The task is not to dramatise suffering or turn victims into exhibits but – quite the opposite – to validate the emotional responses to change, explain how they are normal responses to extraordinary circumstances and then to give space to people to support their processing of them, individually and together with others (Doppelt 2016). More than camera spotlights and news features, communicators of transformative change and those going through it (which may be one and the same) will need to create or seek out calm spaces and times for rejuvenation and healing.

7.4.4 Helping People Resist the Habit of Acquiescing to Going Numb

The American writer Terry Tempest Williams admitted once, 'I cry every day, and not because I'm sad but because I feel.' She went on to explain, 'I think in many ways that's our most important task at this moment in time: to not avert our gaze, to not allow ourselves to be numb to the world. I think being numb is another form of suicide' (Williams 2014). She experienced it as 'daunting' to witness countless 'feeling' advocates for a better world to be heartbroken *and* continue to sing about the beauty of the world, all at once. Joanna Macy addresses just this heartbreak of so many activists. Having dedicated decades of her life to offering a space for people to express their own and witness others' feelings about the world so as to be reinvigorated in their dedication to the work of 'The Great Turning', she believes, 'It is o.k. for our hearts to be broken over the world. What else are hearts for? There is great intelligence in that' (Macy 2016).

Communicators amidst a profound societal transformation must understand that while the world is breaking, hearts are breaking. People will be tempted to go numb to both, within themselves and in others, because experiencing or witnessing such deep emotion can be unbearable. This is why so many have written about apathy, numbness and 'climate or disaster fatigue' (Kerr 2009:926; see also Nordhaus and Shellenberger 2009; Cafaro 2005; Lertzman 2008; Slater 2008). Moreover, when long-suppressed feelings are freed and old habits are broken, intense energies are being set free. Communicators increasingly acquainted with and better prepared for the emotional responses to (climate) change can help to hold spaces for this release – a counter-cultural act in service to transformation – and direct that emotional energy towards contributing constructively towards a safer and more desirable future (emotion as E-motion or energy-in-motion).

Importantly, because of the counter-cultural and powerful nature of working with intense emotions unleashed by transformative forces, communicators must help create safe spaces, foster curiosity rather than allow judgement and create opportunities for people to connect so that they support each other throughout the process.

7.4.5 *Distinguishing (and Deconstructing) Valuable (Un)certainties*

Another way in which communicators can serve the transformation process constructively is to actively engage both certainties and uncertainties. The issue here is not, however, merely an extension of the long-standing interest in the (science and climate change) communication field of how to communicate uncertainties (Webster 2003; Patt and Dessai 2005; Marx *et al.* 2007; Morgan *et al.* 2009; Pigeon and Fischhoff 2011; National Research Council 2016). Rather, in the journey into *terra quasi incognita*, uncertainty is the fundamental condition for being, while alleged certainties can be a hindrance to reckoning with what was and fully exploring the possibilities of what may be.

Thus, a task of transformative communication is to foster curiosity rather than reinforce biases and simplistic answers. This may entail asking more and better questions and giving fewer answers. It might well mean getting away from the widespread practice of 'messaging', and instead become far better at deep listening. Interestingly, asking questions and listening, and making room for not-knowing and silence, are not the traditional strongholds of the communication field (Moser 2016).

If Ulrich Beck is correct, the metamorphosis that has begun, and that climate change will accelerate, will force us to ask not just what we can do *against* climate change but what climate change will do *to* (and maybe even *for*) us (2016:36–39). It will place pressures not only on our institutional structures but also on our ways of thinking, being and doing. As he argues, 'the main source of climate pessimism lies in a generalized incapacity, and/or unwillingness, to rethink fundamental questions of social and political order in the age of global risks' (2016:37). Engaging these fundamental questions inevitably throws us into a place of 'not-knowing'. It demands deep reckoning with what has long been problematic with the many dimensions of our socially, environmentally and economically unsustainable ways of being without having immediate, alternative answers.

Transformative communication must help hold a space for and facilitate these difficult conversations, rather than provide or propagate superficial or easy answers. The task is not to be agnostic but to hold a critical stance towards any prematurely offered compass for moving forward, and to help people (re)learn how to critically examine the directions advanced by others.

7.4.6 Orienting and Course-Correcting towards the Difficult

One certainty to live by was offered more than 100 years ago by the German poet Rainer Maria Rilke, who said, 'We know little, but that we must trust in what is difficult is a certainty that will never abandon us.' In his seventh *Letter to a Young Poet*, written in 1904, he maintained, 'Most people have (with the help of convention) oriented their solutions toward the easy, and often to the easiest of the easy, and yet it is clear, that we must trust in that which is difficult' (Rilke 2000).

What would this mean for communicators? It would, at the very least, demand that we model and stimulate (self-)reflection about our role and participation in the transforming world. It would further demand that we speak that which is uncomfortable to say and help make it possible for others to participate in difficult dialogues (Moser and Berzonsky 2015). It would mean to repeatedly bring the focus to that which is unjust, and put our finger on that which is at risk of staying invisible. We would also need to confront fabricated un-knowing about climate risks. We would need to listen for that which is un-said and help give voice to those who are silenced. And we would need to insist on deep, systemic solutions, not merely participate in the chorus that celebrates quick fixes and then turns away to more pleasant topics.

Between the tasks already laid out and the one added here, it becomes clear that transformative communication would need to engage in a constant balancing act of providing comfort and making people uncomfortable, of offering reprieve and causing trouble. It would need to be at the forefront of what Donna Haraway calls 'staying with the trouble' (Haraway 2016) – a complex concept involving profound questioning, holding still in not-knowing and becoming practiced in undoing that which has brought us the present slate of existential threats (Moser and Berzonsky 2015).

7.4.7 Sense- and Meaning-Making of Difficult Change through Story (Not Facts)

The tasks proposed so far make clear that transformative (climate change) communication is neither a special case of science communication nor is it a special case of communicating with the media or advocacy or a hand-maiden to a well-designed behaviour change program. All those forms and purposes of communication have their place and importance but fail to meet the demands of a transformation process.

Rather, much of the task of transformative communication instead is sense- and meaning-making amidst a change that we cannot fully see, comprehend or control and whose outcome is entirely unpredictable – at least from here. This is why many of those cited in the discussion on naming and framing (Section 7.4.1) draw on mythology and archetypes.

The American physician and writer Rachel Remen is one among many (Bontje and Slinger 2017; Brown 2017; Malone *et al.* 2017; Veland *et al.* 2018) who argues for the importance of story. While fully acknowledging the factual world, she says, 'the facts are the bones of the story ... [But] the important knowledge is passed through stories. It's what holds a culture together. ... And so story, and not facts, are the way the world is made up' (Remen 2010).

Story is also the way to hold together a world that seems to (or actually does) disassemble. But what then are the stories needed amidst a transformation? Communicators might tell stories not about heroes who succeed in their conquest of others but about protagonists of (self-)transformation; they might tell stories that delineate the search for or spread the word about known pathways of transformation (Dahle 2007; Kapoor 2007; Westley *et al.* 2011; Roggema *et al.* 2012; Tippett 2016; Werbeloff *et al.* 2016; O'Neill *et al.* 2017). Other stories people will need to hear and learn from are stories – and examples – of endings and renewal

(Berzonsky and Moser 2017), of difficulties overcome, of redemption and of positive, dignified and just solutions. Importantly, as many will wish to tell stories of comparatively painless solutions, communication in support of transformative change must help discern whether these stories actually withstand scrutiny, and where not, to resist the salesmanship and course-correct towards the difficult, but possible, as Rilke suggests (Section 7.4.6).

7.4.8 Fostering Authentic and Radical Hope

The difficult, however, will have few takers if there is not also hope. In a world at once falling apart and reconstituting in ways we do not yet know, cannot see or maybe do not like, heartbreak and despair, and resulting numbness, will be among the greatest challenges for climate change communicators. Already, American journalist, writer and public radio host Krista Tippett argues, 'the discourse of our everyday lives tends toward despair' (Tippett 2016:4). Transformative changes – maybe to the better and possibly to the worse – will require us to hold the tension between what is and what could be, and from it generate a reliable, authentic hope (Solnit 2004; Lear 2006; Bell 2009; Orr 2011; Macy and Johnstone 2012; Ojala 2012; Sharpe 2013; Stoknes 2015; Hathaway 2017).

Echoing many recent writers on the topic, Tippett maintains that hope is 'not feeling, but choice and practice, becoming spiritual muscle memory, to go through life as it is, not as we wish it to be' (2016:233). She suggests, 'hope is brokenhearted on the way to becoming wholehearted. Hope is a function of struggle' (2016:251). Clearly, such wisdom challenges the shaky promises of wishful thinking for happy endings and easy outcomes. Rather, it suggests that transformative communication must set an expectation of difficulty to build up the willingness and capacity to be vulnerable. It suggests that communicators wishing to support transformative change must remind their audiences (and likely themselves) that hope is daily work, hard-won and probably never fully enshrined and protected against the spectre of loss and seeming futility. Surely, it is not something 'given' to others, once and for all, but more safely shored up in connection with others. It is in this daily struggle for hope that communicators will find more in the wisdom traditions and the humanities (e.g. literature, theology, history) than in the vaults of social science research.

7.4.9 Fostering Generative Engagement in Building Dignified Futures for All

If authentic and radical hope equips us to face the world, then communicators must next play a role in enabling effective, constructive engagement in building a world in which we can live safe and dignified lives. To do so, we need each other. But how do we reach and engage each other constructively after years, decades and sometimes lifetimes of being everything but unified behind a cause?

Being professionally dedicated to being in conversation with others no matter how large the distance to bridge, Tippett claims, 'We hunger, and are ready, for a fresh language to meet each other' (2016:x). If that is true, what responsibility do communicators have in helping this fresh meeting to occur? What language do we choose to open doors and minds, rather than slam them shut? How do we weed out the subtle and not-so-subtle judgements in our writing and speeches to and about those who think differently about climate change and possible solutions? Clearly, we have made much progress understanding 'denialists' and 'climate sceptics' (Jacques *et al.* 2008; Dunlap and McCright 2011; Norgaard 2011; Dunlap and Jaques 2013; Medimorec and Pennycook 2015; McCright *et al.* 2016), but – if we are honest – over the course of twenty years of climate change communication, positions have hardened; the gulf between opposing viewpoints is rarely if ever bridged. A transformative communication would need to help us heal those rifts. Or, as the philosopher Kwame Anthony Appiah once put it: help us 'sidle up to difference' (Appiah 2011).

Maybe it is more important, however, to go beyond denialism and a focus on those who find it hard to embrace the climate change reality anyway. Transformative communication must find ways to touch people's deep desire to want to be good. 'Climate believers' or not, transformation will place large demands on all of us, and connecting with that deep motivation will help us remain committed to undertaking the necessary changes in the face of great difficulty. Communicators can also help people to step out of fear and into care for themselves and others as we move into growing difficulties and disruptions. And finally, transformative communication must foster risk readiness, i.e. a willingness to step not into careless action but instead out of the familiar and onto the transformative path. It must learn to convey – with conviction – that doing so is ennobling not because of the promise of success, but because of the pledge to the greater good. There is nothing easy about this, yet 'we are made by that which could break us' (Tippett 2016:13).

7.4.10 Promoting and Actively Living a Public Love

None of this can be accomplished alone or from a place of adversity. In fact, (climate) communication itself must transform and review and correct any past communication that has hardened societal divisions. It must embody and model a communication that comes from a profound commitment to a *common* future. Promoting and actively embodying such a 'public love' would mean communicating in a way that creates a community-of-solidarity, one that can face the challenges of transformation together.

Promoting and actively living such a public love would entail helping to define a species 'I' and fostering constructive deliberation about the meaning of a collective 'we'. Against the legacies of past adversities and systemic divisions, biases and injustices, this is an extraordinarily difficult task. Communicators would need to learn from any past experience of overcoming hardened divisions, not to wipe out difference – an impossible and undesirable goal – but to find grounds on which to stand shoulder to shoulder against unfavourable odds for survival and well-being. Living a public love would place those facing the greatest risks with the fewest means in the centre of that community of solidarity. It would help community members to learn to go through transformation with an open hand, i.e. to approach each other from a stance of giving instead of taking; from a place of gifting instead of expecting. The transformation gearing up to engulf us demands – if we wish to survive it – that we engage in conversation and dialogue instead of verbal or physical attacks; that we make lasting connections rather than launch persuasion campaigns and that we work to embody in words and deeds that which we wish to become.

This task, maybe more than all others, may seem lofty and unspecific. And yet, we occasionally get to observe it when someone practices it. Such moments are notable and memorable for the rupture in cultural norms that they constitute. Typically, they begin or are made up of simply stopping whatever we have done to date. Stop talking and start listening; stop defending and start hearing; stop taking and start giving; stop pursuing and start holding still; stop striving for more and start being with less or with loss; stop guarding the 'I' and start making space for 'more than I', which over time may become 'we'.

7.5 Conclusion: The (Counter-)Cultural Work of Transformation

The tasks outlined in this chapter tend to – and maybe should – make us uncomfortable. To start, they entail work that many climate change communicators are yet unfamiliar with and communication science does not give us great confidence in 'best practices'. There are no theories to guide the way and there is no laboratory experimentation that could handle the complexity of transformative change. Communication studies have a history of not looking longitudinally at change processes and the role of communication within them (Moser 2016). And if as communication researchers we know little more than a non-researcher, then who are we? Who is a communication expert in a transformation?

Similarly, if there is little guidance from science, then where – as practicing communicators – do we begin? How do we launch a conversation about

Table 7.2 *Challenges posed by transformation to communication research and praxis*

Communication research	Communication praxis
• No theories • No singularly responsible discipline • No experimentation • Isolating influential factors is feasible, but untenable • Little precedence with longitudinal studies of long-term change • Transformation may well happen faster than scientific progress • What kind of science? • Who is 'an expert'?	• Where are the opportunities for getting started if caught between day-to-day pressures and disinterest? • Where is the compass, where are the visions amidst increasing crises? • Need to fulfil the expectations of the past and of the future • To be a pioneer without much guidance from experience or science • Who then are the 'managers', the 'guides', and the leaders?

transformation when overwhelmed and faced with apathy? What compass do we go by and how do we hold all the tensions illustrated above and not become paralysed along with our audiences? The main challenges are set out in Table 7.2.

Maybe the greatest discomfort, however, comes from the fact that – regardless of which role we hold – there will be no 'transformation spectators' or 'bystanders' in a rapidly transforming world. Given the global nature, rapidity and profundity that climate change, and the associated global changes to the land, life support systems and economic activities, will bring to every part of the world (albeit in geographically contextualised ways), none of us will escape witnessing, experiencing and being impacted by these transformative changes. All of us will be implicated and affected somehow, even if we simultaneously wish to study or practice communicating about them. This has several important implications for research and practice.

First, whether we are researchers or practitioners, we must build our own capacities and skills to meet the psychological challenges associated with climate change and society's response to it. This is inward and interpersonal work, often essential, so we can meet a challenging situation from a deeper grounding and more centred place of self-knowledge and ongoing reflexivity. Next, we must identify and tend opportunities – wherever they present themselves – for deeper engagement. A keynote address can become a dialogue; a classroom can become a meeting; a townhall can become a visioning exercise; a circle of strangers can become a council of elders; a heated debate can turn into abiding curiosity. We should not be experts in climate science or communication science first, but human first, and connect with others as human beings. Furthermore, we must realise that climate change is not the only force of transformation, and none of us will experience the future merely as a climate change–impacted future. Rather, the experience will be of multiple, coalescing transformations – climatic, social, economic, technological, environmental, cultural and more. Communicators interested in supporting the transformative work of society must be curious and seek to understand these connections rather than try to artificially isolate climate change from other types of changes. How do these coalescing changes and crises shape the understanding of those involved? How is the totality of changes constellated and experienced in people's lives? Where are the opportunities and openings when everything seems to close in on people?

We must recognise that in these kinds of coalescing transformations lies an opportunity to break out of the climate change trap. In lived reality, not everything is about climate change per se or foremost, but it is part of a lived reality that is dynamic, profound, disturbing, exhilarating and sometimes devastating. And, of course, therein also lies the challenge of rethinking our organisations' or professions' roles and missions. If we are not just *climate* communicators or *climate* scientists or *climate* activists, but members of a community undergoing complex transformations, then what is important? Who are we then? What do we most want, together? And thus, what is most needed of us now? To consider these questions deeply, and to reckon with the challenges such reflection will undoubtedly surface, is part of the work that communicators themselves must do.

Communication of and for societal transformation in the face of climatic and related global changes is thus an unprecedented challenge. It demands that we ask and be in dialogue about where we – as individuals, as communities, as a global 'we' and as a species – are going and where we want or should be going. It thus requires that we grapple with how we got here, what cultural norms we have followed (and maybe still follow), what practices and interactions we have favoured over alternatives and whether the dominant norms and practices will get us safely to a sustainable future, for ourselves and all, human and non-human. At our best, we would initiate, facilitate and take an active role in difficult, yet invigorating dialogues about past, present and future, about direction, purpose and meaning, about losses, gains and treasures to behold. We would help each other navigate the wild seas of questions, impatience, frustration and yearning – not because any of us have the answers, but because we are committed to the necessary work of transformation.

This chapter is all but the beginning of a conversation about this necessary work. Communicators, scientists, activists and other change agents must engage in it. This conversation and the first ten tasks of transformative communication presented here are not just part of, but at the very heart of, the work society must undertake to navigate the metamorphosis now underway. Where it will lead is unknowable from here. Rather, as Ulrich Beck argued, 'the emergence of a compass for the 21st century ... is the result of cultural work' (Beck 2016:118). A transformative communication is essential to discovering that compass.

Acknowledgements

This chapter is based on a keynote address given at the 'K3 – Kongress zu Klimawandel, Kommunikation und Gesellschaft' conference (trans. 'Conference on climate change, communication and society') on September 25–26, 2017 in Salzburg, Austria. I would like to thank Heinz Gutscher and the conference organisers for offering me that first opportunity to think about climate change communication amidst a societal transformation.

References

Adam, B., and Groves, C. 2011. Futures tended: Care and future-oriented responsibility. *Bulletin of Science, Technology & Society*, 31(1), 17–27.
Amer, M., Daim, T. U., and Jetter, A. 2013. A review of scenario planning. *Futures*, 46(0), 23–40.
Ames, S. C. (ed.) 2001. *Guide to Community Visioning*. Chicago and Washington: Planners Press.
Appiah, K. A. 2011. Sidling Up to Difference. In: Tippett, K. (ed.), *On Being*. Available at: https://onbeing.org/programs/kwame-anthony-appiah-sidling-difference-social-change-moral-revolutions/.
Baker, P. M. A., Mitchell, H., and LaForce, S. 2017. Inclusive connected futures: Editorial introduction to special section on 'Envisioning Inclusive Futures'. *Futures*, 87, 78–82.
Beck, U. 2016. *The Metamorphosis of the World*. Cambridge: Polity Press.
Bell, M. M. 2009. Can we? The audacity of environmental hope. *Nature and Culture*, 4(3), 316–323.
Berzonsky, C. L. 2016. Towards *homo sapiens sapiens*: Climate change as a global rite of passage. Dissertation in Depth Psychology, Pacifica Graduate Institute, Carpinteria, CA.
Berzonsky, C. L., and Moser, S. C. 2017. Becoming *homo sapiens sapiens*: Mapping the psycho-cultural transformation in the Anthropocene. *Anthropocene*, 20 (Supplement C), 15–23.
Biggs, R., Diebel, M. W., Gilroy, D., Kamarainen, A. M., Kornis, M. S., Preston, N. D., ... and West, P. C. 2010. Preparing for the future: Teaching scenario planning at the graduate level. *Frontiers in Ecology and the Environment*, 8(5), 267–273.
Bonnano, G. A. 2004. Loss, trauma, and human resilience: Have we underestimated the human capacity to thrive after extremely adversive events? *American Psychologist*, 59 (1), 20–28.
Bontje, L. E., and Slinger, J. H. 2017. A narrative method for learning from innovative coastal projects – Biographies of the Sand Engine. *Ocean & Coastal Management*, 142, 186–197.

Brown, P. 2017. Narrative: An ontology, epistemology and methodology for pro-environmental psychology research. *Energy Research & Social Science*, 31, 215–222.

Burch, S. L. M., Sheppard, S. R., Pond, E., and Schroth, O. 2013. Climate change visioning: Effective processes for advancing the policy and practice of local adaptation. In: Moser, S.C., and Boykoff, M.T. (eds.), *Successful Adaptation to Climate Change: Linking Science and Policy in a Rapidly Changing World*. London, New York: Routledge, pp. 270–286.

Cafaro, P. 2005. Gluttony, arrogance, greed and apathy: An exploration of environmental vice. In: Sandler R., and Cafaro, P. (eds.), *Environmental Virtue Ethics*. Lanham: Rowman and Littlefield, pp. 135–158.

Corner, A., and Clarke, J. 2014. *Communicating Climate Change Adaptation: A Practical Guide to Values-based Communication*. Edinburgh: Climate Outreach & Information Network and Sniffer Centre for Carbon Innovation.

Dahle, K. 2007. When do transformative initiatives really transform? A typology of different paths for transition to a sustainable society. *Futures*, 39(5), 487–504.

Davidson, J. 2016. Plenary address – A year of living 'dangerously': Reflections on risk, trust, trauma and change. *Emotion, Space and Society*, 18(2), 28–34.

Diamond, J. 2005. *Collapse: How Societies Choose to Fail or Succeed*. New York: Penguin.

Dominey-Howes, D. 2015. Seeing 'the dark passenger' – Reflections on the emotional trauma of conducting post-disaster research. *Emotion, Space and Society*, 17, 55–62.

Doppelt, B. 2016. *Transformational Resilience: How Building Human Resilience to Climate Disruption Can Safeguard Society and Increase Wellbeing*. Sheffield: Greenleaf Publishing.

Dunlap, R. E., and Jacques, P. J. 2013. Climate change denial books and conservative think tanks: Exploring the connection. *American Behavioral Scientist*, 57(6), 699–731.

Dunlap, R. E., and McCright, A. M. 2011. Climate change denial: Sources, actors and strategies. In: Lever-Tracy, C. (ed.), *Routledge Handbook of Climate Change and Society*. London: Routledge, pp. 240–259.

Eriksen, C., and Ditrich, T. 2015. The relevance of mindfulness practice for trauma-exposed disaster researchers. *Emotion, Space and Society*, 17, 63–69.

Feola, G. 2015. Societal transformation in response to global environmental change: A review of emerging concepts. *AMBIO*, 44(5), 376–390.

Few, R., Morchain, D. Spear, D., Mensah, A., and Bendapudi, R. 2017. Transformation, adaptation and development: Relating concepts to practice. *Palgrave Communications*, 3, 17092.

Field, C. B. *et al.* (eds.) 2014. *Climate Change 2014: Impacts, Adaptation, and Vulnerability, Part A: Global and Sectoral Aspects*. Working Group II Contribution to the Fifth Assessment Report of the Intergovernmental Panel on Climate Change. Cambridge: Cambridge University Press.

Foust, C. R., and O'Shannon Murphy, W. 2009. Revealing and reframing apocalyptic tragedy in global warming discourse. *Environmental Communication: A Journal of Nature and Culture*, 3(2), 151–167.

Gillespie, S. 2009. Descent in the time of climate change. In: Marshall, J. (ed.), *Depth Psychology: Disorder and Climate Change*. Sydney: Jung Downunder Books, pp. 395–413.

Haraway, D.J. 2016. *Staying with the Trouble: Making Kin in the Chtuhulucene*. Durham and London: Duke University Press.

Hathaway, M. D. 2017. Activating hope in the midst of crisis. *Journal of Transformative Education*, 15(4), 296–314.

Hoggett, P. 2011. Climate change and the apocalyptic imagination. *Psychoanalysis, Culture Society*, 16(3), 261–275.

Intergovernmental Panel on Climate Change (IPCC). 2014. *Synthesis Report of the Fifth Assessment of the Intergovernmental Panel on Climate Change*. Geneva: IPCC.

Izdebski, A., Holmgren, K., Weiberg, E., Stocker, S. R., Buentgen, U., Florenzano, A., ... and Masi, A. 2016. Realising consilience: How better communication between archaeologists, historians and natural scientists can transform the study of past climate change in the Mediterranean. *Quaternary Science Reviews*, 136, 5–22.

Jacques, P. J., Dunlap, R. E., and Freeman, M. 2008. The organisation of denial: Conservative think tanks and environmental skepticism. *Environmental Politics*, 17(3), 349–385.

Jones, C., Hine, D. W. and Marks, A. D. G. 2017. The future is now: Reducing psychological distance to increase public engagement with climate change. *Risk Analysis*, 37(2), 331–341.

Kapoor, R. 2007. Transforming self and society: Plural paths to human emancipation. *Futures*, 39(5), 475–486.

Kerr, R.A. 2009. Amid worrisome signs of warming, 'climate fatigue' sets in. *Science*, 326 (5955), 926–928.

Kunseler, E.-M., Tuinstra, W., Vasileiadou, E., and Petersen, A. C. 2015. The reflective futures practitioner: Balancing salience, credibility and legitimacy in generating foresight knowledge with stakeholders. *Futures*, 66, 1–12.

Lear, J. 2006. *Radical Hope: Ethics in the Face of Cultural Devastation*. Cambridge: Harvard University Press.

Lertzman, R. 2008. The myth of apathy. *The Ecologist Blog*, pp. 16–17.

Macy, J. 2009. Three dimensions of the great turning. In *Joanna Macy and her work*. Berkeley, CA. Available at: www.joannamacy.net/thegreatturning/three-dimensions-of-the-great-turning.html.

Macy, J. 2016. A Wild Love for the World. Tippett, K. (ed.), *On Being*. Available at: https://onbeing.org/programs/joanna-macy-a-wild-love-for-the-world/.

Macy, J., and Johnstone, C. 2012. *Active Hope: How to Face the Mess We're in Without Going Crazy*. Navato: New World Library.

Malone, E., Hultman, N. E., Anderson, K. L., and Romeiro, V. 2017. Stories about ourselves: How national narratives influence the diffusion of large-scale energy technologies. *Energy Research & Social Science*, 31, 70–76.

Marchais-Roubelat, A., and Roubelat, F. 2011. Futures beyond disruptions: Methodological reflections on scenario planning. *Futures*, 43(1), 130–133.

Markusson, N., Ginn, F., Singh Ghaleigh, N., and Scott, V. 2014. 'In case of emergency press here': Framing geoengineering as a response to dangerous climate change. *Wiley Interdisciplinary Reviews: Climate Change*, 5(2), 281–290.

Marx, S. M., Weber, E. U., Orlove, B. S., Leiserowitz, A., Krantz, D. H., Roncoli, C., and Phillips, J. 2007. Communication and mental processes: Experiential and analytic processing of uncertain climate information. *Global Environmental Change*, 17(1), 47–58.

McAlpine, C.A., Seabrook, L. M., Ryan, J. G., Feeney, B. J., Ripple, W. J., Ehrlich, A. H., and Ehrlich, P. R. 2015. Transformational change: Creating a safe operating space for humanity. *Ecology and Society*, 20(1), 56.

McCright, A. M., Charters, M., Dentzman, K., and Dietz, T. 2016. Examining the effectiveness of climate change frames in the face of a climate change denial counter-frame. *Topics in Cognitive Science*, 8(1), 76–97.

McDonald, R. I., Chai, H. Y., and Newell, B. R. 2015. Personal experience and the 'psychological distance' of climate change: An integrative review. *Journal of Environmental Psychology*, 44, 109–118.

Meadows, D. H. 1994. Envisioning a Sustainable World. Third Biennial Meeting of the International Society for Ecological Economics. San Jose, Costa Rica.

Medimorec, S., and Pennycook, G. 2015. The language of denial: Text analysis reveals differences in language use between climate change proponents and skeptics. *Climatic Change*, 133(4), 597–605.

Morgan, G., Dowlatabadi, H., Henrion, M., Keith, D., Lempert, R., McBrid, S., Small, M., and Wilbanks, T. 2009. Best practice approaches for characterizing, communicating, and incorporating scientific uncertainty in decision-making. A Report by the U.S. Climate Change Science Program and the Subcommittee on Global Change Research. *Synthesis and Assessment Product 5.2*. Climate Change Science Program. Washington, DC: NOAA.

Moser, S. C. 2012. Getting real about it: Navigating the psychological and social demands of a world in distress. In: Rigling Gallagher, D., Andrews, R. N. L., and Christensen, N. L. (eds.), *Sage Handbook on Environmental Leadership*. Thousand Oaks: Sage, pp. 432–440.

Moser, S.C. 2013. Navigating the political and emotional terrain of adaptation: Community engagement when climate change comes home. In: Moser, S. C., and Boykoff, M. T. (eds.), *Successful Adaptation to Climate Change: Linking Science and Policy in a Rapidly Changing World*. London, New York: Routledge, pp. 289–305.

Moser, S. C. 2014. Whither the heart(-to-heart)? Prospects for a humanistic turn in environmental communication as the world changes darkly. In: Hansen, A., and Cox, R. (eds.), *Handbook on Environment and Communication*. London: Routledge, pp. 402–413.

Moser, S. C. 2016. Reflections on climate change communication research and practice in the second decade of the 21st century: What more is there to say? *Wiley Interdisciplinary Reviews: Climate Change*, 7(3), 345–369.

Moser, S. C., and Berzonsky, C. 2015. There must be more: Communication to close the cultural divide. In: O'Brien, K., and Selboe, E. (eds.), *The Adaptive Challenge of Climate Change*. New York: Cambridge University Press, pp. 287–310.

National Research Council 2016. *Completing the Forecast: Characterizing and Communicating Uncertainty for Better Decisions Using Weather and Climate Forecasts*. Washington: National Academy Press.

National Research Council. 2015. Climate Intervention: Reflecting Sunlight to Cool Earth. Report by the Committee on Geoengineering Climate: Technical Evaluation and Discussion of Impacts. Washington, DC: National Academies Press.

Nisbet, M. C. (ed.) 2017. *Oxford Encyclopedia of Climate Change Communication*. Oxford, New York: Oxford University Press.

Nordhaus, T., and Shellenberger, M. 2009. Apocalypse fatigue: Losing the public on climate change. Yale Environment 360, http://www.e360.yale.edu/content/feature.msp?id=2210.

Norgaard, K. M. 2011. *Living in Denial: Climate Change, Emotions, and Everyday Life*. Cambridge: MIT Press.

Nurmis, J. 2016. Visual climate change art 2005–2015: Discourse and practice. *Wiley Interdisciplinary Reviews: Climate Change*, 7(4), 501–516.

O'Neill, B. C., Kriegler, E., Ebi, K. L., Kemp-Benedict, E., Riahi, K., Rothman, D. S., ... and Levy, M. 2017. The roads ahead: Narratives for shared socioeconomic pathways describing world futures in the 21st century. *Global Environmental Change*, 42, 169–180.

O'Brien, K. 2012. Global environmental change II: From adaptation to deliberate transformation. *Progress in Human Geography*, 36(5), 667–676.

O'Brien, K., and Selboe, E. 2015. Social transformation: The real adaptive challenge. In: O'Brien, K., and Selboe, E. (eds.), *The Adaptive Challenge of Climate Change*. New York: Cambridge University Press, pp. 311–324.

Ojala, M. 2012. Hope and climate change: The importance of hope for environmental engagement among young people. *Environmental Education Research*, 18(5), 625–642.

Orr, D. W. (ed.) 2011. *Hope is an Imperative: The Essential David Orr*. Washington: Island Press.

Ossewaarde, M. 2017. Unmasking scenario planning: The colonization of the future in the 'Local Governments of the Future' program. *Futures*, 93, 80–88.

Pasztor, J., Scharf, C., and Schmidt, K.-U. 2017. How to govern geoengineering? *Science*, 357(6348), 231.

Patt, A., and Dessai, S. 2005. Communicating uncertainty: Lessons learned and suggestions for climate change assessment. *Comptes Rendus Geosciences*, 337(4), 425–441.

Pearce, W., Brown, B., Nerlich, B., and Koteyko, N. 2015. Communicating climate change: Conduits, content, and consensus. *Wiley Interdisciplinary Reviews: Climate Change*, 6 (6), 613–626.

Pearson, L. J. and Pearson, C. J. 2012. Societal collapse or transformation, and resilience. *Proceedings of the National Academy of Sciences*, 109, E2030-E2031.

Peterson, G. D., Cumming, G. S., and Carpenter, S. R. 2003. Scenario planning: A tool for conservation in an uncertain world. *Conservation Biology*, 17(2), 358–366.

Pichler, M., Schaffartzik, A., Haberl, H., and Görg, C. 2017. Drivers of society-nature relations in the Anthropocene and their implications for sustainability transformations. *Current Opinion in Environmental Sustainability*, 26–27:32–36.

Pigeon, N., and Fischhoff, B. 2011. The role of social and decision sciences in communicating uncertain climate risks. *Nature Climate Change*, 1, 35–41.

Priest, S. H. 2014. Climate change: A communication challenge for the 21st century. *Science Communication*, 36(3), 267–269.

Rabitz, F. 2016. Going rogue? Scenarios for unilateral geoengineering. *Futures*, 84, 98–107.

Raven, P. G. 2017. Telling tomorrows: Science fiction as an energy futures research tool. *Energy Research & Social Science*, 31, 164–169.

Remen, R. N., 2010. Listening generously. In: Tippett, K. (ed.), *On Being*. Available at: https://onbeing.org/programs/rachel-naomi-remen-listening-generously/.

Rilke, R. M. 2000 (1929). *Letters to a Young Poet*. Navato: New World Library.

Roggema, R., Vermeend, T., and Dobbelsteen, A. 2012. Incremental change, transition or transformation? Optimising change pathways for climate adaptation in spatial planning. *Sustainability*, 4(10), 2525–2549.

Scheffer, M., 2016. Anticipating societal collapse: Hints from the Stone Age. *Proceedings of the National Academy of Sciences*, 113(39), 10733–10735.

Schellnhuber, H. J. 2009. Terra quasi-incognita: Beyond the 2°C line. In: *4 Degrees and Beyond: International Climate Conference*. Oxford University, Oxford, UK, September 28–30, 2009. Available at: http://www.eci.ox.ac.uk/events/4degrees/programme.php.

Sharpe, B. 2013. *Three Horizons: The Patterning of Hope*. Fife: International Futures Forum.

Shein, P. P., Li, Y.-Y., and Huang, T.-C. 2015. The four cultures: Public engagement with science only, art only, neither, or both museums. *Public Understanding of Science*, 24(8), 943–956.

Slater, G. 2008. Numb. In: Marlan, S. (ed.), *Archetypal Psychologies: Reflections in Honor of James Hillman*, New Orleans: Spring Journal, Inc., pp. 351–368.

Solnit, R., 2004. *Hope in the Dark: Untold Histories, Wild Possibilities*. York: Nation Books.

Spector, B. 2010. *Madness at the Gates of the City: The Myth of American Innocence*. Berkeley: Regent Press.

Spence, A., Poortinga, W., and Pidgeon, N. 2012. The psychological distance of climate change. *Risk Analysis*, 32(6), 957–972.

Stoknes, P. E. 2015. *What We Think About When We Try Not to Think About Global Warming: Toward a New Psychology of Climate Action*. White River Junction: Chelsea Green Books.

Swyngedouw, E. 2010. Apocalypse forever? Post-political populism and the spectre of climate change. *Theory, Culture & Society*, 27(2–3), 213–232.

Tàbara, J. D., St. Clair, A. L., and Hermansen, E. A. T. 2017. Transforming communication and knowledge production processes to address high-end climate change. *Environmental Science & Policy*, 70, 31–37.

Tarnas, R., 2001. Is the modern psyche undergoing a rite of passage? Available at: www.jung2.org/ArticleLibrary/Tarnas.pdf, also at: https://cosmosandpsyche.files.wordpress.com/2013/05/revision-rite-of-passage.pdf.

Tevis, R. E. 2010. Creating the future: Goal-oriented scenario planning. *Futures*, 42(4), 337–344.

Thomassen, B. 2014. *Liminality and the Modern: Living Through the In-Between*. Farnham: Ashgate Publishing.

Tippett, K. 2016. *Becoming Wise: An Inquiry into the Mystery and Art of Living*. New York: Penguin.

Turner, G. M. 2012. On the cusp of global collapse? Updated comparison of the limits to growth with historical data. *GAIA – Ecological Perspectives for Science and Society*, 21(2), 116–124.

Veland, S., Scoville-Simonds, M., Gram-Hanssen, I., Schorre, A. K., El Khoury, A., Nordbø, M. J., ... and Bjørkan, M. 2018. Narrative matters for sustainability: The transformative role of storytelling in realizing 1.5°C futures. *Current Opinion in Environmental Sustainability*, 31, 41–47.

Waddell, S., Waddock, S., Cornell, S., Dentoni, D., McLachlan, M., and Meszoely, G. 2015. Large systems change: An emerging field of transformation and transitions. *Journal of Corporate Citizenship*, 58, 5–30.

Webster, M. 2003. Communicating climate change uncertainty to policy-makers and the public. *Climatic Change*, 61(1–2), 1–8.

Werbeloff, L., Brown, R. R., and Loorbach, D. 2016. Pathways of system transformation: Strategic agency to support regime change. *Environmental Science & Policy*, 66, 119–128.

Westley, F., Olsson, P., Folke, C., Homer-Dixon, T., Vredenburg, H., Loorbach, D., ... and Banerjee, B. 2011. Tipping toward sustainability: Emerging pathways of transformation. *AMBIO*, 40(7), 762–780.

White, B. 2015. States of emergency: Trauma and climate change. *Ecopsychology*, 7(4), 192–197.

Wiek, A., and Iwaniec, D. 2014. Quality criteria for visions and visioning in sustainability science. *Sustainability Science*, 9(4), 497–512.

Wilenius, M. 2014. Society, consciousness and change: An inquiry into Pentti Malaska's futures thinking. *Futures*, 61, 58–67.

Williams, T. T. 2014. We Are Still Singing, Radio interview, *24th Anniversary Bioneers Conference,* San Rafael, CA. Available at: www.bioneers.org/terry-tempest-williams-still-singing/.

8

At the Frontline or Very Close
Living with Climate Change on St. Lawrence Island, Alaska, 1999–2017

IGOR KRUPNIK

8.1 Introduction

As the planet's climate keeps changing and triggers rapid transition in many terrestrial and marine habitats, some places are visibly altering faster than others. The Arctic, the world's northern polar region, is one such place where the pace of environmental change is twice the global average (Larsen *et al.* 2014; Richter-Menge *et al.* 2016; Taylor *et al.* 2017). Since the 1990s, the Arctic has often been called the 'canary in the coalmine' of global warming and, together with certain other habitats, such as low-lying tropical islands and high mountain areas, has received a high level of scholarly and public attention (Orlove *et al.* 2014). And yet, even across the Arctic, some areas are known as 'frontlines' for their exposure to particularly visible impacts of climate change, such as coastal erosion, flooding, sea ice retreat and permafrost loss.

Alaska, the only Arctic state of the United States, is one of these climate change 'frontlines'. During the past fifty to sixty years, it has warmed more than twice as quickly as the rest of the US territory (Chapin *et al.* 2014:516). Changes to the Alaskan environment are highly visible as the ongoing warming of the land surface and rising ocean temperatures have diminished sea ice and extended the ice-free season, increased the incidence of storms and summer fires, pushed the northward advance of tree and shrub vegetation and brought about shifts in ranges of terrestrial and marine species (Shulski and Wendler 2007; Wood *et al.* 2015; Druckenmiller *et al.* 2017; EPA 2017; Taylor *et al.* 2017). Yet even within the State of Alaska, certain locations are particularly vulnerable to the impacts of warming. For example, a recent US Army Corps of Engineers (2009) study identified twenty-six Alaskan communities that should be considered for immediate action to mitigate the negative effects of climate change.

The notion of increased climate vulnerability is commonly attributed to small communities in the Arctic and elsewhere that are short of economic and financial resources and developed infrastructure (Ford and Smit 2004; Hovelsrud and Smit 2010; Himes-Carnell and Kasperski 2015; McDowell *et al.* 2016; Ristrophe 2017). Many of these communities are situated along the western and northern Alaskan shorelines, where exposure to the progressively ice-free ocean, severe storms and floods and coastal erosion threatens local economy and daily life. This chapter tells the story of what it means to live in a small Alaskan community 'at the frontline' of climate change, for two decades and counting. It is based on data collected in two Native villages, Savoonga and Gambell, on the remote St. Lawrence Island in the northern Bering Sea during 1999–2017. The research started when Arctic climate change was still a novelty anxiously debated by local residents (Huntington 2000; Krupnik 2000), yet cautiously approached by climate scientists, who were uncertain about the warming trend (Serreze 2008/2009). Since then, the perspectives of both climate scholars and Native Alaskans have transitioned from initial puzzlement to widespread agreement on the rapid pace of change and to acceptance of a 'new normal' (Sheffield Guy *et al.* 2016) in both natural and social domains.

My early research on St. Lawrence Island in 1999 was a collaborative study of its indigenous history and heritage (Krupnik *et al.* 2002). Shortly after it started, a local man suggested that I switch its focus when he stated, 'Don't you see that we have other things that bother us more than your "grandfathers" stories?' When asked what really bothered local people he responded that 'Something strange is happening to our environment.' Soon after, we launched a collaborative effort to document local knowledge about the changing ice and weather conditions on the island (Krupnik 2002; Oozeva *et al.* 2004). That pilot study produced a starting baseline to assess how people observed and internalised rapid environmental change in the late 1990s, when it was already a matter of concern in many places across the Arctic (McDonald *et al.* 1997; Ford N. 2000; Fox 2002; Jolly *et al.* 2002; Krupnik 2000; Krupnik and Jolly 2002; Nickels *et al.* 2002).

Throughout the 2000s we kept documenting local response to sea ice and weather change on the island (Krupnik 2009; Krupnik *et al.* 2010). Other scientists working on scores of collaborative studies of people's environmental knowledge, climate observations and the use of subsistence resources confirmed that climate change remained high on local agendas (Noongwook *et al.* 2007; Huntington *et al.* 2013; Gadamus and Raymond-Yakoubian 2015; Rosales and Chapman 2015). Hence, a recent visit to the island in February 2017 offered a chance to record several interviews with local residents and to assess how Arctic warming, now in its third decade, has affected people's activities and their sense of belonging.

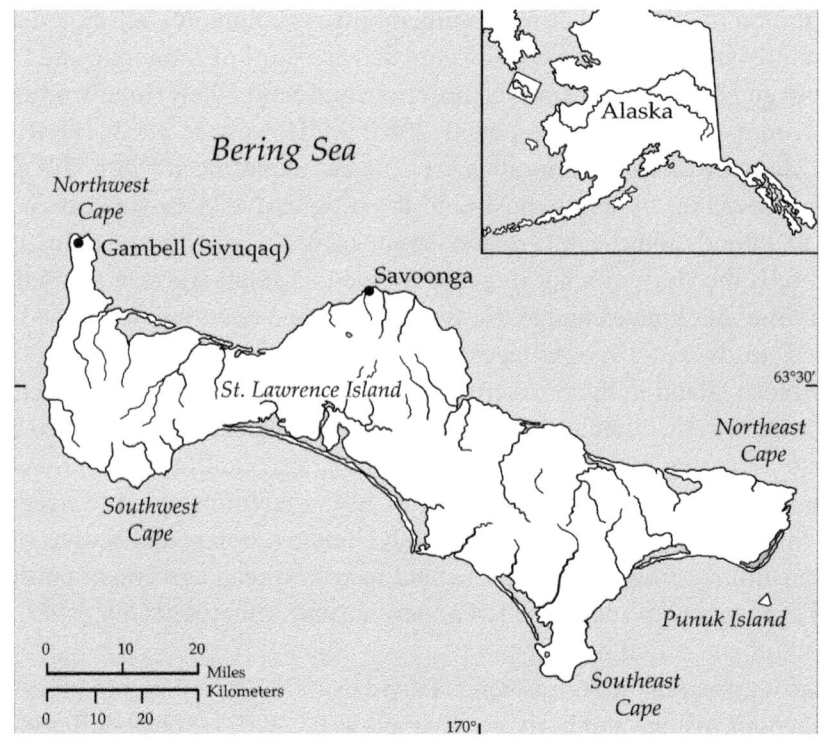

Figure 8.1 Map of St. Lawrence Island and the Bering Strait area (produced by Marcia Bakry, Smithsonian Institution)

8.2 Savoonga: A Story of Change, 1999–2017

The native village of Savoonga (*Sivungaq*, in St. Lawrence Yupik) is a small town of 700+ residents located on the northern shore of St. Lawrence Island facing the Bering Sea. It takes an hour-long flight by small plane to cross the 200 miles separating Savoonga from the mainland hub of Nome (Figure 8.1) (population 3,800; 2016), which has most of the economic, institutional and administrative facilities serving the larger northern Bering Sea-Bering Strait region. A twenty-minute flight connects Savoonga to Gambell (Yupik name, *Sivuqaq*), a sister community of the same size at the north-western edge of St. Lawrence Island facing the Russian Chukotka Peninsula 58 km (36 mi) away. In both Savoonga and Gambell, 95 per cent of residents belong to a group called the St. Lawrence Island Yupik, with deep historical ties to the Yupik people on the Asian side (Krupnik and Chlenov 2013). The two towns are closely related and share the same language and identity. About 400–500 Yupik people with their roots in Gambell or Savoonga now reside in Nome, Anchorage, elsewhere in Alaska and across the continental United States.

Figure 8.2 New public buildings in Savoonga, with the recently erected wind turbines in the background. Photo by Igor Krupnik, February 2017

Stepping off the plane in February 2017, it was hard to miss how profoundly Savoonga had changed in the eighteen years since my first visit. It is now a bustling town with signs of a construction boom – a new large high-school, a new well-stocked village store, rows of modern family homes and administrative buildings with several offices, recently erected wind turbines at the edge of town and many other structures (Figure 8.2). It looks as if Savoonga has gradually moved away from its historical centre of gravity built in the 1930s along the ocean front (Figure 8.3) to higher ground about half a mile inland. Because of that, its vulnerability to the forces of the ocean, such as floods, large waves brought by powerful storms and the crushing impacts of the piling ice, has diminished. However, that slow migration farther inland took place not as a thought-through adaptation, but rather because it was the only direction in which the town could expand. Savoonga leaders talk about the next round of construction farther away from the beach, where a few aging private houses still remain.

Savoonga is intense, crowded and friendly – as it has always been in people's memories. Much of the personal interaction takes place in the street, where on a cold winter's day residents move around on four-wheelers and snow machines. This creates an atmosphere of one big family, which rests well with the history of the community formed a century ago by a tightly related network of nomadic reindeer herders' camps (Krupnik *et al.* 2002). In that sense, Savoonga has preserved its social core, despite its increasingly modern face.

Figure 8.3 Savoonga 'traffic jam'. People's four-wheelers parked next to the public building during community meeting. Photo by Igor Krupnik, February 2017

8.3 Northern Bering Sea Record: 2000–2017

Since the people of St. Lawrence Island first reported changes in their home environment in the late 1990s (Krupnik 2000, 2002; Pungowiyi 2000), the habitats of the northern Bering Sea have been visibly transformed by the forces of Arctic warming. The change has affected all seasons but it is most pronounced from October until May each year, a period that has historically featured temperatures well below freezing and extensive, if not always persistent, sea ice cover (Oozeva *et al.* 2004; Krupnik *et al.* 2010; Ray *et al.* 2016). Today, the fall freeze-up takes place almost two months later and the spring breakup occurs a full month earlier than in the 1980s and 1990s. Since the start of the era of the satellite observations in 1979, the number of days without *any* sea ice in the northern Bering Sea has increased by eighteen days per decade, adding almost three and a half months of open water (Frey *et al.* 2015; Wood *et al.* 2015; Druckenmiller *et al.* 2017).

A longer ice-free season means warmer air and ocean temperatures, more unstable weather, stronger waves and winds and fewer of the quiet cold days that are favoured by hunters. In 2008, a local hunter explained that it was now 'not cold enough' in the winter to build good-quality, solid ice. In the early 2000s, hunters started to complain that the ice was not safe anymore, that it had become 'flimsy', and was not good for animals to use and hunters to walk on (Oozeva *et al.*

2004; Krupnik *et al.* 2010). Since then, several independent studies have confirmed that the ice, wind and weather regime around St. Lawrence Island and, more widely, in the northern Bering and southern Chukchi Sea has been profoundly altered since the early 2000s, corroborating local observations (Kapsch *et al.* 2010; Huntington *et al.* 2013; Frey *et al.* 2015; Wood *et al.* 2015; Ray *et al.* 2016; Huntington *et al.* 2017).

Overall, change in the northern Bering Sea environment over the past two decades has been visible, massive and indisputable. Climate and ocean warming, a novelty in the late 1990s, is now a fact of people's lives that they have to cope with. They hardly question the general forces that have caused this environmental shift, and there are no climate change 'deniers' on the island, at least none that I met in 2017.

8.4 Talking Change in Savoonga

The winter of 2017 has been, again, a remarkable season in a string of several remarkable seasons and years, one after the other. The sea ice did not form in front of Savoonga until mid-to-late January, at least a full month later than in the early 2000s. The thin sliver of shore ice was visible by mid-February but was considered to be unstable and constantly shifting. The new ice that formed in February 2017 was what people call 'local ice', as this was built up by the freezing of local slush ice rather than the winter pack ice, which never arrived from the North. In fact, the winter pack ice stopped coming in the early 2000s. This local ice was barely 5 cm thick, and it was dangerous to walk on it. In Gambell, the town had very little shorefast ice by mid-February, something that people are now well familiar with.

There was also little snow by mid-February, not including the packed cover within town limits. Snow covered the ground very late that year, perhaps a full month later than in the 2000s. When the ice finally formed off Savoonga by the end of January 2017, it was to stay for less than three to four months before breaking up in late April or May the following year. Thus, it was neither 'cold enough', as in the 2000s, nor 'long enough' to build solid winter ice that people and animals rely on.

In many focused and spontaneous talks with people in Savoonga and Gambell in winter 2017, change was on everyone's mind, but more as a matter of curiosity, often of bewilderment. It looked as if everybody could speak about the warming climate at a moment's notice, yet often with the kind of amusement that might mask people's true concerns:

The ice condition [this winter] is very troubling. We didn't have any ice here till late January, almost till early February. Then it finally piled up, but this ice is thin, not good. It is also moving and keeps piling up. It is no good for hunting or ice fishing. And it is built very late in the season; so, it doesn't have much time to grow to get thicker.

Figure 8.4 Gambell beach zone showing little snow and almost no ice in the middle of winter. It has become a typical condition in the past few years. Photo by Igor Krupnik, February 2017

People keep saying that there is no good ice around the island and it scares me. Hunters killed two walruses lately, right in front of the village and they couldn't butcher them on ice, like normally, because the ice kept breaking. So, they had to pull them all the way to the shore and butcher them onshore, which is kind of strange to us in the middle of the winter. (Savoonga, February 14, 2017)[1]

Savoonga residents were well aware that the ice in front of their town was perhaps the *only* ice around and that open water or loose floating ice surrounded their island on all sides. This meant there was no place to go for good winter hunting. None of the usual groups of winter male walruses, *angleghaq*, that live on dense solid ice and that people relied on in the olden days were spotted. The two walruses killed off Savoonga in early February were 'stranders', animals separated from the main group that remained beyond hunters' reach. Instead, Savoonga residents woke up one morning in January to find a stranded sea lion on the beach. It was trapped by the forming ice and had nowhere to go; it was eventually killed despite people's

---

[1] To preserve speakers' anonymity, only the date and the town where the interview was recorded are listed henceforth.

effort to chase it back into the water. Sea lions are familiar to St. Lawrence Islanders, but they are usually present in October and leave by November or early December, before the arrival of the moving ice from the North. Seeing a sea lion in mid-January – and no walruses – was another sign of Mother Nature going 'wild'.

With walruses nowhere to be found, bowhead whales are now abundant off the island in mid-winter, when they are supposed to be hundreds of miles south, closer to the southern edge of the pack ice in the Bering Sea. Island residents believe that, because of weaker ice and more open water, bowhead whales stopped going south to their usual habitat:

With bowhead whales, ... they are being seen in increasing numbers and many are now overwintering here ... There is plenty of open water for them. The ice is thin now and they can break it easily.

Also, between here [Gambell – IK] and Russia there is a large polynya, called *kelligheneq* in our language. It used to be only seasonal, usually in the month of May, but nowadays it is a permanent fixture. ... That is where the whales are wintering now, especially the young ones. But the older ones with very young [females with newborns] still like to go down south. (Gambell, February 17, 2017)

This view has been confirmed by scientists studying bowhead whales' changing distribution due to decreased winter ice in the Bering and Chukchi Seas (Druckenmiller *et al.* 2017).

Another common theme is the early *spring break up* that now happens at least two to three weeks earlier than fifteen to twenty years ago. In many places on the island, like off Gambell, hardly any shore-fast ice forms during the wintertime. The Savoonga spring whaling camp at Pughughileq, on the southern shore, used to have solid shore-fast ice, onto which hunters pulled their boats and butchered the whales that they killed in the moving ice. But that does not happen anymore, as the ice is usually gone by early-to-mid April, and all hunting and butchering now proceeds on open water or on the shore.

The early spring break up and unstable weather in April and May coincide with the main walrus hunting season. Since 2013, hunting in both Savoonga and Gambell has been very poor and the overall walrus catch has dropped to a third of what it used to be ten to twenty years ago (Krupnik and Benter 2016). People are quick to make the connection between these changes, to acknowledge uncertainty and to point to increased risk to their livelihoods and sources of income:

Poor ice affects our walrus hunting. If we don't have good ice here in spring, we won't have enough walrus on the northern side that we normally hunt, like in May or June. People are already moving to the east side [of the island], so that they could catch walrus calves and females in spring. It also scares me, because it means a lot of traveling, hauling boats to the eastern side and then hauling walrus we kill there back to the village. It will be after we finish whaling at Pughughileq in April and we are getting very little snow to last long enough to keep the trails . . .

The ice condition is affecting our walrus hunting in the past few years and we are getting very few walruses, so very few. In my old days, when I worked in the village store [in the 1990s], we always purchased raw ivory and some aged ivory from hunters . . . But today the store is not buying any ivory at all and people do not sell it, because they keep it for themselves. They have very little from hunting these days. (Savoonga, February 13, 2017)

People can talk for hours, mulling over the 'strange things' brought by unusual weather, ice and climate. Some residents, reportedly, have observed birds in the middle of winter, including snow buntings (*Plectrophenax nivalis*) that are not supposed to be seen before April. Snow buntings are the first birds to arrive in spring and are viewed as the sign of a new season. A common explanation is that the buntings now do not depart at all and are overwintering on the island on patches of open ground. 'We used to have only ravens here during the wintertime', one man said. 'Now we see these buntings, also gulls, like never before.'

Another hunter referred to shooting seals from shore blinds in the fall, a common practice. 'Guess what – these seals are supposed to float (on water surface) when killed. Now they most often sink and even when we retrieve some, they have a very thin layer of blubber. They just won't float.' Hunters agreed that this was happening because of the layer of lower-salinity water on top of the sea. 'Normally we have [dense] salt water on top, in which they [seals] remain floating. So, something is wrong with the seals and top water as well.'

The increased occurrence of very strong winds is another noticeable change, particularly in late fall and winter, with strong winds now coming from the south instead of the north, as in the olden days. These winds now present a threat to local infrastructure. The winter storm of December 2010 was exceptionally damaging in Savoonga, as winds of up to 50 mph and sub-zero temperatures contributed to a salt spray freeze on all electrical equipment. The wind knocked off town electric lines and nearly three-quarters of residents lost power for six days and more. Many homes experience burst pipes and flooding, and gusting winds prevented the authorities in Nome from shipping in food, supplies and repairmen. As one resident recalled:

We had that bad storm in 2010, when over 25 houses were left without power and water and we had to set a rescue area (shelter) at school. The school housed almost 200 people for several days. We had it again this year, though not like in 2010. We do not recall such heavy winds in the past; this is new to our memory. Perhaps not new to the Elders, but the infrastructure is very different today and it could easily knock us off for many days. (Savoonga, February 13, 2017)

Melting permafrost is a growing concern in Savoonga, akin to what happens in many other Alaskan communities, though new to the island. Many houses built on once solid ground are affected. Some places closer to the ocean front act like sinkholes and have destroyed older homes:

We lost one house recently – complete collapse; we had to tear it apart and start new construction. It is also happening along the shore, cliffs are falling down.

Our emergency 'shelter' area is at [the village] school. We used the old school building near the shore [as shelter] before that. But it is not functional anymore. It also suffered from erosion [permafrost melting], because it started curving down. Same with the old store building near the shore. It may fall down any time and make trouble to many houses nearby, particularly if it happens under gusty winds. (Savoonga, February 13, 2017)

The list of 'strange' things that people associate with climate change is remarkably diverse and it keeps growing (Krupnik *et al.* 2010; Rosales and Chapman 2015). Yet it is primarily the weak and unsafe winter ice, not the warming temperatures, that is on everyone's mind as a major development introduced by climate change.

8.5 Discussion: How People Adapt

The stories recorded on St. Lawrence Island over the last two decades (Krupnik 2000, 2002, 2009; Pungowiyi 2000; Oozeva *et al.* 2004; Noongwook *et al.* 2007; Krupnik *et al.* 2010; Rosales and Chapman 2015) portray a vibrant discourse, as many thoughtful local observers, including hunters, community leaders and elders, try to take stock of the alarming pace of transition. People are not passive witnesses of change; instead, they seek explanations and look for connections and solutions. Yet certain factors, whether local and of a more general nature, may increase or lessen the chances of small northern towns, like Savoonga and Gambell, successfully adapting to change.

8.5.1 Preserving the 'Core' of Life (and Culture)

Despite the rows of modern houses, brand-new office buildings, snowmobiles and cell phones used by adults and many children, Savoonga and Gambell have successfully sustained their core identity, which is based around subsistence hunting, particularly communal bowhead whaling. Savoonga captures three to four bowhead whales every year from its annual quota of eight 'strikes'. This supports 28 family whaling crews, about 120 to 150 people altogether, who operate for 4 to 5 weeks each spring out of the town's whaling camp at Southwest Cape. Gambell has the same number of strikes and operating crews, although its catch is lower, at one to two whales per year. Receding ice and warmer waters have now added a new whaling season in late fall, particularly in Savoonga, usually in November–December, even in January. Over the past twenty years, twenty-two whales were killed in Savoonga during this late fall season, once unheard of, 38 per cent of the total catch (Nongwook et al. 2007; Savoonga Whaling Captains Association 2017). There is a strong cadre of determined young and middle-age hunters eager to keep this vibrant cultural practice alive into the twenty-first century.

Killing, towing, butchering, storing and sharing meat of a 20- to 40-ton whale requires a host of *collective* actions and thus increases community cohesion. The collective identity generated and sustained by whaling and by maritime hunting more generally is a remarkably strong adaptation asset. Hunting of whales and, to a lesser extent, walruses and seals provides a solid base for a shared identity and common response and continues to bind people together. It looks like an unexpected source of strength for twenty-first-century adaptation and so far it is holding.

8.5.2 New Technology Helps

On recent visits to Savoonga, Gambell and other Alaskan communities, an unfamiliar element of the local landscape was quite visible, the white towers of new wind turbines (Figure 8.2). Since 2008, under Alaska's Renewable Energy Program, small rural communities have received wind turbines installed as a source of electricity, the only local alternative to imported gasoline and diesel, for their electric grid and utilities (DeMarban 2016). It is a small operation – two turbines in Savoonga, three in Gambell, two in Wales – but people rightly view it as their first step in 'energy independence'. It is also the most visible sign of modern-era technology in town (besides the now ubiquitous smart phones, personal computers and the Internet) that may eventually increase people's options as they attempt to cope with climate change.

Other newly introduced or, often, reintroduced technologies come from the twentieth century. In 1994, Savoonga started its first commercial halibut fishing operation via a regional company, Norton Sound Seafood Products (NSSP – Anonymous 2015). Today, ten to fifteen boats with three-to-four-men crews regularly fish for halibut in the summertime using longline hooks. Some thirty to forty Savoonga men are engaged in commercial fishing and are being paid upon delivery of their catches to a small processing facility that NSSP built in town. The facility supplements their family incomes and keeps people busy in the summer months. Yet Savoonga fisheries are a far cry from the more advanced Native commercial fishing operations elsewhere in Alaska that support village flotillas of mid-size boats with Native owners and crewmen. It could be years before Savoonga arrives at this stage, if ever, and Gambell has no commercial summer fisheries so far.

After 2010, Savoonga, in cooperation with Gambell, has restarted small-scale reindeer herding, once a productive operation on the island. The herd now has about 3,000 animals, which allows a sustainable harvest of 600 to 800 reindeer each year (Caldwell 2016). This enterprise relates well to the early 1900s herding tradition of Savoonga and creates a mixture of pride and amusement among local people. Nonetheless, it relies on technology that belongs to the (early) twentieth century and therefore has little room to expand:

> The herd is growing ... but we have neither a good corral nor a meat packing operation here. They don't sell reindeer meat at the store. They [the village corporation] just sell 'licenses' to individual families if they want to have reindeer, so that they can go and kill it (!). Yeah, they kill it with a rifle, like caribou. They only sell meat to other communities ... mostly ship them by plane as bulk [carcasses]. It is growing, but we still do not know how to use it as a good economy. (Savoonga, February 13, 2017)

It is unclear whether these or other new and reintroduced activities, such as birdwatching and rapidly growing Arctic cruise tourism via the Northwest or Northeast Passage, may develop into stable components of the island economy. There are neither mining nor suitable anchoring sites on the island to attract the passing boats. But the fact that the two communities are looking to diversify their use of resources is welcoming as it indicates that they are seeking new technologies to help them cope with rapid environmental change.

8.5.3 It's the Economy, Stupid!

As the attitudes, outlook and living conditions of northern villages become modernised, the gap between the high cost to support new infrastructure and the meager base of the local economy becomes obvious. In many places, various transfers of cash, equipment and personnel from government, state and district-based bodies constitute the prime, if not the only, way to maintain or improve standards of living. As one thoughtful person in Gambell called it, 'As long as we remain a welfare community we will keep our welfare mentality.' Another local leader in Savoonga offered a dire summary of the town's economic prospects:

> Our economy here has two legs only – halibut fishing and ivory carving. Nothing else that we produce and we can sell. So, if we are short of ivory for carving everybody gets hurt. We don't have tourists here, no cruise ships like in Gambell, and very few bird watchers. But even if you have tourists here, particularly from other countries, they cannot purchase our ivory [because of the U.S. Marine Mammal Protection Act]. They just look but don't buy it. So, one of the legs of our economy is really suffering. And the halibut 'plant' is really not a plant; it's just a division of the Norton Sound Sea Food Products with the main processing plant in Nome. They only flense the fish they caught here, mix it with ice and ship it by (small) planes to Nome to the main plant where it gets processed for sale. ... It's a small

operation, but it helps us a lot, particularly in the summertime, when there is not much employment in town.

... This halibut fishing plant is not going to grow much; so, it is some addition to our economy, but it doesn't keep it running fully. We depend on the money we get from the State (of Alaska), from federal agencies, from Kawerak, from the housing administration. Carving and fishing covers just a small part of what we need. (Savoonga, February 14, 2017)

Scholars of climate change impacts commonly point to the limitations of small-scale rural economies, their lack of resources and high costs of infrastructure improvement as the main impediments to successful adaptation in the Arctic and elsewhere (Henriksen 2007; Morton 2007; Salick and Byg 2007; Hovelsrud and Smit 2010; Williams 2012; Rasmussen *et al.* 2015; Nakashima *et al.* 2018). Community leaders on St. Lawrence Island understand these limitations perfectly well. Economic marginalisation thus becomes a vicious circle that pushes communities on the frontline of climate change into reactive mode rather than fostering proactive thinking to address their future.

8.5.4 What Else Is on People's Mind?

Three topics were most prevalent in the interviews, as well as in the spontaneous conversations, phone calls and email communications, with the people of St. Lawrence Island over the past years: whaling, weird weather and recent deaths (or sickness) in the community. Whereas the two former themes are mostly seasonal, death and ill health are ongoing issues that affect every family. Native Alaskan death rates have shrunk or stabilised in the past decades but nonetheless remain substantially higher compared to the overall state population: 1,169.7 versus 736.2 per 100,000 in 2013–2015 (National Center for Health Statistics 2017:118). For the 1,500 people living on the island and additional 500 on the Alaskan mainland and elsewhere, these statistics transform into roughly twenty to twenty-five deaths each year. Passing of family members, old and young, is widely perceived as a loss for the entire community ('a blow to our common body'), which is mourned profoundly and by many people.

To village residents, who essentially belong to clusters of intermarried families, death always has a personal face. Whether coming as the passing of a respected elder, or loss of a young soul to suicide, injury or accident, it generates long grieving and deep personal sadness that clouds almost every conversation. Ill health and prolonged visits to hospitals on the mainland for medical examination or treatment are constant themes and sources of stress, emotional as well as economic. In this regard, the threat of climate change often comes as a distant factor – unless it is pushed to the forefront by yet another natural phenomenon to cope with, talk about and leave behind. And then the cycle and ranking of people's concerns is restored, until it is interrupted once again.

Many other topics are high on islanders' agenda, often with indirect or distant connections to environmental change, such as skyrocketing gasoline prices, a proposed ban on the sale of walrus ivory (reportedly to protect 'other ivories' from poached African elephants) or a widely discussed attempt to put the Pacific walrus on the Endangered Species list that did not materialise (Anonymous 2017). There are also themes that have *no* relation to climate change whatsoever – like basketball games and other popular sport competitions, youth sub-culture, rampant drug use (Weingarten 2005; cf. Collings 2014) and illegal sales of carved ivories to visiting collectors. It is difficult to cover the complexity of village discourse, even to enumerate its ever-morphing topics. These topics can be artificially dissected for an interview or invited conversation with a visiting scientist or journalist, but are not constant themes when people talk among themselves.

My latest visit to St. Lawrence Island in October 2017 was for a memorial service for a late partner in many years of research on St. Lawrence Island. A day prior, an elderly woman passed away in Gambell; so it was also the funeral service at the local Presbyterian Church. In short encounters in Savoonga and in several longer talks in Gambell, people touched upon many familiar themes – the weak local economy, the prospective ban on the sale of walrus ivory, overcrowded housing conditions, sombre lists of those who passed away – but not on weather or climate change.

Snow has already covered the ground (which was called 'kind of normal') and Troutman Lake near Gambell was already hardened by a thin crust of newly formed ice, like in the 2000s (Oozeva *et al.* 2004:182–183). There was not a bit of ice in the ocean nor on the beach, again a 'new normal' in this time of warming. In the morning, hunters with rifles were riding their four-wheelers out of town to shoot seals from shore blinds, again, a normal activity for the season. There was nothing unusual anymore in the weather, climate or the lack of sea ice – so nobody talked about it.

8.6 Diverging Discourses of Climate Change

Indigenous Arctic residents and outsiders differ in how they talk about what should be done to address rapid environmental change. The key terms coming out of a myriad of scholarly and agency meetings that discuss climate change include 'adaptation', 'mitigation', 'increased resilience', 'sustainability' of communities and infrastructure, 'relocation costs', and, now increasingly, 'community actions'. When people on St. Lawrence Island talk about what they need to face climate change with, their first call is 'emergency assistance'. Then they go into specific details on topics related to compensation claims, disaster relief, dedicated shelter areas and the timing of emergency response. In this case, Yupik town leaders in Gambell or Savoonga in 2017 do not differ much from local authorities in other US localities, such as Texas, Florida or Puerto Rico, that have to face the brunt of the changing planetary climate.

With respect to extreme and unusual weather, and emergencies, there are two substantial aspects in which the responses of local leaders in isolated Alaskan communities, like on St. Lawrence Island, vary from those of local authorities elsewhere. First, people in the North have to 'weather in', since there is no place to which they can move or be evacuated to during disasters, except to other buildings within the same community. Second, and more importantly, external state assistance will come only days or weeks *after* the event has occurred, when the damage is already done. This assistance will mostly be in the form of money and equipment to rebuild what has been destroyed. To illustrate, from December 28, 2016, to January 4, 2017, Savoonga was severely affected by a powerful storm. However, the two-person team from the State of Alaska Emergency Response Office did not arrive in the town to examine the damage until a full six weeks after the event has taken place – mostly to take pictures of damaged houses and boats, and to process family claims for state-based disaster compensation. As the local residents explained to me, for the first several weeks they were left to cope with the outcomes of the storm entirely on their own and they took all responsibility for the initial response, damage assessment and community clean-up.

Fortunately, the December 2016 storm was not as destructive as the one in December 2010, when the damage was substantially higher. If the combined, monetary damage from the storm is below a certain level, it is usually covered by compensation from the State of Alaska Disaster Relief Fund, which is made of 75 per cent federal and 25 per cent state monies. If the claimed damage is above a certain threshold, everything is covered by federal money administered by FEMA (Federal Emergency Management Agency) as was the case following major national disasters, such as Hurricane Katrina in 2005, or Hurricanes Harvey and Irma in 2017.

Unfortunately, state and federal agencies generally have few resources and little vision beyond disaster relief, and mostly talk about the relocation of the most affected communities. In places like Savoonga and Gambell, adaptation and disaster mitigation are being left primarily to local leaders to fathom, with hardly any available resources to spare. After twenty years of dramatic changes in the Arctic, and in the absence of any coordinated actions to combat climate change, the least helpful public message is to continue showcasing people's suffering inflicted by climate warming. What people at the 'frontlines' need is a vastly improved emergency response, not a continuous drumbeat of public anxiety and talks of impending 'climigration'. Therefore, scientists' findings about progressing environmental change bolstered by models with various levels of confidence have little relevance to how people actually face climate change.

8.7 'Climigration': Is It Working?

The term 'climigration' was coined by Alaskan human rights lawyer Robin Bronen to describe the 'forced permanent migration of communities due to climate change' (Bronen 2009:68). 'Climigration' results from ongoing climate-induced environmental change and occurs when a community is no longer sustainable for ecological reasons (Bronen 2009:68). Evidently, climigration means organised or spontaneous relocation. It affects the most vulnerable communities and differs from temporary migration caused by catastrophic environmental events, because it means a permanent displacement with no way back.

If we are to follow this definition literally, then no Alaskan community has yet been subjected to forced 'climigration', although at least four, Kivalina, Newtok, Shaktoolik and Shishmaref, have been living with the threat of pending relocation for the past fifteen to twenty-five years (Marino 2015). Both St. Lawrence Island communities, Savoonga and Gambell, are *not* on this list.

Hamilton *et al.* (2016) applied the paradigm of 'climigration' to test whether there is an evident trend of people moving out of at-risk Alaskan communities to safer destinations (see also Hamilton and Mitiguy 2009). The population data for the last twenty-five years (1990–2014) illustrate that most, though not all, rural Alaskan communities are actually growing. This includes all four of the most climate-affected communities mentioned above where families continue to grow, despite a well-known threat of prospective relocation. It is obvious that people do not want to leave their homes or, at the very least, the number of those staying is growing faster than those who have preferred to leave.

Both Savoonga and Gambell have experienced dramatic growth since the past twenty-five years. Savoonga population grew from 519 in 1990 to 740 in 2016 (mayor's office data); the number of residents in Gambell increased from 526 to over 700 during the same period. Both towns have high birth rates, so that children and babies are present in almost every household; twenty to twenty-five babies are being born in Savoonga every year, according to the mayor's office. The town has a new K-12 school with 240 students and rows of modern houses now filled to capacity by growing extended families. No one in Savoonga mentioned the possibility of moving en masse to another place; it is clear that the climate is not yet considered to be a push factor. Evidently, the two communities are preparing to stay, and are literally digging in, in terms of their investments in local infrastructure and new housing.

Yet at least one Alaskan community of Diomede on Little Diomede Island is seemingly shrinking. Little Diomede is a tiny island, with almost no place to go or space to expand. Its population was once about 150 to 180 residents but had dropped to 100 people by 2010. In 2015, according to local schoolteachers, the actual population was around eighty and the island school was on the verge of closing if the number of students did not increase. Several former Diomede residents who recently moved to Nome and Anchorage reported that the main reason for them leaving was the change in sea ice due to the warming climate. It made spring hunting for walruses and connections to the outside world via small ice-landing planes increasingly unreliable. Diomede, thus, illustrates that at least *some* communities may indeed face population decline due to the pressure of climate change.

8.8 Conclusions: Talking Past Each Other?

The changes imposed by Arctic warming during the past twenty years have been profound and undeniable. Neither scientists nor local residents can now dispute the arrival of the 'new normal' (Larsen *et al.* 2014; Richter-Menge *et al.* 2016; Sheffield Guy *et al.* 2016; Taylor *et al.* 2017) which, among other things, means warmer and more unstable weather, stronger winds, a shorter ice season and riskier and more unpredictable sea ice. Whereas scientists primarily point to a dramatic loss of Arctic sea ice, particularly in the summertime (Frey *et al.* 2015), local residents single out how different their 'weird' winter weather is to what it used to be twenty to thirty years ago. Their own interpretations – 'it's not cold enough', 'it's not long enough' (for the winter) or even 'the Earth is faster now' (Krupnik 2002, 2009) – are not on the scientists' lexicon.

Locals and scientists also differ in the ways in which they look into the future of the warming Arctic. Local people have learned to live with climate change and they have gradually pushed it into the realm of 'village talk', even sombre jokes, so that they can cope with it. They certainly have other issues to worry about. They have no means but to accept the new normal, even if they openly lament the predictability of the cold days, solid ice, large herds of walruses and the stable seasonal round of hunting in the 'olden days' which it never was (Krupnik 1993).

I dare to argue that, in this sense, life in Savoonga and Gambell, as well as in other small northern communities at the frontline of climate change, has somehow 'normalised'. True, the Alaskan town of Kivalina now stands protected solely by rows of sandbags; but so does New Orleans with its levees, and the entire nation of the Netherlands with its system of canals and dikes. The village of Shishmaref, another well-known Alaskan case story of rapid climate change, has been literally falling off the coastal cliff for the past thirty years (Marino 2015); nevertheless, life goes on. The process of 'normalisation' for many of these communities means that concerns about climate change have been woven into the fabrics of people's lives, in which other matters – health and death, economy, daily sustenance, the monthly pay cheque or lack thereof and the price of gasoline – are as much, if not more, important. Nobody seems scared of climate change and nobody is leaving. It looks as if the 'love for one's homeland', a factor rarely considered in climatologists' models, keeps people resilient, industrious and often plain stubborn.

I also concur with my several colleagues in the studies of Arctic climate change (Marino and Schweitzer 2008; Tejsner 2013; Collings 2014; Marino 2015) who argue that northern residents, like people in small communities elsewhere, do not like to be viewed as 'canaries in the coalmine'. It is scientists and, even more commonly, visiting journalists who dramatise people's precarious existence in remote northern communities and talk about climate change as an existential threat to their lives and cultures (e.g. Albeck-Ripka 2017; Demer 2017a, 2017b; Goode 2016). This by no means implies that local people do not perceive acute threats from a warming climate. Rather, it means that Nature may shift faster than Culture, a conclusion which is at odds with the common view of rapidly changing indigenous cultures and lifeways, and that a strong Culture is often people's best way to face rapid environmental change (cf. Demer and Lester 2017).

In Gambell and Savoonga, by the end of a second decade of Arctic warming, the towns' residents have obviously learned how to live with climate change and how to factor it into their plans for the future. It is an open question, however, whether these plans will work if global change progresses in the manner predicted by modern climatological models, or on an even faster track.

Acknowledgements

I am grateful to the volume editors, Giuseppe Feola, Hilary Geoghegan and Alex Arnall, who encouraged me to write this chapter to summarise the experience from years of research and interactions with the people of St. Lawrence Island, Alaska. My colleagues – Shari Fox, Jon Rosales, Hajo Eicken, Carleton Ray and Jeffrey Stine – read the first draft and offered valuable comments and criticism; Marcia Bakry produced the map. Warmest thanks to many friends and partners in Savoonga, Gambell and other Native Alaskan communities, who generously shared their knowledge, observations, concerns and a great sense of humour that helped grasp the strength of their culture and dedication to their life at the frontlines of global change.

References

Albeck-Ripka, L. 2017. Why Lost Ice Means Lost Hope for an Inuit Village. *The New York Times*, November 25, 2017 www.nytimes.com/interactive/2017/11/25/climate/arctic-climate-change.html

Anonymous 2015. Savoonga Fishermen Look Forward to Commercial Halibut Fishing Season. *NSEDC Newsletter* Summer:1–2.

Anonymous 2017. After Comprehensive Review, Service Determines Pacific Walrus Does Not Require Endangered Species Act Protection. U.S. Fish and Wildlife Service (October 4, 2017) www.fws.gov/news/ShowNews.cfm?ref=after-comprehensive-review-service-determines-pacific-walrus-does-not-&_ID=36158

Bronen, R. 2009. Forced migration of Alaska indigenous communities due to climate change: Creating a human rights response. In: Oliver-Smith, A. and Shen, X. (eds.), *Linking Environmental Change, Migration and Social Vulnerability*. Bonn: UNU Institute for Environment and Human Security, pp. 68–73.

Caldwell, S. 2016. Savoonga steps up its game in the reindeer meat market. *Alaska Dispatch News* September 26, 2016; see: www.adn.com/rural-alaska/article/rural-village-savoonga-gets-serious-when-it-comes-getting-reindeer-meat-game/2015/04/13/

Chapin, F. S., III, Trainor, S. F., Cochran, P., Huntington, H., Markon, C., McCammon, M., McGuire, A. D., and Serreze, M. 2014. Alaska. In: Melillo, J. M., Richmond, T. and Yohe, G. W. (eds.), *Climate Change Impacts in the United States: The Third National Climate Assessment*, Washington, DC: U.S. Global Change Research Program, pp. 514–536.

Collings, P. 2014. *Becoming Inummarik: Men's Lives in an Inuit Community*. Montreal: McGill-Queen's University Press.

DeMarban, A. 2016. Wind power fuels Alaska's push for rural renewable energy sources. *Alaska Dispatch News* September 27, 2016; www.adn.com/energy/article/wind-power-fuels-alaskas-push-rural-renewable-energy-sources/2012/05/03/

Demer, L. 2017a. For two Alaska villages, walruses remain essential. As sea ice disappears, can it last? *Alaska Dispatch News*, May 26, 2017– www.adn.com/st-lawrence-island/

Demer, L. 2017b. St. Lawrence Island tribal groups tried to protect walruses. Now the animal they rely on faces a threat they cannot control. *Alaska Dispatch News*. Published May 27, 2017 – www.adn.com/features/alaska-news/rural-alaska/2017/05/27/st-lawrence-island-tribal-groups-tried-to-protect-walruses-now-the-animal-they-rely-on-faces-a-threat-they-cannot-control/

Demer, L., and Lester, M. 2017. What subsistence looks like at the Apassingok family table: Platter for all with walrus, whale and more from the Bering Sea. *Alaska Dispatch News*. May 31, 2017 – www.adn.com/features/alaska-news/rural-alaska/2017/05/31/what-subsistence-looks-like-at-the-apassingok-family-dinner-table-platter-for-all-with-walrus-whale-and-more-from-the-bering-sea/

Druckenmiller, M. L., Citta J. J., Ferguson M. C., Clarke J. T., George J. C., and Quakenbush, L. 2017. Trends in Sea-Ice Cover within Bowhead Whale Habitats in the Pacific Arctic. *Deep Sea Research Part II*.

EPA (U.S. Environmental Protection Agency). 2017. Climate impacts in Alaska. https://19january2017snapshot.epa.gov/climate-impacts/climate-impacts-alaska_.html

Ford, J. D., and Smit, B. 2004. A framework for assessing the vulnerability of communities in the Canadian arctic to risks associated with climate change. *Arctic*, 57(4), 389–400.

Ford, N. 2000. Communicating climate change from the perspective of local people: A case study from arctic Canada. *Journal of Development Communication*, 1(11), 93–108.

Fox, S. 2002. These are things that are really happening: Inuit perspectives on the evidence and impacts of climate change in Nunavut. In: Krupnik, I., and Jolly, D. (eds), *The Earth*

Is Faster Now: Indigenous Observations of Arctic Environmental Change. Fairbanks: ARCUS, pp. 12–53.

Frey, K. E., Moore, G. W. K., Cooper, L. W., and Grebmeier, J. M. 2015. Divergent patterns of recent sea ice cover across the Bering, Chukchi, and Beaufort seas of the pacific arctic region. *Progress in Oceanography*, 136, 32–49.

Gadamus, L., and Raymond-Yakoubian, J. 2015. A Bering strait indigenous framework for resource management: Respectful seal and walrus hunting. *Arctic Anthropology*, 52(2), 87–101.

Goode, E. 2016. A wrenching choice for Alaska towns in the path of climate change. *Arctic Now* June 26, 2017 www.arcticnow.com/arctic-news/2016/11/29

Hamilton, L. C., and Mitiguy, A. M. 2009. Visualizing population dynamics of Alaska's arctic communities. *Arctic*, 62(4), 393–398.

Hamilton, L. C., Kei, S. Loring, P. A., Lammers, R. B., and Huntington, H. P. 2016. Climigration? Population and climate change in arctic Alaska. *Population and Environment*, 38(2), 115–133.

Henriksen, J. B. 2007. Report on Indigenous and Local Communities highly vulnerable to Climate Change inter alia of the Arctic, Small Island States and High Altitudes, with a focus on causes and solutions. Convention on Biological Diversity, 2007. UN Document symbol: UNEP/CDB/WG8J/5/INF/.

Himes-Cornell, A., and Kasperski, S. 2015. Assessing climate change vulnerability in Alaska's fishing communities. *Fisheries Research*, 162, 1–11.

Hovelsrud, G. K., and Smit, B. (eds.) 2010. *Community Adaptation and Vulnerability in Arctic Regions*. Dordrecht: Springer.

Huntington, H. P. (ed.) 2000. *Impacts of Changes in Sea Ice and Other Environmental Parameters in the Arctic. Background Report to the International Arctic Sea-Ice Change Workshop*. Bethesda, MD: Marine Mammal Commission.

Huntington, H. P., Noongwook, G., Bond, N. A., Benter, B., Snyder, J. A., and Zhang, J. 2013. The influence of wind and ice on spring walrus hunting success on St. Lawrence Island, Alaska. *Deep-Sea Research II*, 94, 312–322.

Huntington, H. P., Quakenbush, L. T., and Nelson, M. 2017. Evaluating the effects of climate change on indigenous marine mammal hunting in northern and western Alaska using traditional knowledge. *Frontiers in Marine Science*, 29 https://doi.org/10.3389/fmars.2017.00319

Jolly, D., Berkes, F., Castleden, J., Nichols, T., and the Community of Sachs Harbour. 2002. We can't predict the weather like we used to: Inuvialuit observations of climate change, Sachs Harbour, western Canadian arctic. In: Krupnik, I., and Jolly, D., (eds.), *The Earth Is Faster Now: Indigenous Observations of Arctic Environmental Change*. Fairbanks: ARCUS, pp. 92–125.

Kapsch, M. L., Eicken, H., and Robards, M. 2010. Sea ice distribution and ice use by indigenous walrus hunters on St. Lawrence Island, Alaska. In: Krupnik, I., Aporta, C., Gearheard, S., Laidler, G., and Kielsen Holm, L. (eds), *SIKU: Knowing Our Ice. Documenting Inuit Sea Ice Knowledge and Use*. Dordrecht: Springer, pp. 115–144.

Krupnik, I. 1993. *Arctic Adaptations. Native Whalers and Reindeer Herders of Northern Eurasia*. Hanover and London: University Press of New England.

Krupnik, I. 2000. Native perspectives on the climate and sea-ice change. In: Huntington, H. P. (ed.), *Impacts of Changes in Sea Ice and Other Environmental Parameters in the Arctic. Background Report to the International Arctic Sea-Ice Change Workshop*. Bethesda, MD: Marine Mammal Commission, pp. 25–39.

Krupnik, I. 2002. Watching ice and weather our way: Some lessons from Yupik observations of sea ice and weather on St. Lawrence Island, Alaska. In: Krupnik, I. and Jolly, D., (eds.), *The Earth Is Faster Now: Indigenous Observations of Arctic Environmental Change*. Fairbanks: ARCUS, pp. 156–199.

Krupnik, I. 2009. 'The Way We See It Coming': Building the legacy of indigenous observations in IPY 2007–2008. In: Krupnik, I., Lang, M., and Miller, S. (eds.), *Smithsonian at the Poles: Contributions to International Polar Year Science*. Washington, DC: Smithsonian Institution Scholarly Press, pp. 129–142.

Krupnik, I., Apangalook, L., Sr., and Apangalook, P. 2010. 'It's Cold, but Not Cold Enough': Observing ice and climate change in Gambell, Alaska, in IPY 2007–2008 and beyond. In: Krupnik, I., Aporta, C., Gearheard, S., Laidler, G., and Kielsen Holm, L. (eds.), *SIKU: Knowing Our Ice. Documenting Inuit Sea Ice Knowledge and Use*. Dordrecht: Springer, pp. 81–114.

Krupnik, I., and Benter, B. 2016. A disaster of local proportion: Walrus catch falls for three straight years in the Bering Strait region. *Arctic Studies Center Newsletter*, 23, 34–36.

Krupnik, I., and Chlenov, M. 2013. *Yupik Transitions: Change and Survival at Bering Strait, 1900–1960*. Fairbanks, AK: University of Alaska Press.

Krupnik, I. and Jolly, D., (eds.). 2002. *The Earth Is Faster Now: Indigenous Observations of Arctic Environmental Change*. Fairbanks, AK: ARCUS.

Krupnik, I., Walunga, W., and Metcalf, V. K. (Comps.). 2002. Akuzilleput Igaqulghet/Our Words Put to Paper. Sourcebook in St. Lawrence Island Heritage and History. *Contributions to Circumpolar Anthropology*. 3. Washington, DC: Arctic Studies Center.

Larsen, J. N., Anisimov, O. A., Constable, A., Hollowed, A. B., Maynard, N., Prestrud, P., Prowse, T. D., and Stone, J. M. R. 2014. Polar regions. In: *Climate Change 2014: Impacts, Adaptation, and Vulnerability*. Part B: Regional Aspects. Contribution of Working Group II to the Fifth Assessment Report of the Intergovernmental Panel on Climate Change. Cambridge: Cambridge University Press, pp. 1567–1612.

McDonald, M., Arragutainaq, L., and Novalinga, Z. (comps.) (1997). *Voices from the Bay: Traditional Ecological Knowledge of Inuit and Cree in the Hudson Bay Bioregion*. Ottawa: Canadian Arctic Resources Committee.

McDowell, G., Ford, J., and Jones, J. 2016. Community-level climate change vulnerability research: Trends, progress, and future directions. *Environmental Research Letters*, 11 (033001).

Marino, E. 2015. *Fierce Climate Sacred Ground*. Fairbanks: University of Alaska Press.

Marino, E., and Schweitzer, P. 2008. Talking and not talking about climate change in northwestern Alaska. In: Crate, S. A. and Nuttal, M. (eds.), *Anthropology and Climate Change. From Encounters to Actions*. Walnut Creek, CA: Left Coast Press, pp. 209–217.

Morton, J. F. 2007. The impact of climate change on smallholder and subsistence agriculture. *PNAS*, 104(50), 19680–19685.

Nakashima, D., Krupnik, I., and Rubis, J. (eds.). 2018. *Indigenous Knowledge for Climate Change Assessment and Adaptation*. Cambridge: Cambridge University Press.

National Center for Health Statistics. 2017. Health, United States, 2016: With Chartbook on Long-term Trends in Health. *DHHS Publication 2017–1232*. Hyattsville, MD.

Nickels, S., Furgal, C., Castleden, J., Moss-Davies, P., Buell, M., Armstrong, B., Dillon, D., and Fonger, R. 2002. Putting the human face on climate change through community workshops: Inuit knowledge, partnerships, and research. In: Krupnik, I., and Jolly, D. (eds.), *The Earth Is Faster Now: Indigenous Observations of Arctic Environmental Change*. Fairbanks: ARCUS, pp. 300–333.

Noongwook, G., The Native Village of Savoonga, The Native Village of Gambell, Huntington, H. P., and George, J. C. 2007. Traditional knowledge of the bowhead whale (Balaena Mysticetus) around St. Lawrence Island, Alaska. *Arctic*, 60(1), 47–54.

Oozeva, C., Noongwook, C., Noongwook, G., Alowa, C., and Krupnik, I. 2004. *Watching Ice and Weather Our Way/Sikumengllu Eslamengllu Esghapalleghput*. Washington, DC: Arctic Studies Center.

Orlove, B., Lazrus, H., Hovelsrud, G. K., and Giannini, A. 2014. Recognitions and responsibilities: On the origins and consequences of the uneven attention to climate change around the world. *Current Anthropology*, 55(3), 249–275.

Pungowiyi, Caleb. 2000. Native observations of change in the marine environment of the Bering Strait region. In: Huntington, H.P. (ed.), Impacts of Changes in Sea Ice and Other Environmental Parameters in the Arctic. Background Report to the International Arctic Sea-Ice Change Workshop. Bethesda, MD: Marine Mammal Commission, pp. 18–20.

Rasmussen, R. O., Hovelsrud, G. K., and Gearheard, S. 2015. Community viability and adaptation. In: Larsen, J. N., and Fondahl, G., (eds.), Arctic Human Development Report: Regional Processes and Global Linkages. TemaNord, pp. 423–473.

Ray, G. C., Hufford, G. L, Overland, J. E., Krupnik, I., McCormick-Ray, J., Frey K., and Labunski, E. 2016. Decadal Bering Sea seascape change: Consequences for pacific walruses and indigenous hunters. *Ecological Applications*, 26(1), 24–41.

Richter-Menge, J., Overland, J. E., and Mathis, J. T. (eds.). 2016. *Arctic Report Card 2016*. www.arctic.noaa.gov/Repoart-Card

Ristroph, E. B. 2017. Presenting a picture of Alaska native village adaptation: A method of analysis. *Sociology and Anthropology*, 5, 762–775.

Rosales, J., and Chapman, J. L. 2015. Perceptions of obvious and disruptive climate change: Community-based risk assessment for two native villages in Alaska. *Climate*, 3, 812–832.

Salick, J. and Byg, A. 2007. *Indigenous Peoples and Climate Change*. Report of a symposium held on 12–13 April 2007, Oxford, Tyndall Centre for Climate Change Research. www.ecdgroup.com/docs/lib_004630823.pdf

Savoonga Whaling Captains Associations. 2017. Data on whale catches by Savoonga whaling crews, 1972–2017. Document on file at the Office of the Savoonga Whaling Captains Association, Savoonga, Alaska. Cited with the Association's permission (February 13, 2017).

Serreze, M. C. 2008/2009. Arctic climate change: Where reality exceeds expectations. *Witness the Arctic*, 13(1), 1–4.

Sheffield Guy, L. S., Moore, S. E., and Stabeno, P. J. 2016. What does the pacific arctic's new normal mean for marine life?. *EOS Trans. Am. Geophys. Union*, 97, 14–19.

Shulski, M., and Wendler, G. 2007. *The Climate of Alaska*. Fairbanks: University of Alaska Press.

Taylor, P. C., Maslowski, W., Perlwitz, J., and Wuebbles, D. J. 2017. Arctic changes and their effects on Alaska and the rest of the United States. In Climate Science Special Report: Fourth National Climate Assessment Vol. 1. Washington, DC: U.S. Global Change Research Program, pp. 303–332.

Tejsner, P. 2013. Living with uncertainties: Qeqertarsuarmiut perceptions of changing sea ice. *Polar Geography*, 36(1–2), 47–64.

U.S. Army Corps of Engineers. 2009. Alaska Baseline Erosion Assessment. Study Findings and Technical Report. http://climatechange.alaska.gov/docs/iaw_USACE_erosion_rpt.pdf

Weingarten, G. 2005. Snowbound. *Washington Post Magazine*. May 1, 2005. Available at: www.washingtonpost.com/wp-dyn/content/article/2005/04/26/AR2005042601144.html

Williams, J. 2012. The impact of climate change on indigenous people – the implications for the cultural, spiritual, economic and legal rights of indigenous people. *The International Journal of Human Rights*, 16(4), 648–688.

Wood, K.R., Bond, N.A., Danielson, S.L., Overland, J.E., Salo, S.A., Stabeno, P.J., and Whitefield, J. 2015. A decade of environmental change in the Pacific Arctic region. *Progress in Oceanography*, 136, 12–31; https://doi.org/10.1016/j.pocean.2015.05.005

9

Localising and Historicising Climate Change
Extreme Weather Histories in the United Kingdom

GEORGINA ENDFIELD AND LUCY VEALE

9.1 A Cultural History of Extreme Weather in the United Kingdom

As Mike Hulme has recently noted, 'there is something very personal ... about the way our minds work with the idea of climate. While meteorologists define climate using the statistical average of weather measured over a period of at least thirty years, for most people climate becomes reified through a rather unstructured assemblage of *remembered* weather' (Hulme 2016b:160). Accepting that there is a need to 're-culture' discourses around climate (Hulme 2008; Livingstone 2012), it has been suggested that we 'think more directly about weather' (Hulme 2015:2).

Instrumental and descriptive observations of weather in meteorological registers, diaries, letters, administrative records and many other documents have enabled the comparison of climates between places for several hundred years. Today, these archival documents allow us to explore how people experienced climatic changes in the past and to assess and model the impacts of past climates upon past societies based on 'detailed record of past changes which is in fact a history of weather' (Pfister 2003:118).

The research for our AHRC-funded project *Spaces of Experience and Horizons of Expectation: Extreme Weather in the UK, Past, Present and Future* has focused on the collection of narrative accounts of extreme weather events from original documents located in archival repositories in five case study regions of the United Kingdom (Central England, East Anglia, South-West England, Wales and North-West Scotland). In particular, we have focused on how and why extreme weather events become inscribed into the memory of a community or an individual in the form of oral history, ideology, custom, behaviour, narrative, artefact, technological and physical adaptation, including adaptations to the working landscape and built environment. We have also been keen to explore how these different forms of remembering and recording represent central media through which information on past events is curated, recycled and transmitted across generations. A key purpose of the work was not to interrogate the archives in a targeted way but to let the archives speak for themselves. We have thus searched archival collections not by year or known event but through 'weather words'.[1] This has allowed us to retrieve information on past weather events and associated repercussions.

With the rest of our team, we have collated this information in the TEMPEST (Tracking Extremes of Meteorological Phenomena Experienced in Space and Time) database, a freely available online resource comprising over 18,000 weather event narratives. These span over 500 years of weather history and are searchable by weather type, place, date, impact and response type, document type, author and keyword (see Endfield *et al.* 2017; Veale *et al.* 2017). The value of TEMPEST is in bringing together previously disparate and often little-consulted documentary sources on UK weather history. By searching TEMPEST by weather type or time period, it is possible to quickly assess where multiple accounts of the same weather event, as recorded by different people, perhaps in different places, have been captured and also to ascertain whether a particular event entered the popular memory (i.e. if it is recalled by those living after its occurrence). Even single word or line entries in a diary, when combined with other sources, can help us to piece together the anatomy of particular events. In some cases TEMPEST adds new events to established chronologies of extreme weather in Britain, and in all, it adds new, personalised and localised understandings of the relationships between extreme weather events, people and places.

Working with these materials, it is, and it seems always has been, much more common for people to talk (and to write) about weather than climate. Indeed, in the c. 18,000 entries in our database of historical extreme weather event narratives only sixteen feature the word 'climate'. Although small in number, amongst these entries are accounts of three truly extreme episodes in the country's weather history: the winters of 1739–1740, 1783–1784 and 1794–1795. We reproduce extracts from these events below alongside findings from recent research on the meteorological context to introduce our source material, how combining sources can help to construct a story of a particular event, and to suggest that extreme events (and perhaps severe 'freezes' in particular) have, at least at specific times in history, been key moments when we have talked about and questioned climate or reassessed our expectations for weather. The materials provide insight into the relationship that people have with weather, at particular moments in their lives and in history. These

---

[1] Search terms included: 'drought', 'flood', 'heat-wave', 'storm' and 'weather'.

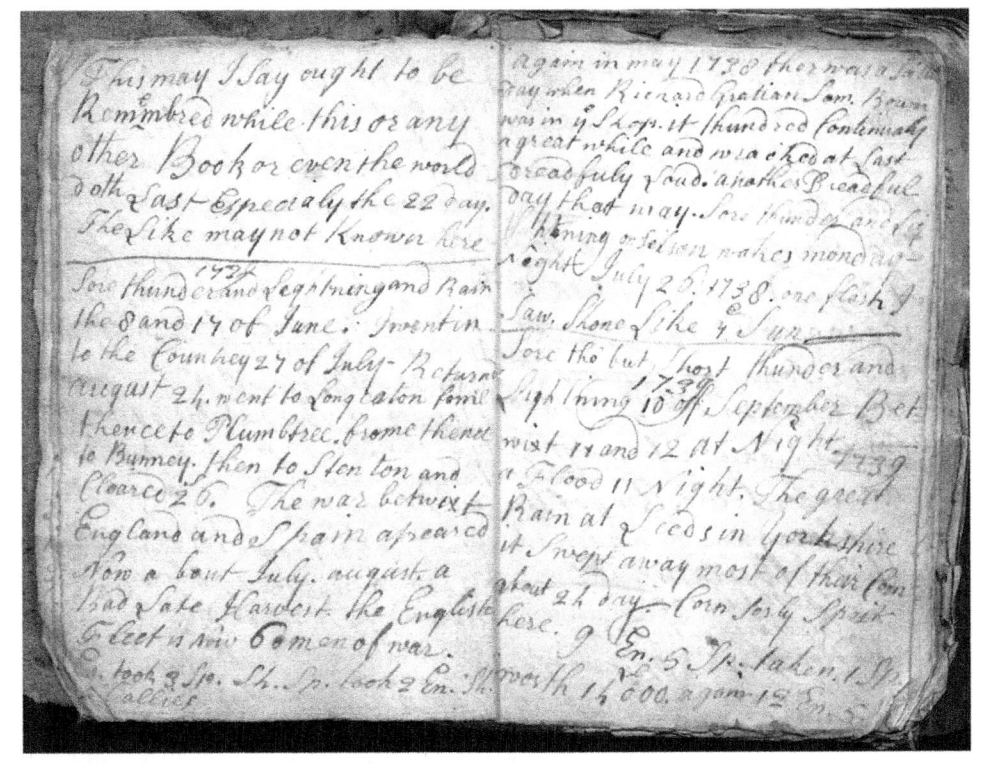

Figure 9.1 John Harrison's 'Book of Remarks' © Derbyshire Record Office

are peopled and geographically specific narratives that tell us something of what it was like to live through particular weather conditions.

9.1.1 1739–1740

John (also known as James) Harrison, from Belper, Derbyshire, was twenty years old when he began his Hemerologium or 'Book of Remarks' in 1736, a record he kept until 1746 (Figure 9.1). Harrison's notebook captured much of the military conflict with Spain (War of Jenkins' Ear 1739–1748) and the very cold winter of 1739–1740. The entries Harrison made for that winter tell us of the local conditions in Derbyshire, where at Christmas the milk had to be 'cut out of the pannetions [pails] with knives tho' close by the fireside', and the nearby river 'Trent was so hard that a wagon and 8 horses and 9 quarters of malt went over'.[2] He was able to play football with twenty of his friends on the frozen River Derwent on 2 January. On 8 February he writes 'the frost was so strong in Holland that boiling punch

---

[2] D2912/10, Derbyshire Record Office.

would freeze in 8 minutes' and 'thawed bread one hour to cut it the like is thought never was known in these *climates*'. He then writes that 'nothing had happened like this for 57 years', indicating that the winter of 1683–1684 had left its mark in the cultural memory. Harrison was not alone in recognising the weather of winter 1739–1740 as extreme and unusual throughout Europe. In Scotland, the *Caledonian Mercury* of 17 January 1740 reported: 'We had Tuesday night and yesterday long, the most bitter frost ever known (or perhaps recorded) in this part of the world, a piercing Nova Zembla [an archipelago in the Arctic Ocean] air. Old people aver it was far more violent that any day of the winter 1684' (in Pearson 1973:20).

For twentieth-century meteorologist and geographer Gordon V. Manley, it was research into the 'great winter of 1740' that prompted his assertion that 'In a modern civilisation, the existence of a public memory of the weather is essential' (1958:11). The year 1740 is the coldest in the Central England Temperature (CET) Series (average temperature 6.8°C) (Manley 1974), and Jones and Briffa concluded that the year 1740 'and especially the 1739/1740 winter were exceptionally cold' (2006:361).

9.1.2 1783–1784

Master carpenter and brewer John Clifton maintained 'day books' from 1763 to 1784. The entries begin largely in the form of a business diary but gradually come to include more references to social events and the weather in Oundle, Northamptonshire. The final volume, coinciding with the end of his life, was 1784. As the severity of the winter 1783–1784 increased, Clifton's first thought was to reflect back on the winter of 1740 (a winter we assume he lived through). 'We have had now two months of the most piercing cold weather & severe frost as ever was remembered by the oldest man living, it was as piercing as it possibly could be in January & February 1740 & all business as much at a stand & poor people as near starving.'[3] A month later the entry reads, 'Monday 29 March – One of the most piercing cold snowy days as is possible to happen in this *climate*; it is allowed by everybody to be full as cold as the last day of the old year & nothing but the dismal appearance of the continued winter'. There was no let up and at Easter, just a few days before his death, Clifton writes:

Easter Sunday 11th April – More piercing cold than it was at Christmas & not a leaf to be seen on the trees, nor no vegetables to eat.
   Easter Monday 12th – it snow'd very fast this morning from a quarter after 5 till quarter after 8 & then a piercing cold day afterwards – no gardens stir at all. Instead of the children

---

[3] ZA8746, Northamptonshire Record Office.

going into the fields to get violets & primroses & lords & ladies & hear the birds whistle & see the lambs play, they are forced to keep within doors by the fire.

The extreme cold of winter 1783–1784 and subsequent floods followed the Lakagígar (Laki) eruption in Iceland, and many climatologists are of the opinion that the eruption 'significantly influenced the weather of Europe' (Brazdil et al. 2010:181), although others conclude that a negative phase of the North Atlantic Oscillation (NAO) and an El Nino-Southern Oscillation (ENSO) warm event 'was more fundamentally to blame for the severe conditions over North America and Europe' (D'Arrigo et al. 2011:L05706:1).

9.1.3 1794–1795

Eleven years later, expectations of winter were challenged again, as surgeon, apothecary and midwife Matthew Flinders (1748–1802) of Donington, Lincolnshire, recorded in his detailed record of life and work that covers the period 1775–1802:

This has been the severest winter in these *climates* known in living memory, the severe frost & snow being about 3 months continuance. The frost commenced 2d week in Decr & continuing to the same time in March – the snow began at Xmas Eve – and continued with intervals most of the time. I think I may say more has fallen than in the last 7 years together and several times more on the ground, than has been since the great snow in 1767 when it was a yard deep on the level – We had several partial thaws during the 3 months – but none in wch we could say the Frost and Snow were wholly gone.[4]

Flinders had lived through the winter of 1783–1784 but rated 1794–1795 as the more severe. Contributors to the *Gentleman's Magazine* agreed, detailed in a quotation that hints at the importance of cultural and political context in our judgements of extreme weather events:

The unprecedented inclemency will leave the deeper impression on the mind from having occurred at an era of political history, when men are awefully [*sic*] contemplating an extraordinary and rapid succession of momentous events.[5]

The average temperature of December, January and February 1794–1795 was 0.5° C (Lamb 2002), making it the fifth coldest winter between 1659 and 1979, with January 1795 the coldest month in the CET with an average temperature of $-3.1°C$ (Manley 1974).

A 'national memory of the weather' (Manley 1958:11), or awareness of weather events in history (and specifically we argue knowledge of *extreme* weather events

---

[4] FLINDERS 2, Lincolnshire Archives.
[5] The Chronicles of the Seasons. Winter 1794–1795 *The Gentleman's Magazine* 77 (1795): 181.

in our past), is an important factor in how we interpret and understand current weather and climate. As we face predictions of increased frequency and higher intensity extreme weather events in the United Kingdom, documentary records charting past events not only present an information resource by which to judge the severity of current events but also have very real potential to engage publics with both weather history and futures. The ability to view extreme events in meteorological as well as cultural context is important given that both factors influence how an event is felt, interpreted and subsequently remembered. To date, 'only limited attention has been paid to cultural and meteorological "interconnections" and "imprints" associated specifically with extreme or unusual weather, such as droughts, floods, storm events and unusually high or low temperatures' (Endfield and Veale 2017:2).

In the remainder of this chapter we feature a small selection of examples from our documentary history of weather, all from or with a connection to the English Midlands.[6] Some are local records of what were national-scale events, whilst others capture weather active only over small areas. These indicate the significant potential for the construction of regional and multiregional weather histories and broaden our ideas about the documentary archive of weather to include a much wider range of narratives than is typically imagined, in an area of investigation that has come to focus on the records of 'famed' observers of weather. Largely ordinary documents become extraordinary records of observing and living through particular episodes of weather, together providing insight into the relationship between culture and climate and demonstrating how 'ideas about climate are always situated in a time and a place' and often through weather per se (Hulme 2016:15). Multiple narratives enable confidence in the accounts, whilst human stories, set in particular places, are intensely engaging. The examples also speak to cultures of writing, record keeping, archiving and the place of weather in daily life. We argue that such histories and experiences of weather events could serve a powerful role in informing memory but may also influence popular understanding of both local and global climatic change.

9.2 Observing, Recording and Transmitting Extreme Weather Events

We begin with the everyday experience of extreme events recorded in daily diaries. We then consider weekly summaries of weather as reported by land agents on large estates in correspondence to their employers, and finally events that came to

---

[6] The English Midlands is a region in Central England which is commonly divided into the East and West Midlands. The East Midlands includes Lincolnshire, Northamptonshire, Derbyshire, Nottinghamshire Leicestershire and Rutland whilst the West Midlands includes Staffordshire, Shropshire, Warwickshire, Hertfordshire and Worcestershire.

dominate particular years, marked in parish registers (and in the built environment). The authors of the documents demonstrate that it was common practice to reflect on weather and to draw on personal life experience of weather, and local weather memory, in the interpretation of current weather, especially when it challenged expectations.

9.2.1 Personal Diaries and Everyday Experiences

There is a growing body of research that employs historical documentary and oral history-based approaches to reconstruct regional and local climate change and to investigate how climate variability has been experienced (Culver 2014). Weather diaries are firmly established as key sources in these investigations in that they provide place- and date-specific observations on the weather. Of course, 'compiling a weather diary was not part of anybody's job; the duty was a self-imposed one', and in this regard 'weather journals could be seen as products of a certain kind of exercise of self-formation' (Golinski 2007:84). Diaries do not have to be 'weather diaries', however, to contain information about the weather. As sources, they cut across academic disciplines, owing to the fact that the meteorological data they contain are inseparable from the subjectivities of the author, and, in many cases, from the other information present – details of meetings and activities, ailments, friendships and confessions. Diaries then can provide information about the nature of diary keepers and weather observers as well as past weather and 'uncover the "human" in human-environment relationships exceptionally well' (Adamson 2015:607).

9.2.1.1 Thistlewood's Diary

The Laki eruption, Iceland, began on 8 June 1783 and continued for around six months (Thordarson and Self 1993)[7]. The 'reports from across Europe attest to the presence of atmospheric haze or fog at various times in the summer and autumn of 1783' believed now to have been pollution from the volcano, with possible impacts on vegetation and human health (Witham and Oppenheimer 2005:15). The period July 1783 to June 1784 is recognised as one of mortality crisis in the United Kingdom, with September 1783 being the most fatal month, and the east of England bearing the 'brunt of the increase in mortality' (Witham and Oppenheimer 2005:20). Summer 1783 was notably hot and stormy and, as we have already mentioned above, the winter that followed was severe.[8]

---

[7] It is thought that the eruption lasted until 7 February 1784 (Thordarson and Self 1993).
[8] July 1783 was the hottest recorded in Manley's collection of Central England temperatures, and not surpassed for 200 years (Manley 1974).

Dated diary entries allow us to 'reconstruct a detailed picture of the extreme atmospheric conditions which followed the Laki eruption, but also to evaluate the perception of these events and the responses engendered in those who experienced at first-hand the unusual and alarming episodes of climatic disturbance which then occurred' (Grattan and Brayshay 1995:125; Grattan and Sadler 1999) at a time when the link between volcanic activity and the weather was only just beginning to be the subject of speculation.

The haze and intense lightning storms of that summer are captured in the diary of John Thistlewood of Lincolnshire (1716–1793). Thistlewood's diaries have lain firmly in the shadow of the archive of his younger brother Thomas (1721–1786) whose collections include valuable sources on sugar plantations and slave labour as well as on the eighteenth-century weather and climate of Jamaica, the island to which he emigrated in 1750 (Chenoweth 2003). Much less is known about John, although it seems reasonable to assume that, like Thomas, he would have received a good education (including mathematics) and a practical training in agriculture. As well as farming, he practised as a land surveyor and is named as such on a total of sixteen of the enclosure maps for Lincolnshire, as well as on a number of other plans, maps and surveys held in Lincolnshire Archives. His diary is now catalogued as a series of loose papers detailing daily life in Lincolnshire between June 1779 and his death in June 1793, with entries dominated by reports on the weather.[9] There is also mention of work on his farm, visits to local markets at Horncastle and Bardney and reference to a number of surveying jobs. Each daily entry is usually just a couple of lines long and describes the strength of the wind and its direction, the presence or absence of precipitation (with type), appearance of clouds and temperature, presenting a very similar appearance to Thomas' weather journals. Developments or changes in the weather over the course of a day are usually accompanied by a time reference with descriptions of storms or other notable events demanding lengthier entries. He also includes notes on halos, mock suns, eclipses, comets and earthquakes.

References to people are very few in number and generally without any expression of emotion. There are two mentions of his brother Thomas: the first comes with news of the 1780 hurricane in Jamaica, the second with news of Thomas' death which reached him on 10 April 1787. This 'matter of fact' approach to recording was by no means unique to Thistlewood (and he may, like his brother, have kept a separate private diary), though viewing this source alone places him in

---

[9] MON/31/92 (microfilm MF/1/42), Lincolnshire Archives. The original diary is now held in the Beinecke Rare Book and Manuscript Library and has been digitised, the pages available to view online: John Thistlewood, Notes concerning Lincolnshire (subsequently NL), TTP, Oversize, Box 13, Folders 93–95. http://beinecke.library.yale.edu/collections/highlights/thomas-thistlewood-papers

a different category of diary keeper to those writers who used their diaries for deep emotional reflection and thought sharing.

Thistlewood first notes the presence of a blue haze in Tupholme, Lincolnshire, on 23 June 1783, 'calm & hot, thick blue haze, sun went down very red'.[10] By 26 June it had grown 'more dark & thick the sun scarcely ever appeared'. On 2 July, as well as the continued presence of blue haze, 'at 2PM began to darken (beyond the blue haze) in the West, with far off thunder, the darkness increased all round, with great & almost continual lightning and loud rumbling thunder which continued till 9PM, a pleasant mild shower fell about 6 o'clock, the thunder clouds was a great height & scarcely ever appeared thro' the haze'. From other documentary accounts we know that this storm was particularly severe and memorable, causing loss of life in the east of England.[11]

Thistlewood documents further notable summer storms that struck Lincolnshire on 10–11 July, 20 July, 2 August, 19 August and 25 August. He also describes the intense heat of summer 1783. On 2 August the wind 'blew but warm as the mouth of a furnace, the sun shone extreme hot, in the afternoon wild thunder clouds passed over, with distant thunder, evening some lightning out of the N . . . I don't remember of ever being so hot. Strong blue haze'. Other atmospheric phenomena added to a general sense of foreboding. On 18 August he reports the appearance of the 'great meteor', as captured by Paul Sandby (see Daniels 2011), 'wind low NW, some light clouds, very hot & sultry, a great deal of blue haze (at 10PM a fiery meteor in the NE large as the full moon, broke with a noise like distant thunder and passed swiftly southward with a long train of bright sparkling fire. It was light almost as day and continued for about a minute)'. Mentions of the presence of blue haze become infrequent after 8 September though 'dark', 'thick' air continues and the final mention of blue haze is on 23 March 1784. This fits with previous studies showing 'the earliest references to the appearance of haze over England date to 12 June 1783 and its presence is recorded over periods of days to weeks hence' (King and Ryskamp in Witham and Oppenheimer 2005:15).

John's diary entries go on to detail the harsh winter of 1783–1784. Day after day he makes entries of heavy falls of snow and hard frosts, the entries becoming simply 'do. [ditto]' as the wintery days repeat. The record is objective and generally distanced, the weather frequently described as 'severe', 'extreme', 'excessive', 'violent', 'sharp' and 'dreadful'.

---

[10] This fits with Witham and Oppenheimer's conclusion that, 'the earliest references to the appearance of haze over England date to 12 June 1783 and its presence is recorded over periods of days to weeks hence' (2005:15).

[11] We have collected multiple narratives of the July 1783 storms in TEMPEST (Endfield et al. 2017). The diaries of William Strong of Old Bolingbroke, Northamptonshire stand out as he entered a note on the storm of 2 July into the front of his diary on an annual basis, 'July 2nd Wednesday. A violent storm of thunder & lightning from 4 to 6PM which killed three boys at Wansford, and hurt several others, whilst at school. Miss Cranwell and a man at March were killed; for many days prior to the storm, the weather has been particularly hazy and warm' (William Strong, diaries for 1785–1791, ST 87–93, Northamptonshire Archives).

The Thistlewood diary pairing provides insight into family observation of the weather and suggests that both were encouraged to keep records of the weather at an early age, probably during their formal education. In the main, both Thistlewood brothers 'concealed their feelings and personal qualities' (Golinski 2007:84), rendering them objective observers of the weather. Their weather journals were therefore 'a paradoxical vehicle for self-development – one in which self-effacement and concealment of personal motives were norms of the genre' (Golinski, 2007:84). For much of their lives the two brothers lived in contrasting climates where they respectively experienced very different 'extremes' of weather, yet their diaries are of very similar appearance. Although not unique in capturing the hot, hazy summer of 1783, John's diary adds further detail to any reconstruction of that summer as it was experienced in the east of England, contributing to our memory of that extreme season.

9.2.1.2 Pegge-Burnell's Diary

Farmer and landowner Peter Pegge inherited Winkburn Hall *c*. 1781. On obtaining the estate he took the name and arms of Burnell and lived at the hall in Nottinghamshire until he died, childless in 1836, at the age of eighty-five. Pegge-Burnell kept a diary for much of his life and the multiple volumes afford an opportunity to explore his relationship with the weather.

The final days of 1798 were severely cold, especially the 27 December when 'men engag'd in active and dry Employments, within and without Doors, tho' aware of the impending Danger, and exerted every Effort to repel it, had, notwithstanding, their Hands and Feet froze, in the Middle of the Day' (Hopkinson 1800:6). In the last week of January 1799 there were again severe frosts and on the 31 it 'began to snow small snow but fast about half past three, & continued to do so the whole night, & the wind at times very high & the snow much driftd'.[12] Pegge-Burnell's family were 'almost all ill of colds & bad coughs' and, although continuously pessimistic in the pages of his diaries, he expresses a real anxiety at this time over what lies ahead 'this hath been the coldest month & much fodder consumed & a most severe winter too'. As Oliver acknowledges in relation to Welsh weather diaries, 'Memories are short when it comes to comparative remarks', and 'anxious anticipation often colours the comment on the weather of the moment' (1965:188). Frustratingly, it is impossible to compare Pegge-Burnell's experience of living through the winter of 1798–1799 with the more famously severe season of 1794–1795 as the diary for 1795 is missing, and no written references to that year are present in successive volumes, perhaps erased by the challenges of the present.

---

[12] DD/CW/8c/5/21, Nottinghamshire Archives.

The new month, however, brought more severe weather:

February – This month began in the most winterly & severe stile, frost, snow and a high wind – Lord save us & have mercy on the poor, a worse night & day I have never beheld, froze very hard in the even'g
11 Feb – a most shocking day of wind, snow & rain – there pas't the severest season I ever knew
25 Feb – large flood upon the Trent

There was some fine weather towards the end of February, but by the middle of March conditions had taken a worrying turn for the worse:

14 March – terrible to relate frost & more snow – winter
25 March – showers, the ground perfectly drowned with wet and no seed yet got into the ground
26 March – cold, rain &c months of the worst weather I ever knew, plenty of lambs this season so far but nothing for the ewes to eat
30 March – a very hard frost & flying snow – a dreadful season this – this night the meat & milk frose
31 March – still a most severe frost, wind high – poor lambs dieing from cold & want of suck. Oh what is to become of us this season – god only knows – but his will be done says PB

Entries are very brief in the early part of April as Pegge-Burnell suffered from a cold and gout, but when they resume they only describe a worsening situation:

14 April – a fair day but very cold, the ground a perfect sponge & no seed yet in, sheep almost starving, want of meal – the ground being both black with wet & cold – this the latest season & worst I remember ...
29 April – cold as xmas. This hath been a bitter cold, dismal month, the latest season by far I ever remember, no grass, no cuckoo, nor swallows, in short nothing seasonable.

Evidence suggests that thousands of lambs did indeed perish in the spring of 1799.[13] Agriculturalist Arthur Young (having previously surveyed the state of agriculture in 1795) was prompted to issue a questionnaire to a number of his correspondents in England and Wales. The responses were subsequently printed in the *Annals of Agriculture* and 'give a vivid description of its [the severe season] effects' (Jones 1953:371). Read alongside Pegge-Burnell's diary, the distressing experience of living through this season, even for the relatively wealthy and well fed, becomes starkly apparent.

Summer was no better, generally very wet, with 'outbreaks of elemental fury' (Ashton 1959 in Neumann and Kington 1992:191). A sample of entries from

---

[13] The chronology of weather compiled by amateur meteorologist Martin Rowley (booty.org.uk) states: From records in Kendal (Westmorland / Cumbria), we have .. 'No vegetation in the fields, nor blossoms upon the fruit trees, on the 7th May, 1799. The skins of upwards of 10,000 lambs, which perished in the spring, were sold in this town. The weather was cold and wet all through the year.'

Pegge-Burnell's diary helps us to understand the developing picture of distress and again provides details of the cultural context for this challenging weather:

31 May – cold wind – this hath been a gloomy cold & black month, gardens & pastures the latest ever known

31 July – Rains, the hay & there is a deal down spoiling in my meadow – neither sun or wind but close & wet – oh I never had so little hay upon the ground & that little injured by the rain. Corn improves daily.

17 August – a tremendous flood. God speedily help us. Several lds of hay aflote in the meadow

22 August – very heavy rain – O God what a winter spring & summer – the Trent greatly out – oh me. This a most dismall month of weather. Corn very late & bad – shocking hay season. What a winter, spring & summer . . . this a most dismal month of weather, corn very late & bad – shocking hay season

31 Dec – the severest frost I ever knew Lord have mercy on us, thank God my large family are in health. This hath been the most untoward year, perhaps ever remembered for bloodshed, bad weather &c &c

Looking at the CET record, 1799 was within the 'top-20' of coldest years (Manley 1974). The impact of what was at best unfavourable and at times extreme weather during 1799 had a long lasting impact on society. In Pegge-Burnell's county town of Nottingham, at the beginning of 1800, 'a large public subscription was raised for the relief of the sick poor, and the establishment of a public soup-kitchen' (Sutton 1852:244). On 19 April 'The high prices of provisions at the market, provoked a riot' (Sutton 1852:246).[14] Another riot began the 31 August, and ironically peace was only restored as a consequence of extreme bad weather, 'when one of the most awful storms of lightning, thunder, and rain, ever witnessed in this town, put a final end to the protracted disturbance' (Sutton 1852:247; and see Neumann and Kington 1992). Reverend Samuel Hopkinson of Etton, Northamptonshire, was of the opinion that the period 1798–1800 together represented 'A Series of bad Weather the most memorable perhaps in the History of this Kingdom' (1800:26)[15].

Pegge-Burnell's surviving documentary archive certainly fits with Lejeune's assertion that 'The end of the eighteenth century and the beginning of the nineteenth is . . . the moment when diarists begin to address their diary as though it was an intimate friend in whom they could confide' (2009:7). Now unique records (handwriting and format acting to individualise the pre-printed diaries) of living through episodes of interesting weather diaries are necessarily 'written without the

---

[14] 'The most consistent cause of rioting was an increase in prices, and 'the butchers' stalls near the Market Place were looted on at least five occasions between 1788 and 1800' (Wardle 2010:31).

[15] Hopkinson himself is said to have been constantly attentive to the needs of the local poor, administering charitable donations on behalf of Earl Fitzwilliam, including in direct response to hard winters, and was himself a keen observer of the weather. See FM Misc vols/136, Northamptonshire Record Office for a notebook of accounts of Reverend Samuel Hopkinson of Etton of charitable distributions by Earl Fitzwilliam.

knowledge of where it will end' (Lejeune 2009:170) and express very real fears relating to the weather.

9.2.2 Weekly Letters and Estate Matters

The potential of estate records, in the United Kingdom and elsewhere, as useful sources of information on past weather and climate has not escaped the attention of those interested in the reconstruction of past weather (Chenoweth and Thistlewood 2003; Endfield 2008; Dolak *et al.* 2013). Estate records of weather are (usually) geographically and temporally specific (also covering large areas and with surviving records for lengthy periods of time), often include numerical and narrative weather data and also detail the timescales and processes by which information on the weather (particularly extremes) and responses to it travelled and were exchanged. Many people on the estate were closely connected to, and observant of, the weather: landlords, agents and stewards gardeners, tenant farmers and their families, and the wider population of the estate and its surrounding area. In terms of sources, diaries, correspondence, account books, legal documents, surveys and maps can all be valuable. In this section we focus on the correspondence exchanged between agents and owners.

Land agents were popularly regarded as 'mediators' and 'a conduit not simply for reports about the estate and requests for instructions and decisions, but also for a flow of intelligence about local reactions to national affairs: legislation, taxation, elections, county and borough politics, wages and prices, natural disasters, the impact of disease or the weather, the problem of the poor' (Hainsworth 2008:4). The place of weather is clearest when we look at agricultural and horticultural activity on the estate, most notably the harvest, but also the health, growth and quality of livestock and game. Forestry too was affected by the favourability of growing seasons and also by strong winds capable of uprooting significant numbers of sizeable trees in a matter of hours. Building and improvement works could be hampered by poor conditions for outdoor work and both domestic and working buildings at risk of damage in extreme weather events. Weather affected the outward appearance of the estate, to residents and visitors. In addition, before the introduction of a state welfare system, the landowner was a key figure in providing localised assistance during periods of hardship that were sometimes caused or exacerbated by extreme meteorological conditions.

As a popular conversation starter or topic that often pertained centrally to estate matters, weather features in many letters, written to loved ones or professional contacts alike. Single letters can provide insight into the experience of living through a particular extreme weather event, whilst larger collections and letter books document regular conversations (albeit often one-sided) about the weather, allowing us to not only reconstruct longer periods of weather (and climate variability; see Endfield and Nash 2002) but also to build up pictures of the nature of the weather observation and its assessment as extreme or unusual and to place particular events in context. Our materials come from the (well catalogued) Newcastle estate papers and date from the eighteenth and nineteenth centuries – a time when there was a rapid expansion in the publication of agricultural periodicals, manuals and surveys, as well as interest in studying the effect of weather and climate on cultivation. Keeping weather notes was in many cases encouraged, and any respected landowner would want to keep abreast of weather news.

9.2.2.1 1838

In February 1838, the letters of estate and building agents at the 4th Duke of Newcastle's Hafod estate in Wales (purchased in 1835) relayed the impacts of the severe cold weather to the Duke, detailing the spoilage of fodder crops and the stoppage of building works. This one-sided conversing about weather shows that there was a direct relationship between weather and work, or progress, on the estate, and between weather, employment, and financial well-being for labourers. 'Newcastle, who was away from Hafod much of the time, employed a galaxy of agents, baliffs and surveyors, many of whom he brought from his Nottinghamshire estates' (Colyer 1976/77:279).

1st February 1838

My Lord Duke

... The weather continues so severe that the masons cannot work and very little has been done by the carpenters outside since my last. I have suspended several hands until the weather breaks and the timber is got in that they may be able to work outside at the shed &c.

Several slates were blown off in various parts of the Offices during the late high winds which I have had reinstated ...

S Heath[16]

Feb'y 26th 1838

My Lord Duke

The weather has been very severe this last fortnight, frosts & snows. Your grace the weather has now changed & I fear we are going to have some cold rain, it is very rainy & stormy today.

The turnips your Grace are a good deal injured by the frosts & there is a great complaint of a many potatoes being spoiled in the hills by the severe frosts.

---

[16] Ne C 8428, University of Nottingham Manuscripts and Special Collections.

I have been endeavouring to your grace to sell the fat bullocks, I offered them to an Aberystwyth butcher for 16d pr lb & weight them, he would not give it & bid me 5d per lb. ... It is desirable your Grace to sell them as keep is becoming very scarce ...

John Lown[17]

Entries in the Duke's diary relay his personal fears and indicate that his thoughts at this time were often in Wales. Well-connected, active in politics and a frequent traveller, however, the Duke also had an awareness of weather crises elsewhere:

10 Jan – a great deal of snow has fallen for several successive days
 11 Jan – I have completed the filling of my new ice house today – it holds 326 loads
 21 Jan – the thermometer last night was as low as 8 d – 19 d was the highest temperature in the day & some people have been frozen to death
 1 Feb – the weather still continues with great severity, the rivers are frozen, the Trent & the Thames have been passable for a fortnight
 3 Feb – ... on the 20th Jan'y the thermometer was down to 5d below zero – within 1½ degrees of the lowest temperature ever known in England – This report is from the meteorological journal
 5 Feb – the accounts of the severity of the weather abroad far exceed anything that we have here ...
 8 Feb – ... the frost has lasted almost interruptedly for 5 weeks
 12 Feb – the intense frost continues unluckily without snow & I must fear that the turnips will be injured, therm' last night 18d
 17 Feb – Today the whole country is again veiled in white, we are buried in a heavy snow & what the poor larks are to do I know not if this weather continues *especially in Wales* where with the increased snow there will be the greatest difficulty in providing for the existence of the animals
 25 Feb – our eyes are at last treated with a view of verdure – for nearly two months the country has been enveloped in snow with a continued severity of weather which I do not remember to have experienced above once or twice in my life
 26 Feb – our verdant prospects have been very transitory. It has snowed all day & the ground is again in unbroken white. I fear for the lambs and the vegetable world
 6 March – the accounts from Ireland describe the loss of life by the last snow storm to have been quite awful and the extent ... 11 men ... shortcut over the hills from the town to a fair had been .... Every one perished, they were found all huddled together & quite dead.[18]

Remembered as 'Murphy's winter', the extremely cold winter of 1837–38 and especially the low temperatures of the 20 January, one of the coldest days for generations (as detailed in Newcastle's diary), was predicted by weather prophet Patrick Murphy, Irish-born futurist and author of *The Weather Almanack*. First issued, and a bestseller in 1837, according to its title the almanac showed 'the state of the weather every day of the year 1838'

---

[17] Ne C 8429/1–2, University of Nottingham Manuscripts and Special Collections.
[18] Ne 2F 5/1 (emphasis added), University of Nottingham Manuscripts and Special Collections.

Figure 9.2 Caricature of Patrick Murphy. Courtesy of the Visual Telling of Stories website, author Dr Chris Mullen

(Murphy 1837). Murphy's fame was to be temporary. Future weather and weather fortune tellers like Murphy increasingly became the target for much ridicule among rational, scientific communities. In Figure 9.2 he is shown in one of a number of satirical caricatures depicting him as a potato-headed minstrel.

9.2.2.2 1860

Summer 1860 was intensely stormy, the fourth coldest summer season according to the CET and the fourth wettest with 370.8 mm rainfall, including a very wet June (only surpassed/equalled by June 2012 as the wettest in the England and Wales monthly series (EWP) from 1766 (Wigley *et al.* 1984; Alexander and Jones 2001; HADUKP 2016). For Henry Heming, agent to the 5th Duke of Newcastle on his Nottinghamshire estates from the 1840s to 1860s, the summer storms coincided with the Duke's request of the 9 July to receive a 'weekly report of matters going on at Clumber' (the Duke, at this time Secretary of State for the Colonies, was in Canada). Heming soon found his letters dominated by bad weather and described the practice as 'very painful to write', the weather responsible for 'so much anxiety' by 'all classes of people'.[19]

August 17 1860

My Lord Duke,

The weather is now the all-engrossing matter. Knowing the deep interest your Grace takes in all that concerns the welfare and condition of your dependents I do not refrain from informing you of the state of the weather about which so much anxiety is now felt by all classes of the people although to is very painful to write bad news week after week. There has been more rain since I wrote last week than in any previous week. The lowlands have been submerged and a great deal of the spoilt hay carried away. I have heard from Cromwell this morning that after three days fearful apprehension of a flood the river is going down. This is cause for thankfulness as a flood from the Trent at this time would be most disastrous. On the 11th it rained torrents for 14 hours. It has rained here every day since but the showers have been partial ...

Henry Heming[20]

August 31, 1860

My Lord Duke,

... I have bad news to communicate, especially in reference to the weather which is not only most alarming as regards the food of the people & the welfare of the farming classes, but it hinders every kind of business more or less & causes a great impediment to all building operations. A slight improvement has taken place, yesterday was a fine blowing day after a hard night's rain. This day promised to be fine but a strong shower has just fallen & more threatens to fall ... The wheat is as green as it was two months ago ...

If we only get fine weather, even now, it will compensate for a great deal of the adversity we have thus far experienced; but a continuance of bad weather would be most appalling ... I rec'd a letter from an Irish Landlord this morning in which he says, 'I am sorry to say the prospects of the Harvest are very bad indeed. The Hay actually swimming away with the floods. The turnip crop poor & thin ... in the

---

[19] Ne C 13804/1–2, University of Nottingham Manuscripts and Special Collections.
[20] Ne C 13804/1–2, University of Nottingham Manuscripts and Special Collections.

wheat: the potatoes diseased & God only knows what the consequences will be. What an honourable post his Grace is now filling.'

Henry Heming[21]

Sepr 14 1860

My Lord Duke,

It is with the greatest pleasure & satisfaction I have to inform your Grace of six days fine harvest weather. There have, however, been some severe frosts. Considerable progress has been made in harvesting the crops. Upon your own farm all the oats have been stack'd & more than half the barley in fair condition. The wheat is still green & it will be a week or 9 days before it is fit to cut. In my own small concern I am more than six weeks later than last year ... Great complains are made by the best farmers on all the strong soils of the inferiority of their crops – Mr Denman 'never had worse' ...

It will take full three weeks of fine weather to secure all the corn that remains uncut. Such was the haste of many farmers to cut & carry the crops that several stacks have had to be taken down ...

Henry Heming[22]

Although the British Rainfall Organisation (BRO) was set up in 1859 in response to concerns about perceived decreasing levels of rainfall over the British Isles in the middle of the century (Pedgley 2010), it was the wet summer of 1860 that dominated the first issue of George Symon's *British Rainfall*. This included observations from Welbeck, a neighbouring estate to Clumber (where 30.54 inches of rain were recorded for the year).[23] The year also marked the introduction of Robert Fitzroy's gale warning service (the first forecast published in *The Times* in 1861), following the Royal Charter storm of 25–26 October 1859.

The letters from Heming to Newcastle infer that personal and estate-based experience and memory of weather, and the presence of documentary records of previous summers, available to farmers and land managers, to look back to and compare dates of harvests, helped to place current events in context. Each dated piece of correspondence facilitates a reconstruction of weather and its immediate impact during the summer of 1860.

9.2.3 Parish Registers and Benchmarked Years

Brázdil and co-authors describe European parish registers as 'a special class of weather record' (2006:743). Parish priests or appointed officials felt it important,

---

[21] Ne C 13806/1–4, University of Nottingham Manuscripts and Special Collections.
[22] Ne C 13808/1–2, University of Nottingham Manuscripts and Special Collections.
[23] *British Rainfall. On the Distribution of Rain Over the British Isles During the Years 1860 and 1861*. Compiled by George Symons. 1862. Available online through the Met Office Digital Library and Archive: https://digital.nmla.metoffice.gov.uk/file/sdb%3AdigitalFile%7C8dd7e35e-6f29-4412-9625-6a42ff24cf2c/

and in some cases were actively encouraged, to record events of local or national significance, including weather and weather-related events as they affected parishioners and the church, alongside entries of baptisms, marriages and burials. In the eighteenth century, Bishop Kennett of Peterborough (1718–1728) is said to have encouraged his clergy to note down, 'any notable incident of times and seasons, especially relating to your own parish, and the neighbourhood of it, such as storms and lightning, contagion, and mortality, drought, scarcity, plenty, longevity, robbery, murders, or the like casualties', in order that the registers would become 'chronicles of many strange occurrences that would not otherwise be known, and would be of great use and service for posterity to know' (in Waters 1887:69–70). The memoranda, 'as chronicles of communal memory and experience ... provide a vibrant counterbalance to the laconic style of the bureaucratic purpose of the registers' (Hobbs 2008:95).

Using Tate's (1946) terminology, Morgan (2015) points out that despite being locked away for many years in the 'parish chest', parish registers were generally accessible to those who wished to consult them. Today, they function as a record of the community or a public history and are among the most popular materials in Local Government Record Offices and available through subscription websites. Parish registers contain memoranda of particularly catastrophic or memorable weather events, important in some instances as the main source for otherwise little-documented events that might be missing from existing chronologies and for others in detailing localised impacts. However, as Tufnell found in relation to the Beetham registers, known and important environmental events can be absent, even from the registers kept by those with an obvious interest in the weather, adding 'to the problems of evaluating the significance of those events which actually feature in the registers' (Tufnell 1983:147).

It is widely recognised that members of the clergy have long been important weather observers, and indeed of regional study and natural history more widely. Their observations could serve to raise the profile of their region as 'Local curiosities of nature reinforced regional identity and highlighted the region on the geographic and cultural maps of the country' (Jankovic 2000:78). Clergy famous for their weather observing include Reverend Ralph Josselin (see Macfarlane 1991), Reverend Gilbert White (see White 1836) and Parson James Woodforde (see Woodforde 1998). Clergymen were key figures in the community and many regarded the keeping of records relating to important events in the locality (affecting their parishioners) as part of their duty. Weather also had an impact on the size of the congregation, and, in the case of extremes, sometimes on their ability to conduct services. Extreme weather events and celestial phenomenon would have been viewed by many as 'Acts of God' throughout the eighteenth and nineteenth centuries, another reason why many vicars would have thought it important to make a note of them in the Parish Register.

9.2.3.1 The Windstorm of 1 February 1715

Although absent from existing national storm chronologies (for example, Lamb and Frydendahl 1991), the 1 February windstorm is recorded in a number of parish registers in central and northern England and in these places came to dominate that weather year owing to the extensive damage to the built environment and to trees.[24] At Old Bolingbroke, Lincolnshire, the incumbent recorded that there was 'a remarkable storm of wind which according to common report blew down some thousands of houses in the Kingdom'.[25] Fifty-five miles to the north at Alkborough the wind 'blew down many thousand trees in Lincolnshire', 'a comparable hurricane storm event apparently not recalled in the 'memory of any man living'.[26] At Rolleston, in neighbouring Nottinghamshire, it was noted that 'On wch day was such a violent tempest of wind as was never known in any man's memory, it struck down two pinnacles from the steeple and did great damage to the Church and a good deal more in town'[27]. The churchwardens' accounts give more detail of the damage incurred, the repairs and the costs of clearing the rubble and creating a temporary covering for the church amounting to £27 7s 6d.[28] In Derbyshire the wind blew the weathercock off the steeple at Chapel-en-le-Frith and an ash tree was uprooted in the churchyard whilst a number of houses were destroyed,[29] and at Ashbourne William Johnson was killed when a stable collapsed.[30]

In Wintringham in the north of the county of Yorkshire, Ralph Hodgson recorded in the register of St Peter's, 'this year there happened a greate Intemperance of winde on ye first day of February that it blowed down the West battlements of this Church against the broatch, and the southeast pinacle down to the ground, and did much damage in manie places, and blowed down Part of manie Churches, and manie trees, and many appletrees, and many wind millnes in manie places' (Yorkshire Parish Register Society 1922). Evidence of the storm remains in the built structure of St Peters too, 'One of the shields on the west face of the parapet, which is dated 1715, commemorates either the repair or rebuilding of the parapet almost certainly after storm damage on 1st February 1715' (Dennison and Richardson 2008, executive summary). Churches at Bainton, Huggate and Rillington also lost the tops of their spires in the storm (Dennison and Richardson 2008).

---

[24] Other documents detail extensive damage to trees in Sherwood Forest (DD/FJ/11/1/2/194–5, Nottinghamshire Archives), and to farm barns (see Blundell, 1968).
[25] OLD BOLINGBROKE PAR/1/2, Lincolnshire Archives.
[26] ALKBOROUGH PAR/1/3, Lincolnshire Archives. [27] Rolleston PR, Nottinghamshire Archives.
[28] Rolleston PR 777, Nottinghamshire Archives. [29] D3453/1/2, Derbyshire Record Office.
[30] D662/A/PI/1/7, Derbyshire Record Office.

9.2.3.2 The Register for Droitwich St Peters

Reverend Lea's entries in the register for the parish church of St Peters, Droitwich, Worcestershire, demonstrate how closely the weather was intertwined with the local industries of salt mining and fruit growing in the latter half of the nineteenth century.[31] They also provide important information concerning the aid provided to local poor during periods of hardship and distress, and capture a number of seasons or years subsequently 'benchmarked' for their notable or extreme weather events or conditions.

1864 – We have had the driest season known since 1826 – The grass fields are all [...] The turnip crop has generally failed though several may recover [?] Hay has risen to 6.10 per ton –In some parts, as Sheffield and Derby they have had abundance of ruin – at Malvern the poor are obliged to buy their water.

Drought conditions had prevailed throughout much of the country for the whole summer, prompting Lea to think back to the year 1826. As a year, 1864 was actually drier, sixth driest according to the EWP (1826 is eleventh). The narratives in our database suggest that 1864 then became a 'benchmark' dry year itself, compared to the dry years of 1870, 1887 and 1921.

1872 – This summer has been remarkable for the violence frequency and destructiveness of thunderstorms, much damage has been done to life and property. In the spring Vesuvius was in eruption, weather has been unsettled ever since – the potato disease has set in with greater violence than ever.

Vesuvius erupted on 26 April 1872. Certainly, in Central England, the weather of May and June was very stormy, the storms of 18–19 and the 24–27 June particularly notable with sheep and cattle struck by lightning and several people, and acres of hay swept away in floods.[32] The *Aurora Borealis* were of 'considerable magnitude and rather unusual features' and 'of wonderful proportions' in late June, accompanied by more violent storms of hail and 'thunder and lightning of terrible and destructive energy'.[33]

1879 – A year of disasters and distress – trade bad crops worse than ever was known. The season wet and sunless [...] Farms are being thrown on the Landlord's lands & farmers ruined, which is the result of 4 bad seasons, this last the worst of all. The winter of 1878–79 was the longest and hardest for many years past. The thermometer sometimes fell so low as to show 28 degrees of frost – in 79 – 80 it was also severe and [...] I have not heard of more than 20 degrees of frost being registered in this district.

1880 – This has been a most disastrous year for farmers – the season was very wet, the sheep died in flocks – the harvest was not half the average. The potatoes were badly

---

[31] 850DROITWICHSPA/1/a/iii, Worcestershire Record Office.
[32] THORPE MALSOR 3229/4, Northamptonshire Record Office; 899/1576/2–6, Worcestershire Record Office.
[33] 899/1576/7–10, Worcestershire Record Office.

diseased. The worst year for farmers ever known, the crops were poor and spoilt by the rain; many farms are thrown on the landlords' hands, especially on the clay soils ...

The last quarter of the nineteenth century has been termed an agricultural crisis, and agricultural depression was particularly intense during 1879–1880. 'The depression was associated with a prolonged run of disastrously unfavourable seasons, with delayed growth in the spring, and with wet and late harvests, which themselves had a cumulatively adverse effect by producing late and difficult ploughing and sowing conditions' (Crittall 1959). However, the picture was complicated and increasing costs of production and decreasing prices were also major factors. 'The meteorological explanation for the depression certainly does not add up in all parts of the country' (Crittall 1959). Reverend Lea was very interested in agriculture and in 1872 had published a book on small farms detailing how to make them pay by fruit growing (Lea 1872).

Throughout these almost annual weather summaries, Lea draws on his personal weather memory in making his entries and compiling a history for the parish, characterising years by their dominant weather events and types.

9.3 The Utility of Localised Weather Memory for a Climate Change World

Twentieth-century climatologist Gordon V. Manley regarded a national memory of the weather as 'essential' for any civilised society. As Eden (2008:4) has suggested, however, with the exception of the most extreme or unusual events, 'once a weather phenomenon has reached two years old it seems to fall out of the human memory bank'. Jones agrees, stating that 'The recollection of past weather, however unusual, rarely lasts more than a lifetime' (1953:371). Documentary narratives, however, can enrich our awareness of weather history, triggering personal memories of living through connected events, as well as extending our weather awareness to include events from a much longer timeframe, contributing to this national memory. This chapter has referred to the writings of a wide (though by no means inclusive) range of people, which can be used to reconstruct particular extreme weather events as well as longer periods of weather. The records reveal extreme weather and weather-related events to be 'hybrid phenomenon', constructed through 'sensory experiences, mental assimilation, social learning and cultural interpretations' (Hulme 2009:197), and in most cases based on personal experience and recourse to secondhand testimony.

It is important to note that regional circumstances, particular biophysical and environmental conditions, an area's social and economic activities and embedded cultural knowledges, norms, values, practices and infrastructures, including those associated with human adaptations, will have affected the nature of the experiences of and reactions and responses to weather events (Oliver Smith and Hoffman 1999). Unusual, anomalous or extreme events were judged against what was perceived to be a normal range of variation, which itself was a function of the nature and span of an individual's experience and the average range of variation communicated through oral histories or historical knowledge. There are, therefore, multiple opportunities for bias and subjectivity. These biases are of interest of and in themselves for the perspective they provide on how different people understood and responded to weather variability and indeed extreme or unusual weather events in the past. Equally TEMPEST allows for the assembly and comparison of different accounts in place and/or time which can act as a means of confirming the scale, longevity or severity of an event. In interpreting these events, we also of course have the benefit of hindsight and access to other research that enables us to see the broader context and to recognise the 'event' within everyday observations.

As we saw in the opening section, extreme weather events in the present are moments when we question our expectations of climate. Looking to the future, 'the idea of climate change disorientates our memories, many of which are profoundly attached to past weather, and unsettles our expectations about the future. It begins to unsettle our belief that we know how the weather of the future *should* be' (Hulme 2016a:161). Yet equally, 'today's weather can only be made sense of through reference to the past' (Hulme 2016a:161). The documentary record is a tremendous resource to recover this past, and the recollection of past weather exerts powerful influences on our interpretation of present weather. It also influences our apprehensions about future changes in climate (Hulme 2016b). Indeed, local, experiential weather stories might also serve a powerful role both not only in informing memory but also in influencing judgement and popular understanding of both local and global climatic changes (Marx *et al.* 2007; Scruggs and Benegal 2012). Historical, experiential narratives of the type we have collected may provide 'grounded' evidence of the potential societal implications of future climate change and reassurance for our unsettled minds. Today, therefore, the kinds of narratives and stories which have furnished our study have great potential not only to improve 'weather memory' but to engage people in broader debates about climate change.

Acknowledgements

This chapter is one outcome of the project 'Spaces of Experience and Horizons of Expectation: The Implications of Extreme Weather in the UK, Past, Present and Future', funded by the Arts and Humanities Research Council (AHRC) through grant number AH/K005782/1. We would like to acknowledge the help and support

References

Adamson, G. 2015. Private diaries as information sources in climate research *WIRES Climate Change*, 6, 599–611.

Alexander, L. V. and Jones, P. D. 2001. Updated precipitation series for the UK and discussion of recent extremes *Atmospheric Science Letters*, 1(2), 142–150.

Blundell, N. 1968–1972. *The Great Diurnall of Nicholas Blundell of Little Crosby, Lancashire Vol 2 1712–1719* Transcribed and annotated by F. Tyrer, edited for the Record Society by J.J. Bagley, The Record Society of Lancashire and Cheshire

Brazdil R., Demaree, G. R., Deutsch, M, Garnier, E., Kiss, A., Luterbacher, J., Macdonald, N., Rohr, C., Dobrovolny, P., Kolar, P., and Chroma, K. 2010. European floods during the winter 1783/1784: Scenarios of an extreme event during the 'Little Ice Age' *Theoretical and Applied Climatology*, 100, 163–189.

Brazdil, R., Kundzewicz, Z. W., and Benito, G. 2006. Historical hydrology for studying flood risk in Europe *Hydrological Sciences Journal*, 51, 739–764.

Chenoweth, M. 2003. *The 18th Century Climate of Jamaica Derived from the Journals of Thomas Thistlewood, 1750–1786*. Philadelphia: American Philosophical Society.

Chenoweth, M. and Thistlewood, T. 2003. The 18th century climate of Jamaica: Derived from the journals of Thomas Thistlewood, 1750–1786. *Transactions of the American Philosophical Society*, 93(2), i–153.

Colyer, R. 1976/77. The Hafod estate under Thomas Johnes and Henry Pelham, fourth duke of Newcastle *Welsh History Review*, 8, 257.

Crittall, E. 1959. Agriculture since 1870. In: Crittall, E. (ed.), *A History of the County of Wiltshire: Volume 4*. London: Victoria County History, pp. 92–114.

Culver, L. 2014. Seeing climate through culture. *Environmental History*, 19(2), 311–318.

D'Arrigo, R., Seager, R., Smerdon, J. E., LeGrande, A. N., and Cook, E. R. 2011. The anomalous winter of 1783–1784: Was the Laki eruption or an analog of the 2009–2010 winter to blame? *Geophysical Research Letters*, 38: L05706.

Daniels, S. 2011. Great balls of fire: Envisioning the brilliant meteor of 1783. In: Daniels, S., DeLyser, D., Entrikin, N., and Richardson, D. (eds.), *Envisioning Landscapes, Making Worlds: Geography and the Humanities* London and New York: Routledge, pp. 155–169.

Dennison, E. and Richardson, S. 2008. *St Peter's Church, Wintringham, North Yorkshire Architectural and Archaeological Recording of the Tower Parapet* Ed Dennison Archaeological Services Ltd.

Dolák, L., Brázdil, R., and Valášek, H. 2013. Hydrological and meteorological extremes derived from taxation records: the estates of Brtnice, Třebíč and Velké Meziříčí, 1706–1849. *Hydrological Sciences Journal*, 58(8), 1620–1634.

Eden, P. 2008. *Great British Weather Disasters*. London and New York: Continuum.

Endfield, G. H. 2008. *Climate and Society in Colonial Mexico: A Study in Vulnerability.* Oxford: Blackwell.

Endfield, G. and Nash, D. 2002. Drought, desiccation and discourse: Missionary correspondence and nineteenth-century climate change in central Southern Africa. *Geographical Journal*, 168, 33–47.

Endfield, G. and Veale, L. 2017. Climate, culture, weather. In: Endfield, G. and Veale, L. (eds.), *Cultural Histories, Memories and Extreme Weather: A Historical Geography Perspective*. Abingdon: Routledge, pp. 1–15.

Endfield, G. H., Veale, L., Royer, M. J., Bowen, J. P., Davies, S., Macdonald, N., Naylor, S., Jones, C., and Tyler-Jones, R. 2017. Tracking Extremes of Meteorological Phenomena Experienced in Space and Time (TEMPEST) dataset http://dx.doi.org/10.5285/d2cfd2af036b4d788d8eddf8ddf86707

Golinski, J. 2007. *British Weather and the Climate of Enlightenment* Chicago and London: The University of Chicago Press.

Grattan, J. and Brayshay, M. 1995. An amazing and portentous summer: Environmental and social responses in Britain to the 1783 eruption of an Iceland volcano. *Geographical Journal*, 161, 125–134.

Grattan, J. and Sadler, J. 1999. Regional warming of the lower atmosphere in the wake of volcanic eruptions: The role of the Laki fissure eruption in the hot summer of 1783. In: Firth, C. R. and McGuire, W. J. (eds.), *Volcanoes in the Quaternary Geological Society Special Publication No 161*. London: Geological Society, pp. 161–172.

HadUKP. 2016. England and Wales precipitation data (www.metoffice.gov.uk/hadobs/hadukp/data/download.html)

Hainsworth, D. R. 2008. *Stewards, Lords and People: The Estate Steward and His World in Later Stuart England* Cambridge: Cambridge University Press.

Hobbs, S. 2008. The abstracts and brief chronicles of the time: Memoranda and annotations in parish registers 1538–1812. *The Local Historian*, 38, 95–110.

Hopkinson, S. 1800. *Causes of the Scarcity Investigated: Also an Account of the Most Striking Variations in the Weather, October, 1798, to September, 1800*. Stamford: Printed and sold by R. Newcomb.

Hulme, M. 2009. *Why We Disagree about Climate Change. Understanding Controversy, Inaction and Opportunity*. Cambridge: Cambridge University Press.

Hulme, M. 2015. Climate and its changes: A cultural appraisal. *Geo: Geography and Environment*, 2, 1–11.

Hulme, M. 2016a. Climate change and memory. In: Groes, S. (ed.), *Memory in the Twenty-First Century*. London: Palgrave Macmillan, pp. 159–162.

Hulme, M. 2016b. *Weathered: Cultures of Climate*. London: Sage.

Jankovic, V. 2000. *Reading the Skies: a Cultural History of English Weather, 1650–1820*. Manchester: Manchester University Press.

Jones, G. E. 1953. The winter of 1798–1789. *Weather*, 8, 371–375.

Jones, P. D., and Briffa, K. R. 2006. Unusual climate in Northwest Europe during the period 1730 to 1745 based on instrumental and documentary data. *Climatic Change*, 79, 361–379.

Lamb, H. 2002. *Climate, History and the Modern World*. Abingdon: Routledge.

Lamb, H. and Frydendahl, K. 1991. *Historic Storms of the North Sea, British Isles and Northwest Europe*. Cambridge: Cambridge University Press.

Lea, W. 1872. *Small Farms: How They Can Be Made to Answer by Means of Fruit Growing*. London: Journal of Horticulture and Cottage Gardener Office.

Lejeune, P. 2009. *On Diary*. Honolulu: University of Hawaii Press.

Livingstone, D. 2012. Reflections on the cultural spaces of climate. *Climatic Change*, 113, 91–93.

Macfarlane, A. (ed.) 1991. *The Diary of Ralph Josselin, 1616–1683*. Oxford: British Academy.

Manley, G. 1958. The great winter of 1740. *Weather*, 13, 11–17.

Manley, G. 1974. Central England temperatures: Monthly means 1659–1973. *Quarterly Journal of the Royal Meteorological Society*, 100, 389–405.

Marx, S. M., Weber, E. U., Orlove, B. S., Leiserowitz, A., Krantz, D. H., Roncoli, C., and Phillips, J. 2007. Communication and mental processes: Experiential and analytic processing of uncertain climate information. *Global Environmental Change*, 17, 47–58.

Morgan, J. E. 2015. Understanding flooding in early modern England. *Journal of Historical Geography*, 50, 37–50.

Murphy, P. 1837. *The Weather Almanack* (on Scientific Principles, Showing the State of the Weather for Every Day of the Year of 1838). London: Whittaker & Co.

Neumann, J. and Kington, J. 1992. Great historical events that were significantly affected by the weather: Part 10. Crop failure in Britain in 1799 and 1800 and the British decision to send a naval force to the Baltic early in 1801. *Bulletin of the American Meteorological Society*, 73, 187–199.

Oliver, J. 1965. Problems in agro-climatic relationships in Wales in the eighteenth century. In: Taylor, J. A. (ed.), *Climatic Change with Special Reference to the Highland Zone of Britain*. Aberystwyth.

Oliver Smith, A. and Hoffman, S. (eds.) 1999. *The Angry Earth: Disaster in Anthropological Perspective*. New York: Routledge.

Pearson, M. G. 1973. The winter of 1739–40 in Scotland. *Weather*, 28, 20–24.

Pedgley, D. E. 2010. The British rainfall organisation, 1859–1919. *Weather*, 65, 115–117.

Pfister, C. 2003. Monthly temperature and precipitation in central Europe 1525–1979: quantifying documentary evidence on weather and its effects. In: Bradley, R. S. and Jones, P. D. (eds.), *Climate Since AD 1500*. London: Routledge pp. 118–142.

Scruggs, L. and Benegal, S. 2012. Declining public concern about climate change: Can we blame the great recession? *Global Environmental Change*, 22(2), 505–515.

Sutton, J. F. 1852. *The Date Book of Remarkable and Memorable Events Connected with Nottingham and Its Neighbourhood*. Nottingham: Simkin and Marshall.

Tate, W. E. 1946. *The Parish Chest: A Study of the Records of Parochial Administration in England*. Cambridge: Cambridge University Press

Thordarson, T. and Self, S. 1993. Atmospheric and environmental effects of the 1783–1784 Laki eruption: A review and reassessment. *Journal of Geophysical Research*, 108, AAC 7-1–AAC 7-24.

Tufnell, L. 1983. Environmental observations by the rev. William Hutton of Beetham, Cumbria. *Transactions of the Cumberland and Westmorland Antiquarian and Archaeological Society*, 83, 141–150.

Veale L., Endfield G., Davies S., Macdonald N., Naylor S., Royer M. J., Bowen J., Tyler-Jones R., and Jones C. 2017. Dealing with the deluge of historical weather data: The example of the TEMPEST database. *Geo: Geography and Environment*, 4: e00039.

Wardle, D. 2010. *Education and Society in Nineteenth-Century Nottingham*. Cambridge: Cambridge University Press.

Waters, R. E. C. 1887. *Parish Registers in England: Their History and Contents, With Suggestions for Securing their Better Custody and Preservation, a New Edition*. London: Longmans, Green, and Co.

White, G. 1836. *The Natural History of Selbourne*. New Edition London.

Wigley T. M. L., Lought, J. M., and Jones, P. 1984. Spatial patterns of precipitation in England and wales and a revised, homogenous England and Wales precipitation series. *Journal of Climatology*, 4, 1–25.

Witham, C. S. and Oppenheimer, C. 2005. Mortality in England during the 1783–4 Laki craters eruption. *Bulletin of Volcanology*, 67, 15–26.

Woodforde, J. 1998. *The Diary of James Woodforde edited with introduction and notes by Winstanley, R. L.* Parson Woodforde London: The Folio Society.

Yorkshire Parish Register Society. 1922. *The Parish Register for Wintringham, 1558–1812* Wakefield: Privately printed for the Yorkshire Parish Register Society.

Part III

Doing in a Climate Change World

10

From Denial to Resistance
How Emotions and Culture Shape Our Responses to Climate Change

ALLISON FORD AND KARI MARIE NORGAARD

10.1 Introduction

The basic premise of an enlightened, democratic society is that scientific information will lead to public concern and institutional response. This premise fits into models of human behaviour that centre individuals as 'rational actors' maximising clearly defined interests (Becker 1993). These logics undergird a worldview that has dominated Euro-American society since the Enlightenment; celebrating logic, deliberation, critical reasoning and rational thought as the pinnacle of human achievement. It is this mode of thinking that leads us to expect that, when faced with evidence that climate change is real, human-caused and problematic, most people would accept this new information, formulate a decisive, rational response and act upon the danger. Instead, despite widespread awareness, the crisis of climate change has largely failed to register on the political radar of Western citizens (Jacques 2006; Jacques *et al.* 2008; Norgaard 2011; Capstick *et al.* 2015). Rather, in the United States, only 17 per cent of Americans are 'alarmed' about climate change, and actively acting to address it (Leiserowitz *et al.* 2009; Roser-Renouf *et al.* 2016a). Even for the 28 per cent of Americans who were concerned about global warming, the issue placed number 14 in a list of political priorities during the 2016 American national election (Roser-Renouf *et al.* 2016b). And while public perception has increased globally since early efforts by scientists to communicate about it, parts of Europe (such as the United Kingdom), Australia and the United States have also experienced significant growth in scepticism (Capstick *et al.* 2015).

There is a profound misfit between public climate engagement and the seriousness of this problem. Public awareness of a social problem does not necessarily translate to active problem-solving (Blake 1999; Shove 2010a; Norgaard 2011; O'Brien 2012). This gap between the climate crisis and public response raises a series of questions: Why aren't *knowledge of* and *concern about* climate change good predictors of action? How might public mobilisation be achieved? Under what cultural conditions do climate actions take place?

Adopting new practices and policies is not simply a matter of rational deliberation. Emotions and culture are important factors in shaping how people experience information and adopt sets of environmental practices. New information must be interpreted through cultural worldviews that are constantly being affirmed, challenged, critiqued or bolstered (Zerubavel 1997). Emotions regulate our attachments to cultural worldviews, signalling when things are amiss, or on track. In this chapter, based on the comparison of four case studies, we argue that culture operates in multiple and dynamic ways to shape social reality and that the role of emotion in the operation of culture must be accounted for if we are to understand responses to climate change to date, now and in the future.

10.1.1 Theorising Climate Responses across Social Domains

Most climate change researchers share a broad sense that climate change is a structural problem that is symptomatic of global macroeconomic and political cultures (Rosa and Dietz 1998; Stern 2008; Lazarus 2009; Leiserowitz *et al.* 2009). There is also a sufficient body of work on individual attitudes, beliefs, values, concern and levels of knowledge to suggest that these aspects of culture are significant to our understanding of climate change as a social problem (Dietz *et al.* 2005; Kellstedt *et al.* 2008; Clements *et al.* 2015; Leiserowitz *et al.* 2016). However, we have largely side-stepped the meso-level of climate change culture, in which we try to understand how the macro-level, political economic cultures of contemporary globalised capitalism, intersect with individual, everyday lived experience.

Understanding how people respond to climate change is virtually impossible without a meso-level cultural analysis that links lived experience to macro-structural phenomena. All groups interpret knowledge about climate change through cultural processes that take their own worldviews and values as a starting point. New information is interpreted through prior beliefs and values (Kunda 1990; Petty and Wegener 1999); when worldviews are put in opposition, people will engage in motivated reasoning to justify their own worldview (Kunda 1990). Cultural frames, selective interpretive schema that simplify and condense information, filter out information that is incompatible with familiar ways of seeing the world (Snow and Benford 1992; Zerubavel 1997; Benford and Snow 2000). This renders certain pathways of action viable, and others incomprehensible (Swidler 1986; Kunda 1990; DiMaggio 1997). Culture shapes how individuals see the world, and how they feel about what they see. And culture is anything but rational! Emotional relationships to culture can help us understand why some people outright deny climate change, while others live in fear of it, but do nothing to change its trajectory, and still others challenge the social systems that cause climate change, fighting passionately against injustices to overturn inequality.

10.1.2 Emotions and Social Structure

The study of emotions is often individualised as biological or psychological phenomena. Emotions are also deeply social. We learn how to interpret and make sense of our internal states (feelings), and read those of others, through cultural categories (Shott 1979; Turner and Stets 2005). Rather than being idiosyncratic, cultures of emotion are structured by social norms and expectations (Hochschild 1979; Hochschild 2012). For some theorists of emotion and culture, emotions are essentially internalised culture (Illouz 2007:3). Emotion serves as the glue that binds us to norms, the performance of socially sanctioned roles and the maintenance of status hierarchies (Collins 1990; Illouz 2007; Hochschild 2012). When we stray from social norms and our proscribed roles, emotions signal to us through feelings of discomfort that we are out of cultural bounds. The material structures that produce climate change are reinforced by cultural narratives of justification, norms, habits and taken-for-granted assumptions about the world. Emotions regulate our attachments to these practices.

10.1.3 Culture, Social Order and Climate Change

While key theorists have considered the importance of culture in understanding our responses to climate change (Shove 2010b; Urry 2011; Smith and Howe 2015), a sustained and complex engagement that considers how culture shapes climate action on a structural level is missing from dominant discourse on climate change. Instead, the most widely cited social scientific engagements with climate change ostensibly address culture by measuring people's environmental knowledge, attitudes, concern, values and behaviours. These measures lack a theoretical framework beyond the assumption that if individuals know about, understand and care about climate change, they will act to address it though (Shove 2010a).

This research fails to chart the complexity of how people make sense of climate change by incorporating new knowledge into already existing cultural worldviews. Individual attitudes, values, beliefs, concern and knowledge are units of meaning, all of which are part of cultural systems that make up what we here refer to as *worldviews*. Yet these variables are often handled independently; without a unifying cultural theory to explain their role in social change, they tell us little about how people make sense of the world around them in ways that perpetuate, ignore or challenge climate change. The study of culture is particularly concerned with meaning; how people make sense of the world around them and enact their roles in it according to what certain practices, relationships and circumstances mean to their adherents (Sewell 1999). Units of meaning are important constituent elements of this, but do not alone explain the complexity of how people respond to socio-environmental circumstances. Cultural worldviews are not necessarily cohesive (Swidler 1986; Bourdieu 1977; DiMaggio 1997), but they are less disjointed and individual than the current literature on climate culture would attest.

Further, this literature tends to privilege cognitive processes. Even variables like values and concern, which are infused with affect, may be reduced to their role in producing deliberative, rational responses to climate change. Values and attitudes are thought to drive behaviours that individuals choose (Shove 2010:1272); however, these variables are all bits of culture held by individuals, contained in mental structures that DiMaggio calls 'schematic representations of complex social phenomena' (1997:273).

Meaning is not strictly cognitive but also organised by symbolic systems external to individuals (Alexander and Smith 1993; DiMaggio 1997; Patterson 2014). People retain and store vast amounts of cultural information, uninterpreted; yet much remains unknown about which cues determine which cultural tools we use in which situations. We contend that emotions – the affective interpretations attached to sensations – are central to the interpretation of cultural cues, signalling which frames are compatible with deeply held, often embodied beliefs and habits, and which ones would require the undertaking of what Swidler calls 'a drastic and costly cultural retooling' (1986:277).

The meaning of climate change must be filtered through cultural frames, eliciting emotions of varying strength, endurance and valence (Thoits 1989; Robnett 2004). Emotional resonance influences the adoption of cultural frames (Robnett 2004) as individuals respond to surrounding culture and situations based on their own biographical histories, 'mak[ing] certain feelings salient, certain beliefs plausible, certain moral principles more important than others' (Jasper 1997:13). Emotional and cultural resonances are distinct, but connected (Robnett 2004; Gould 2009), and both must be understood to make sense of group emotion cultures that shape responses to paradigm shifting informative-frames, such as the concept of human-caused climate change.

10.2 Methods

To understand how emotions and culture interact to shape group level responses to climate change, we explore four cases in which groups perceive and respond to climate change based on the alignment of their pre-existing cultural worldview and their perception of climate change as a social problem. This chapter is based on ethnographic studies on (1) Norwegians who accept climate science but implicitly deny its implications in everyday life, (2) American homesteaders who see climate change as one of many socio-environmental risks they must protect themselves from and (3) members of the Karuk tribe, for whom climate change is yet another symptom of ongoing colonial oppression. The three cases pull from original ethnographic research conducted by the authors on a Norwegian village (Norgaard), including participant observation, interviews and media analysis, American homesteaders (Ford), including participant observation and interviews, and members of the Karuk tribe (Norgaard), including interviews, observation and analysis of comments submitted in policy documents. We also include, as our fourth case, a review of research on overt climate scepticism, based on sociological studies of climate denialists, because this movement has played such a central role in framing the public discourse around climate change in the United States and throughout the world.

10.2.1 Common Cultural Frames

In each of these sections, we consider how pre-existing cultural worldviews interact with the concept of climate change to generate complex emotional states. Response to climate change information is shaped by emotional management in interaction with available culture. Based on the above studies, we identify four particularly salient cultural frames that contribute to a group's collective response to climate change:

1. Relationship to the natural world
2. Beliefs in dominant institutions such as the state and market and commitment to the project of modernity, to which climate change is inextricably linked
3. Political values and level of political engagement
4. Religious/spiritual beliefs and moral values

These cultural frames act as mirrors in a kaleidoscope that reflect onto each other in different configurations. They filter information, shape beliefs and tacitly confirm ideological assumptions. When they are challenged, emotional response is invoked. Emotional states will vary in content and intensity depending on the relationship between existing cultural beliefs and the assumptions necessary to support the dominant framing of climate change (i.e. the assumption that it is possible to know about abstract, material processes through scientific empiricism). These elements of a cultural worldview may be explicit, as in beliefs, which we tend to know we hold, or implicit, as in habitus – sets of dispositions that establish taken-for-granted understandings of what is natural and real (Bourdieu 1977). Both implicit and explicit facets of worldviews interact with the dominant diagnostic

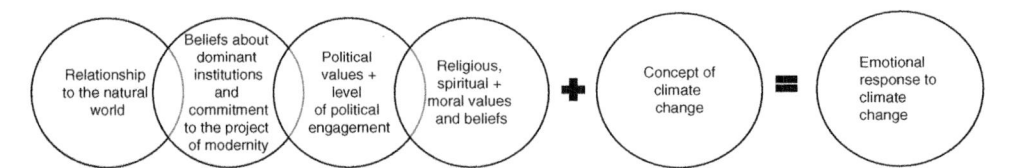

Figure 10.1 The relationship between pre-existing cultural worldviews and the concept of climate change generates emotional responses that affect how individuals and groups will respond to the issue
The linear diagram reflects the limits of the authors' graphic design tools and skills; in reality, these facets of a cultural worldview are overlapping, simultaneous and may not always be equally weighted in each social circumstance.

Figure 10.2 The relationship between emotions and culture in shaping responses to climate change

framing of the problem. Local feeling rules may diminish or magnify the intensity of response, which, combined with available cultural alternatives that are not incompatible with existing worldviews, render some responses to climate change more viable than others (Figure 10.1).

Climate change is more than just an environmental process; in the words of sociologist Philip Smith, 'it is also a new and interesting concept, signifier and drama in a surprisingly complex cultural field' (Smith 2017:746). The power of culture to shape the way we act arises from the way it makes us feel. But the expression of feelings is also structured; as people manage their emotions within the context of their cultural worldviews, they must work with available 'chunks of culture' that can be organised into coherent 'strategies of action' (Swidler 1986:273, 283). In the following case studies, we show how cultural worldviews influence the formation of a narrative about climate change as they modulate emotional responses to environmental risk in ways that make sense to different groups of people (Figure 10.2).

10.3 Case Study Results

10.3.1 Outright Rejection: The Deep Anthropocentricism of Climate Sceptics

Often referred to as climate scepticism or denial, the outright rejection of climate change is closely associated with conservative political organising and the financial resources of the American right (Jacques 2006; Goldenberg 2013; Mccright and Dunlap 2013). Climate denialists have received significant attention in part for their success in blocking proactive attempts by the state to reduce or replace fossil fuel consumption.

At stake in the political fight over climate change is the legitimation of society's dominant core social values (Jacques 2006). Environmental sceptics, including climate denialists, operate from an ethical foundation of 'deep anthropocentricism', a cultural worldview that 'believes humanity is utterly independent of non-human nature, and moral obligation is dependent upon strict relative and immediate human benefit, otherwise the ethic sees no obligation for human concern' (Jacques 2006:85). Deep anthropocentricism supports a strong belief in dominant institutions such as the state and market.

Influenced by conservative political philosophy, denialists tend to highly value the perceived freedom of the market to allocate resources and wealth, and the underlying structure of state power that supports it. Climate denial is thus closely linked to the protection of corporate interests to continue to use and develop the fossil fuel–based infrastructure upon which the modern economy depends. McCright and Dunlap (2013) show that conservative think tanks linked to powerful corporate interests mobilised to construct the 'non-problematicity' of climate change, in a direct challenge to the work scientists and environmental organisations had done to establish climate change as a legitimate social problem. The effects of this have been twofold; firstly, a politically active minority of conservatives have denied relationships between economic activity and climate change, successfully obstructing political action at the national level, and thus international level (such as the failure of the United States to ratify the Kyoto Protocol, thereby derailing American climate policy); secondly, the campaign to de-problematise climate change has established doubt in the mind of the public, legitimising a cultural narrative of environmental doubt that can be used to justify short-term, status-quo environmental management at the expense of both mitigating and adapting to climate change (Hartter *et al.* 2018).

The cultural work of undermining scientific consensus, done by national conservative elites, offers a legitimised narrative of doubt that can be used to justify continuing with a valued way of life, without having to respond to inconvenient information. Peter Jacques argues that climate scepticism is not really about science, but about the politics of citizenship, and who and what citizens are ultimately responsible to and for (2006). The importance of dominant institutional power structures for denialists is undergirded by unwavering belief in the positive value of modernity and human technological and industrial progress. Conservatives who actively champion industrial progress and the sanctity of the free market are politically situated to resist regulation on behalf of an environment that they do not value.

For conservative American climate sceptics, information about climate change is filtered through a frame of deep anthropocentrism supported by Judeo-Christian values which are frequently interpreted by evangelicals as the creed that God gave humans dominion over nature to use as a tool for human wellbeing (Jacques 2006). This affirms the notion that humans are not interdependent with other species in ecological webs; they are rather superior beings made in the image of God, whose survival is not based on the sustainable management of finite resources, but rests in the hands of an omnipotent being. While not all climate denialists are necessarily evangelical Christians, the ethical underpinnings of environmental scepticism are rooted in these values. Thus, calls to limit human economic development to protect the natural world are viewed by sceptics as unfounded, and even dangerous to the social order they protect. This may generate negative emotions such as anger, frustration or defensiveness, yielding a rejection of cultural narratives that challenge a valued worldview.

If we take conservative cultural worldviews as a starting point, the ferocity and emotional fierceness of the sceptical response to climate change is not a puzzle to be solved, but an obvious reaction to the threat of a strongly valued cultural worldview, with major implications for a re-ordering of the political-economic system in a way that directly contradicts the group's commitment to the project of modernity. Climate sceptics protect their worldview by denying the validity of climate change; resulting in a sustained, aggressive attack on scientists, environmentalists and politicians who take seriously the scientific consensus that climate change is real, serious and human-caused. The ferocity and intensity of this attack suggests the emotional attachment that is threatened by climate change.

Emotions serve an evaluative role in responding to changing circumstances (Schwarz and Clore 1983; Goodwin *et al.* 2002; Hochschild 2012). The ferocity and grandiloquence of climate denial rhetoric suggests more than a rational, measured disagreement of a scientific nature, but an emotional reaction grounded in threat, and the fear, anxiety and unease that such a threat might engender. Kemper theorises that power and status are universally linked to the elicitation of emotions; when power and status are threatened, negative emotions motivate attempts to restore them (Kemper 1990). The development of powerful political campaigns and social movements to deny the threat of having to take climate change seriously may serve to manage and release emotion, affirming the collective identity of supposedly modern, rational, free individuals, ingenious enough to survive any environmental conditions nature/God creates.

10.3.2 Ignoring: Holding on to Continuity and the Reproduction of Everyday Life

For many who accept the scientific consensus on climate change, thinking seriously about the matter evokes a series of troubling emotions (Norgaard 2011, 2014). This was the case in Norway, where Norgaard spent ten months in 2001, engaging in ethnographic research that asked why Norwegians did not actively respond to climate change. Norwegians seemed particularly likely to respond to the threat of climate change for several reasons. Firstly, climate change was covered regularly and seriously by the Norwegian media and scepticism of climate change was much lower than in the United States, where a successful right-wing campaign challenged its scientific framing (Norgaard 2006; McCright and Dunlap 2013). Secondly, the environment and humanitarian ethics are both central to Norwegian national identity (Reed and Rothenberg 1993). When asked, Norwegians in Norgaard's study identified environmental protection as important. They identified with moral values that emphasised equality, and a connected, humanitarian view of the world. They possessed the political tools for informed action. However, when it came to action, the majority were at a loss. 'Despite my knowledge of the wider climate issues, I am still living the same life', one student reported.

The cultural worldviews of climate sceptics directly contradicted the possibility of human-caused climate change, generating momentum for a concerted backlash against environmental regulation. For these Norwegians however, things were not as simple (see Table 10.1 for comparison). Even as people valued the environment, and believed they had a responsibility to tread lightly upon it, they upheld a taken-for-granted commitment to the project of modernity, to which climate change is inextricably linked. While they knew about and felt concerned about climate change, the material reality of Norway's political economy was that their high quality of life depended upon the export of oil to the rest of the world. Norwegians were deeply implicated in causing climate change and they both knew it and refused to know it. Robert Lifton (1982) calls this the *absurdity of double life* and shows that living with conflicting, incompatible realities profoundly challenges our thinking, feeling, identity, morality and sense of personal and political empowerment. Unlike American conservatives, who have a long history of distrust of academic and state institutions where most climate science happens, Norwegians had no reason to distrust the claims of scientists, so they took scientific

Table 10.1 *Case comparison by kaleidoscopic cultural frames*

	Relationship to Nature	Commitment to modernity	Political commitments and engagement	Religious/spiritual beliefs and moral values
American climate-sceptics	Deep anthropocentricism (Jacques 2006)	Strong and explicit (Jacques 2006)	Conservative, active engagement, well-funded (McCright and Dunlap 2013)	Fundamentalist Judeo-Christian (Jacques 2006)
Norwegians engaged in implicit denial	Environmentalist	Strong and implicit	Active locally	None discussed
American Homesteaders	Environmentalist	Ambivalent	De-politicised	Various
Indigenous Karuk Tribe members	Indigenous	None	Strong and active radical resistance	Traditional indigenous religious and spiritual practices

information at face value. They did have a reason to avoid the information emotionally though; they were implicated in the production of climate change through their economically privileged way of life.

Because they accepted the scientific framing, Norwegian interpretation of climate change included complex, negative emotions such as fear, guilt and helplessness. In addition to fear of a future with more heat waves, droughts and increased storm intensity, the Norwegians interviewed feared that present political and economic structures are unable to effectively respond to what climate change will bring. This fear challenged an implicit commitment to modern globalised capitalist society, despite its connection to climate change, deep social inequality and other major global environmental problems, all of which allow people in wealthy nations to live comfortably at the expense of marginalised people in majority world nations. Many Norwegians also experienced guilt over their complicity in this system and its role producing climate change.

When climate change incited difficult emotions such as fear, anxiety, guilt and despair, Norwegians adopted tactics for normalising awkwardness about socially unaccepted feelings about climate change. Local feeling rules that discouraged open expression of intense emotion allowed Norwegians to minimise the threat of climate change, thus upholding the status quo and protecting their privilege. Norgaard observed two broad strategies of denial: *interpretive* and *cultural*, which allowed Norwegians to tacitly *not know* the offending information without explicitly denying it. Both involved emotional management.

Interpretive denial refers to the ways in which information about climate change was interpreted, while cultural denial refers to the ways in which this information was acted upon. *Interpretive denial* allowed people to assign meaning to events in a way that legitimised their response. This took the form of narratives that gave definition and meaning to information in relation to a pre-existing worldview. First, information about the issue was minimised, or backgrounded. Norgaard notes that climate change was not frequently discussed in local newspapers, or at strategy meetings of local, political, volunteer or environmental groups she attended. When climate change was acknowledged, it was often normalised using what Morris Rosenberg calls *perspectival selectivity*, 'the angle of vision that one brings to bear on certain events' (Rosenberg 1991:134). For example, people may manage unpleasant emotions by searching for and repeatedly telling stories of others who are worse off than they are. Three narratives in this category – 'Amerika As A Tension Point', 'We Have Suffered' and 'Norway Is A Little Land' – served to minimise Norwegian responsibility for the problem of global warming by pointing to the larger impact of the United States on carbon dioxide emissions, stressing that Norway had been a relatively poor nation until quite recently and emphasising the nation's small population size. For example, multiple newspaper articles in the national papers in the Winter and Spring of 2001 listed the figure that the United States emits 25 per cent of total greenhouse gas emissions, while accounting for only 4 per cent of the global population visibly in their articles. While obviously the United States must be held accountable for its emissions, framing the figure in terms of total emissions and population made the difference between the United States and 'little Norway' appear greatest. When looking at per capita emissions in each country the contrasts are not so large. Perspectival selectivity was used to create what social psychologists Susan Opotow and Leah Weiss call *denial of self-involvement* (Opotow and Weiss 2000), protecting Norwegians from taking seriously their implication in the problem. This served as a form of emotional management that downplayed a responsibility to act, protecting villagers from having to rethink their commitment to modernity and its privileges.

While interpretive denial shaped how people read and applied meaning to knowledge about climate change; *cultural denial* refers to how they acted upon that knowledge to hold information about global warming at arm's length. By following established cultural norms and feeling rules that guided what they called attention to, felt, talked and thought about in different contexts, information about climate change was absorbed into daily life and minimised in a process of cultural denial. For example, following norms of attention with respect to time encouraged community members to not think too far ahead into the future, hence minimising the extent to which the implications of immediate events are forecasted. Cultural norms of emotion limited the extent to which community members could bring strong feelings they privately held regarding climate change into the public political process, which in turn served to reinforce the sense that everything was fine. Their adherence to these norms supported the dominant social paradigm, a set of core beliefs, values and wisdom about the world that 'forms the core of a society's cultural heritage' (Dunlap and Liere 1984:1013).

Without constructive cultural outlets for feelings that otherwise threatened a valued way of life, Norwegians socially constructed collective, implicit denial of the implications of climate change. This allowed them to maintain their sense of what Giddens (1984, 1991) calls *ontological security*; the sense of continuity in everyday life, without questioning either their commitment to modernity, or their cultural identity as an environmental, humanitarian people. Although politically engaged, Norwegians focused their political efforts on problems other than climate change, which was held outside of the sphere of political responsibility by the narratives of perspectival selectivity (Norway is a Little Land, Amerika as a Tension Point). The dissonance between conflicting cultural values undermined any political commitments to change, and local feeling rules that encouraged individuals to downplay strong negative emotions like guilt, fear and despair made it difficult to apply the offending knowledge of climate change to everyday life without undoing the cultural worldview that the community was invested in maintaining.

10.3.3 I'll Do It Myself: Self-Sufficiency as a Response to Social and Environmental Risk

While many Americans continue to accept environmental risk as the cost of modernity, to which they remain committed, a small but growing movement challenges the assumption that it is worth the cost. 'For me it's the three big things: climate change, peak oil and then the economic collapse', explains Don, a middle-aged white American man. 'I think that our "leaders", and I use air quotation marks around the word "leaders", I think some of them must to some degree know how utterly unsustainable things are ... they're patching the world's economy together with so much chewing gum and duct tape and crazy glue and band aids, but when it finally does fall over, it's just going to go kaboom.'

Don is a part of a loose collective of people who participate in what is often referred to as urban homesteading, which Ford observed through participant observation and interviews between 2013 and 2014 (Ford 2014). Homesteaders seek to live in ways that they describe as 'self-sufficient' by attempting to minimise their reliance on systems and institutions that they distrust, including both public and private entities. For example, rather than rely on a municipal water source, homesteaders may try to provide for their own hydrological needs by collecting rainwater, digging a well, moving to land on a creek or river, or constructing a grey water system that conserves and reuses water. Homesteaders tend to value the environment; many report a feeling of spiritual connection with nature. But their relationship to nature is also instrumental, as they depend upon ecological systems for sustenance as they renegotiate their material flows (Schlosberg and Coles 2015) away from institutional supply chains. Homesteaders strongly believe in a sustainable human–environment relationship as not only ethically desirable, but as necessary for human survival.

Urban homesteaders believe in climate change, and link it to contemporary, global political-economic processes. While they largely accept the scientific framing of climate change, they rarely speak of it in scientific terms. Their recognition of climate change, as inextricably linked to what is often referred to vaguely as 'the system' yields a social framing of climate change that emphasises human need and 'greed' as causal factors in producing it. Unlike members of our previous cases, who both maintain a commitment to the project of modernity (explicit in the former and implicit in the latter), homesteaders have an ambivalent relationship with modernity, which they frame as unstable and replete with environmental and economic risk. The practice of homesteading emphasises pre-industrial human-environmental relations, where time is spent tending to the subsistence needs of a single household, sometimes at the expense of diversified contemporary labour and leisure practices. Homesteaders seek alternative practices beyond business-as-usual, challenging modernity, but their solution is retrograde.

Here emotions play a central role. All the homesteaders interviewed by Ford (13 in 2013–2014) expressed complex, negative emotions about global environmental risk, such as guilt, shame, anger, frustration and hopelessness about the current political-economic system. Homesteaders felt that institutions upon which they relied for their subsistence needs (food producers, manufacturers, utilities and health services) did not have their best interests at heart, but rather operated out of profit motive and greed, the guiding sentiments of modern capitalist states. Noah, a middle-aged man who practices homesteading and trains young people in urban farming, attributes problems like economic inequality, resource depletion, and global warming to an economic system out of alignment with social needs: 'The guiding value of economic development is one of maximizing profit. We have an economy based on greed essentially.' Homesteaders are critical of the dominant social paradigm, sceptical about technological fixes to environmental problems, and wary of market-oriented solutions. While the Norwegians discussed above worked hard to keep their knowledge of risk in the background, urban homesteaders acknowledged it readily, and indeed, often hyper-focused on it.

Although urban homesteaders are critical of dominant cultural worldviews, their cultural worldviews remain limited by them. Even as they distrust state and market institutions, they embody a spirit of individualism associated with traditional American self-identity recently deepened by contemporary neoliberalism, encouraged by both state and market ideologies. Thus, even as they saw climate change and other environmental risks as symptoms of immoral systems, the cultural frames in which they were embedded made it difficult for them to move beyond centring a free, rational, autonomous individual as the solution to the very problems this worldview is implicated in creating. Implicitly libertarian, homesteading pulls from a discourse of individualism that sees individuals as autonomous, rational actors who are responsible for their own wellbeing and success. Individualism is a historic American cultural value (Bellah *et al.* 1985; O'Brien 2015) that has further been reinforced by neoliberal narratives that emphasise the significance of the (rational) individual over all other social units (Brown 2003; Harvey 2005; Carlson 2015). Homesteading and other self-sufficiency movements reflect this narrow focus, encouraging individuals to focus on their own practices, rather than those of the institutions that produce the risk in the first place.

Faced with anxiety, despair and guilt, people who accept the assumption that individuals are ultimately responsible for their experiences and success have nowhere to turn but inwards. Cultural individualism precludes recognition of the structural factors contributing to climate change. For example Don, who acknowledges feeling incredibly guilty about his part in producing climate change attributes the guilt to his own mental imbalance, rather than to a culture that does not provide alternatives to fossil fuel–intensive development, or even outlets for conversations about alternatives: 'Maybe others who are more mentally well balanced would not feel terrible guilt about it, but I do.'

The de-politicisation of American civic life forecloses the possibility of collective action to address a widely shared distrust in institutions; as Nina Eliasoph (1998) shows, even within the context of political activism, individual participants are encouraged to frame their public concerns as private matters, perpetuating the narrative that individuals only act out of self-interest. Some homesteaders were politically active, but many also reported having felt discouraged by participation in collective action that felt neither effective nor satisfying. Parker explained:

When I was younger I used to protest and go do things like this, which is what a lot of people do, but I found that to be really disempowering . . . I found what seemed a better response is to be more constructive, so with the power that I have, it was to engage and then for me it became about food, cause that's something I could . . . have agency into [a] certain scale. And so that's where I began to expend my energy.

Some participants of self-sufficiency movements see their actions as contributing broadly to social change, not directly as part of a social movement, but as part of a renegotiation of the organisation of material life. Such justification resembles Eliasoph's politically concerned subjects who rely upon culturally available political narratives that allowed them to 'translate feelings of impotence into feelings of efficacy' (Eliasoph 1997:614). Self-sufficiency practices serve as emotional management strategies to defray the stress of being dependent on institutions by which individuals feel betrayed.

Being discouraged by the enormity of the social changes necessary, homesteaders turn to personal lifestyle changes as a way of managing negative emotions including guilt, hopelessness and a sense of overwhelm. Annette, who was active in her local community, saw homesteading as an empowering option within a system that otherwise left her feeling powerless:

My number one reason why I homestead is because of the empowerment aspect. Like the personal power that you feel from . . . taking part in your own necessities as a human being. I think it's important for human happiness to participate and to have your hands physically working towards some aspect of your basic needs . . . clothing, food, shelter, you know . . . And that's why I homestead.

Many homesteaders recognised that individual lifestyle change was insufficient to counteract the fossil fuel–dependent system of consumer capitalism. Some joined together in communities of likeminded people to share skills, knowledge and resources. However, the fact that individualism remained the guiding community value limited the collective political potency of the movement.

Even as homesteaders see climate change as a symptom of a broader destructive system that is doomed to collapse, they do not have a strong structural critique that allows them to see the violence of 'the system' as anything more than the composite of selfish individual actions. In comparison, Kyle Powys Whyte (2017) writes that many indigenous people understand climate change as a logical extension of the destructiveness of white, Euro-American settler political overreach; however, urban homesteaders did not offer a structural critique of colonialism, as evidenced by their unproblematised adoption of the name 'homesteaders'. While some expressed discomfort with the moniker, only one did so from the perspective of the racial insensitivity towards Native Americans contained in mythologising the violent Westward expansion of the United States.

Rather than address this violent history, in which people's land claims were legitimised or denied based on race and indigeneity, urban homesteaders focus on the industriousness of the white settlers facing hostile conditions on the untamed frontier; a deep cultural story that retains salience for Americans who are still invested in an American national identity, even as they are critical of some of its manifestations (Dunbar-Ortiz 2018). Some contemporary urban homesteaders proudly identified ancestors who had been 'homesteaders', if not directly part of the wave of settlement following the Homestead Act of 1862, then settlers who moved west at some point in America's westward expansion, and many focused on the capitalist ideal of the significance of owning property of one's own as the hallmark of true self-sufficiency. In common usage, the term 'homesteading' was largely de-historicised, keeping alive an invented tradition of rugged self-sufficiency without a political critique of power (Hobsbawm and Ranger 1983; Norgaard 2011).

10.3.4 Climate Change as a Continuously Unfolding Chapter in a History of Destruction

Our first three cases addressed responses to climate change that centre on privileged western communities, consisting mostly of middle-class, Euro-American, heterosexual, white people. Often missing from the discourse about climate action is consideration of how our ties to histories of colonialism, racial domination, as well as hetero-patriarchal social systems might influence the adoption of new practices. Our last case looks at indigenous responses to climate change to de-centre these taken-for-granted variables and suggest ways in which their invisibility upholds their privilege, and obscures pathways out of political inaction.

Indigenous peoples across North America describe experiencing climate change from a very different set of circumstances and cultural and political reference points than settler-citizens (Norton-Smith *et al.* 2016). Indigenous peoples can often be found leading the way in climate change policy, strategy and resistance by participating in the political process, engaging in sustainable land stewardship and being at the forefront of many direct-action climate activism efforts. As organisations, tribes have been key leaders in responding to climate change through both mitigation – efforts to stop further climate change – and adaptation – developing responsive measures for coping with the unfolding ecological and atmospheric changes. The Karuk Tribe of Northern California exemplifies this commitment to a radically different socio-environmental community than offered by mainstream environmentalisms.

For many Native people – especially those in rural areas – closer social, economic and cultural connections with the natural world shape the immediacy with which people experience the changing climate. Many Karuk people hold traditional beliefs in human–environment relations that transcend western categories of difference between the human and non-human world. For example, in their tribal climate assessment, Karuk leaders write:

As Karuk people we are fortunate to retain relationships with hundreds of species we consider our relations (Lake *et al.* 2010). These foods, medicines and fibers are embedded within cultural, social, spiritual, economic and political systems, and daily life (Lake 2013; Norgaard 2014). Impacts to culturally significant species in the face of climate change have thus more direct impacts on Karuk people than for communities who no longer retain such intimate connections with other beings and places in the natural world. Part of the increased vulnerabilities Karuk people face as the climate changes are a direct result of the strength of these connections. For example, the loss of acorn groves that have been family gathering sites for generations is much more than an economic impact.

While Euro-American cultural worldviews emphasise the role of technology in meeting human needs and valourise individualistic industry in which the environment is largely conceptualised as a natural resource there for human use, Karuk and other indigenous peoples conceptualise other species as relations, and retain an intimate, embodied bond with them. As climate change threatens ecological relationships, individuals and the Karuk tribe *as a whole* experience these changes directly, rather than through the abstract filter of scientific language. Although Norwegians saw themselves as nature-loving people, this characterisation was a stereotypical abstraction, often linked to their love of landscape (the mountains), a recreational activity (skiing) and a mythologised history of daily connection to the land. Life in an economically developed early twenty-first-century, wealthy European nation left less direct connection to the broader ecological community in which they were embedded. Without these connections, climate change was easy to hold at arm's length.

The political positionality of Karuk people in a settler society created through colonisation leads to a different cultural framing of the changing climate. Rather than an anomalous consequence of an otherwise benign and beneficial political economic system, global environmental destruction appears as simply one more dimension of the ongoing destructiveness of a political economic order that has long favoured economic profit over human and environmental well-being. Kyle Powys Whyte writes that indigenous people are living in their 'ancestors' dystopia now', having already experienced the unravelling of the human-animal-plant-environment entanglements that were considered essential to tribal ways of life. Whyte writes, 'The vast majority of our history precedes the campaigns that have established states such as the U.S. and Canada. Our conservation and restoration efforts are motivated by how we put dystopia in perspective as just a brief, yet highly disruptive, historical moment for us – at least so far' (2017:3).

Many indigenous peoples retain value systems that are organised around ethics of responsibility to the natural world and to the human community. Leaf Hillman, Director of the Karuk Department of Natural Resources explains:

We believe that we were put here in the beginning of time, and we have an obligation, a responsibility, to take care of our relations, because hopefully, they'll take care of us. And it's an obligation that we have, and so just like – we say, well, we have to fish. They say, 'Well, there aren't that many fish this year, so I don't think you should be fishing.' That is a violation of our law. Because it's failure on our part to uphold our end of the responsibility. If we don't fish, we don't catch fish, consume fish . . . then the salmon have no reason to return. They'll die of a broken heart.

This understanding of history in relation to land and ecological communities offers a cultural pathway of appropriate behaviour, represented here as a responsibility to maintain specific connections to other species, conceptualised as relations. Fishing is one of the responsibilities people have to the fish. These dimensions of reciprocity and responsibility to other species are key social and cultural values.

Cultural beliefs structure which behaviours and actions feel acceptable, and which do not. Emotions signal when actions are culturally acceptable, in line with a predominant belief or value, and when they are not. In a cultural context where relationship to the environment is marked by responsibility and reciprocity to familial relations, failure to live up to these cultural expectations may be experienced emotionally, as guilt, or shame. The role of community provider is a crucial aspect of a fisherman's identity and a source of pride.

In contrast to beliefs in deep relationality between humans and the environment, mainstream Western environmental ethics emphasise differences between human and environmental 'others', using nature as a boundary-mark against which human civilisation is measured (Merchant 1980; Plumwood 1993; Cronon 1995). The cultural logic of colonialism and environmental resource management was one of utilitarianism, constrained by liberal belief in individual freedom as the pinnacle of human development. The right to make use of the environment, and people associated with it were justified by the cultural logics of various institutions, including religion, the state, science and the market. These cultural logics continue to shape environmental politics, confining even well-intentioned sustainability and social justice initiatives to cultural criteria that uphold a belief in individual freedom as a dominant belief (Alkon and Mares 2012).

While much focus has been placed on the vulnerability of indigenous communities, the traditional knowledge and deep interpersonal and interspecies network ties that form part of indigenous communities are sources of significant resilience. The depth of critique culturally available to the indigenous community translated into an active political engagement, albeit one that has historically been marginalised by dominant settler-political institutions. Historically, tribes have adapted to numerous large-scale changes, be they environmental or social (Turner and Clifton 2009; Whyte 2013). If tribes are more vulnerable today it is not because of an inherent weakness, but because of the way colonialism and Western land management policies and practices have limited – if not outlawed – the ability of indigenous communities to exercise their resilient lifeways (see Vinyeta *et al.* 2015:19 for more discussion). And despite not being major greenhouse gas emitters, many tribes have embarked upon ambitious projects to reduce their emissions through use of renewable energy and transportation sources.

The economic logic of capitalism has prioritised profit over well-being and individualism over community. From a tribal perspective, climate change vividly reveals the flaws of Western economic and environmental principles and practices. Many proposed climate change solutions protect the status quo by prioritising individual responsibility and right to profit. Even more radical environmental movement goals that critique dominant institutions often retain cultural frames that emphasise the cultural values of capitalism and the neoliberal state, such as individual freedom at the expense of collective well-being. Thus, taken together, the contrast between indigenous and mainstream Western climate activism results from different cultural frames of interpretation. Even when their emotional responses to climate change overlap, the difference in structural location and cultural worldviews shape group-based understandings of alternative cultural narratives, and practices. A community's response to information about climate change is a function of the congruency of new information with existing ideological standpoints, and emotional management strategies.

10.4 Conclusion

By comparing the above cases, we see the ways in which social location sets groups up to interpret new information about climate change in vastly different ways. Initial framing of an emerging social problem must be reconciled with beliefs about the current world-order, socio-environmental ethics, political values and other salient aspects of integrated cultural worldviews. As we have shown, emotions play a significant role in mediating between cultural worldviews, new information or knowledge that is framed by an outside group, and desires for the future. Cultural meaning structures are malleable, but they are also durable. Emotional responses to them may vary by intensity, valence and durability, but they will always be present, as we navigate environmental problems in socially structured contexts. Culture and emotion are simultaneous and co-constitutive. Analytically, we may separate out different levels of culture and interaction to make sense of how they work in context; however in practice emotional responses to cultural attachments may be fleeting, unregistered or contradictory.

When we approach the problem of climate change response from a cultural-emotional, as well as a material-structural perspective, responses of resistance, inaction or action that fail to adequately address the scope of the problem are no longer puzzling inconsistencies in otherwise rational human behaviour. They become the anticipated and logical outgrowth of deeply entrenched cultural worldviews that are being challenged. This accounts for tendencies to reduce current social conditions to categories of perception shaped by a limited understanding of the past.

One implication of socially organised denial of climate change is that privileged individuals are separated from the immediate consequences of climate change. Bolstered by predominant cultural narratives insisting on their innocence and right to excessively consume, citizens of wealthy, industrialised nations occupy a position of privilege in the global hierarchy in which the material consequences of their lifestyles are often invisible to them. As manufacturing of products has increasingly moved to less wealthy majority world nations, economies in places like the United States and Europe have shifted towards service work and professional labour. As the labour and environmental resources that sustain materially privileged citizens come from farther away, it is easy to lose sight of the connection between our subsistence and quality of life and the physicality of work and dependence on intact ecological systems. As we see in the cases above, there are various ways to make sense of these transitions culturally; it is important, however, to question the power structures that support and benefit from the invisibility of material realities. Both culture and emotions can be co-opted into supporting the status quo; or, channelled into challenging it.

Acknowledgements

The authors wish to thank the Karuk Tribe and all research participants for allowing them to observe their responses to climate change and environmental risk.

References

Alexander, J. C., and Smith, P. 1993. The discourse of American civil society: A new proposal for cultural studies. *Theory and Society*, 22(2), 151–207.
Alkon, A. H., and Mares, T. 2012. Food sovereignty in US food movements: Radical visions and neoliberal constraints. *Agriculture and Human Values*, 29(3), 347–359.
Becker, G. S. 1993. Nobel lecture: The economic way of looking at behavior. *Journal of Political Economy*, 101(3), 385–409.
Bellah, R. N. Madsen, R., Sullivan, W. M., Swidler, A., and Tipton, S. M. 1985. *Habits of the Heart: Individualism and Commitment in American Life*. Berkeley, CA: University of California Press.
Benford, R. D., and Snow, D. A. 2000. Framing processes and social movements: An overview and assessment. *Annual Review of Sociology*, 26(1), 611–639.
Blake, J. 1999. Overcoming the 'Value-Action Gap' in environmental policy: Tensions between national policy and local experience. *Local Environment*, 4(3), 257–278.
Bourdieu, P. 1977. *Outline of a Theory of Practice*. Cambridge: Cambridge University Press.

Brown, W. 2003. Neo-liberalism and the end of liberal democracy. *Theory and Event*, 7(1), 1–25.
Capstick, S. Whitmarsh, L., Poortinga, W., Pidgeon, N., and Upham, P. 2015. International trends in public perceptions of climate change over the past quarter century. *WIREs Climate Change*, 6, 35–61.
Carlson, J. 2015. *Citizen-Protectors: The Everyday Politics of Guns in an Age of Decline*. Oxford: Oxford University Press.
Clements, J. M., McCright, A. M., Dietz, T., and Marquart-Pyatt, S. T. 2015. A behavioural measure of environmental decision-making for social surveys. *Environmental Sociology*, 1(1), 27–37.
Collins, R. 1990. Stratification, emotional energy, and the transient emotions. In: Kemper, T. D. (ed.) *Research Agendas in the Sociology of Emotions*. Albany, NY: State University of New York Press, pp. 27–57.
Cronon, W. 1995. The trouble with wilderness, or getting back to the wrong nature. In W. Cronon (ed.) *Uncommon Ground: Rethinking the Human Place in Nature*. New York: W. W. Norton and Company, pp. 69–90.
Dietz, T., Fitzgerald, A., and Shwom, R. 2005. Environmental values. *Annual Review of Environment and Resources*, 30(1), 335–372.
DiMaggio, P. 1997. Culture and cognition. *Annual Review of Sociology*, 23(1), 263–287.
Dunbar-Ortiz, R. 2018. *Loaded: A Disarming History of the Second Amendment*. San Francisco, CA: City Lights Publisher.
Dunlap, R. E. and Van Liere, K. 1984. Commitment to the dominant social paradigm and concern for environmental quality. *Social Science Quarterly*, 65(4), 1013.
Eliasoph, N. 1998. *Avoiding Politics: How Americans Produce Apathy in Everyday Life*. Cambridge: Cambridge University Press.
Eliasoph, N. 1997. 'Close to Home': The work of avoiding politics. *Theory and Society*, 26 (5), 605–647.
Ford, A. 2014. *The Emotional Landscape of Risk: Self-sufficiency Movements and the Environment*. Unpublished Master's thesis. Eugene, OR: University of Oregon.
Giddens, A. 1991. *Modernity and Self-Identity: Self and Society in the Late Modern Age*. Cambridge: Polity Press.
Giddens, A., 1984. *The Constitution of Society*. Cambridge: Polity Press.
Goldenberg, S. 2013. Secret funding helped build vast network of climate denial think tanks. *The Guardian*, February 14.
Goodwin, J., Jasper, J. M., and Polletta, F. 2002. Introduction: Why emotions matter. In: Goodwin, J., Jasper, J. M., and Polletta, F. (eds.), *Passionate Politics: Emotion and Social Movements*. Chicago: University of Chicago Press, pp. 1–24.
Gould, D. B. 2009. *Moving Politics: Emotion and ACT UP's Fight against AIDS*. Chicago: University of Chicago Press.
Hartter, J., Hamilton, L. C., Boag, A. E., Stevens, F. R., Ducey, M. J., Christoffersen, N. D., and Palace, M. W. 2018. Does it matter if people think climate change is human caused? *Climate Services*, 10, 53–62.
Harvey, D. 2005. *A Brief History of Neoliberalism*. Oxford: Oxford University Press.
Hobsbawm, E., and Ranger, T. 1983. *The Invention of Tradition*. Cambridge: Cambridge University Press.
Hochschild, A. R. 1979. Emotion work, feeling rules, and social structure. *American Journal of Sociology*, 85(3), 551–575.
Hochschild, A. R. 2012. *The Managed Heart: Commercialization of Human Feeling*. Berkeley, CA: University of California Press.
Illouz, E. 2007. *Cold Intimacies: The Making of Emotional Capitalism*. Cambridge: Polity Press.

Jacques, P. 2006. The rearguard of modernity: Environmental skepticism as a struggle of citizenship. *Global Environmental Politics*, 6(1), 76–101.

Jacques, P. J., Dunlap, R. E., and Freeman, M. 2008. The organisation of denial: Conservative think tanks and environmental skepticism. *Environmental Politics*, 17(3), 349–385.

Jasper, J. M. 1997. *The Art of Moral Protest: Culture, Biography, and Creativity in Social Movements*. Chicago: University of Chicago Press.

Kellstedt, P. M., Zahran, S., and Vedlitz, A. 2008. Personal efficacy, the information environment, and attitudes toward global warming and climate change in the United States. *Risk Analysis*, 28(1), 113–126.

Kemper, T. D. 1990. Research agendas in the sociology of emotion. In: Kemper, T. D. (ed.) *Research Agendas in the Sociology of Emotions*. Albany, NY: State University of New York Press, pp. 207–237.

Kunda, Z. 1990. The case for motivated reasoning. *Psychological Bulletin*, 108(3), 480–498.

Lake, F. K. 2013. Historical and cultural fires, tribal management and research issue in Northern California: Trails, fires and tribulations. *Occasion: Interdisciplinary Studies in the Humanities*, 5(22), 5.

Lake, F. K. Tripp, W., and Reed, R. 2010. The Karuk tribe, planetary stewardship, and world renewal on the middle Klamath river, California. *Bulletin of the Ecological Society of America*, 91, 147–149.

Lazarus, R. J. 2009. Super wicked problems and climate change: Restraining the present to liberate the future. *Cornell Law Review*, 94, 1153–1232.

Leiserowitz, A., Maibach, E., Roser-Renouf, C., and Smith, N. 2016. *Climate Change in the American Mind*. New Haven: Yale Program on Climate Change Communication.

Leiserowitz, A., Maibach, E., and Roser-Renouf, C. 2009. Global Warming's Six Americas: An Audience Segmentation Analysis. In: AGU Fall Meeting Abstracts.

Lifton, R. 1982. *Indefensible Weapons: The Political and Psychological Case against Nuclear Weapons*. New York: Basic Books.

Mccright, A. M., and Dunlap, R. E. 2013. Defeating Kyoto: The conservative movement's impact on U.S. climate. *Social Problems*, 50(3), 348–373.

Merchant, C. 1980. *The Death of Nature: Women, Ecology, and the Scientific Revolution*. New York: HarperCollins.

Norgaard, K. M. 2014. Karuk Traditional Ecological Knowledge and the Need for Knowledge Sovereignty, Available at: http://pages.uoregon.edu/norgaard/pdf/Karuk-TEK-and-the-Need-for-Knowledge-Sovereignty-Norgaard-2014.pdf.

Norgaard, K. M. 2011. *Living in Denial: Climate Change, Emotions, and Everyday Life*. Cambridge: MIT Press.

Norgaard, K. M. 2006. 'People want to protect themselves a little bit': Emotions, denial and social movement nonparticipation. *Sociological Inquiry*, 76(3), 372–396.

Norton-Smith, K., Lynn, K., Chief, K., Cozzetto, K., Donatuto, J., Redsteer, M. H., and Whyte, K. P. 2016. *Climate Change and Indigenous Peoples: A Synthesis of Current Impacts and Experiences*. Portland: US Department of Agriculture, Forest Service, Pacific Northwest Research Station.

O'Brien, J. 2015. Individualism as a discursive strategy of action: Autonomy, agency, and reflexivity among religious Americans. *Sociological Theory*, 33(2), 173–199.

O'Brien, K. 2012. Global environmental change III: Closing the gap between knowledge and action. *Progress in Human Geography*, 37(4), 587–596.

Opotow, S., and Weiss, L. 2000. New ways of thinking about environmentalism: Denial and the process of moral exclusion in environmental conflict. *Journal of Social Issues*, 56(3), 475–490.

Patterson, O. 2014. Making sense of culture. *Annual Review of Sociology*, 40(1), 1–30.
Petty, R., and Wegener, D. 1999. The Elaboration likelihood model; Current status and controversies. In: Chaiken, S., and Trope, Y. (eds.) *Dual-process Theories in Social Psychology*. New York: Guilford Press, pp. 41–72.
Plumwood, V. 1993. *Feminism and the Mastery of Nature*, London: Routledge.
Reed, P., and Rothenberg, D. 1993. *Wisdom in the Open Air: The Norwegian Roots of Deep Ecology*. Minneapolis, MN: University of Minnesota Press.
Robnett, B. 2004. Emotional resonance, social location, and strategic framing. *Sociological Focus*, 37(3), 195–212.
Rosa, E. A., and Dietz, T. 1998. Climate change and society: Speculation, construction and scientific investigation. *International Sociology*, 13(4), 421–455.
Rosenberg, M. 1991. Self-processes and emotional experiences. In: Howard, J. and Callero, P. (eds.) *The Self-Society Dynamic: Cognition, Emotion and Action*. Cambridge: Cambridge University Press, pp. 123–142.
Roser-Renouf, C., Maibach, E., Leiserowitz, A., and Rosenthal, S. 2016a. Global Warming's Six Americas and the Election 2016, Available at: http://climatecommunication.yale.edu/publications/six-americas-2016-election/.
Schlosberg, D., and Coles, R. 2015. The new environmentalismof everyday life: Sustainability, material flows and movements. *Contemporary Political Theory*, 15(2), 160–181.
Schwarz, N., and Clore, G. L. 1983. Mood, misattribution, and judgments of well-being: Informative and directive functions of affective states. *Journal of Personality and Social Psychology*, 45(3), 513–523.
Sewell, W. H. J. 1999. The concept(s) of culture. In: Bonnell, V., and Hunt, V. (eds.) *Beyond The Cultural Turn*. Berkeley, CA: University of California Press, pp. 35–61.
Shott, S. 1979. Emotion and social life: A symbolic interactionist analysis. *American Journal of Sociology*, 84(6), 1317–1334.
Shove, E. 2010a. Beyond the ABC: Climate change policy and theories of social change. *Environment and Planning A*, 42(6), 1273–1285.
Shove, E. 2010b. Social theory and climate change. *Theory, Culture and Society*, 27, 2–3.
Smith, P. 2017. Narrating global warming. In: J. C. Alexander, Jacobs, R. N. and Smith, P. (eds.), *The Oxford Handbook of Cultural Sociology*. Oxford: Oxford University Press.
Smith, P. and Howe, N. C. 2015. *Climate Change as Social Drama: Global Warming in the Public Sphere*. Cambridge: Cambridge University Press.
Snow, D. A., and Benford, R. D. 1992. Master frames and cycles of protest. In: Morris, A. D., and McClurg Mueller, C. (eds.) *Frontiers in Social Movement Theory*. New Haven: Yale University Press, pp. 133–156.
Stern, N. 2008. The economics of climate change. *American Economic Review*, 98(2), 1–37.
Swidler, A. 1986. Culture in action: Symbols and strategies. *American Sociological Review*, 51(2), 273–286.
Thoits, P. A. 1989. The sociology of emotions. *Annual Review of Sociology*, 15, 317–342.
Turner, J. H., and Stets, J. E. 2005. *The Sociology of Emotions*. Cambridge: Cambridge University Press.
Turner, N. J., and Clifton, H. 2009. 'It's so different today': Climate change and indigenous lifeways in British Columbia, Canada. *Global Environmental Change*, 19(2), 180–190.
Urry, J. 2011. *Climate Change and Society*. Cambridge: Polity Press.
Vinyeta, K., Whyte, K. P., and Lynn, K. 2015. Climate Change Through an Intersectional Lens: Gendered Vulnerability and Resilience in Indigenous Communities in the United States. Gen. Tech. Rep. PNW-GTR-923. Portland, OR: U.S. Department of Agriculture, Forest Service, Pacific Northwest Research Station. 72 p.

Whyte, K. P. 2013. Justice forward: Tribes, climate adaptation and responsibility. *Climatic Change*, 120(3), 517–530.

Whyte, K. P. 2017. Our ancestors' dystopia now: Indigenous conservation and the anthropocene. In: Heise, U., Christensen, J., and Niemann, M. (eds.) *Routledge Companion to the Environmental Humanities*. Abingdon: Routledge.

Zerubavel, E. 1997. *Social Mindscapes: An Invitation to Cognitive Sociology*. Boston: Harvard University Press.

11

Effective Responses to Climate Change
Some Wisdom from the Buddhist Worldview

PETER DANIELS

11.1 Introduction

The global-scale environmental issues that now confront humanity are largely of our own making and are consistent with the view that the Anthropocene era prevails. They highlight a situation that has not been well-recognised to date – the extremely complex, highly interconnected, and rather finite nature of the world that nurtures human beings. Reductionist analysis, and the simplistic and narrow application of mechanical causal laws, no longer seem relevant for meeting climate change and other major environmental threats to the welfare, and perhaps the very survival, of humankind.

One of the most relevant worldviews, one that establishes a logic and path for appropriate action in a highly interconnected world, is that of Buddhism. In common with Hinduism and many other philoso-religions, the universe is perceived as an 'Indra's net' – a net of 'jewels' in which every jewel is reflected in all other jewels (Thiele 2011). The significance of the powerful interrelationships between all phenomena underlies their role as the lynchpin or *raison d'etre* for effective thought or action in such worldviews and calls upon systems-based ways of experiencing and viewing reality. Hence, there is a natural complementarity and mutual beneficial exchange between Buddhism and the growing focus on interconnectedness and complex systems in modern science and related policy (Bhaskar 2010; Ledford 2015). This union becomes even more cogent under the threat of major ecospheric disruption from anthropogenic disturbances, such as that of climate change.

Although Buddhism is often perceived as a way of life or a philosophy rather than a religion in the traditional sense, it can still be dismissed as unworthy of the respect and attention that is accorded to secular sciences rooted in objectivity and empirical epistemologies when it comes to the analysis of, and responses to, major problems confronting humanity. All non-positivist scientific worldviews face this critique. Buddhism does encourage a substantive empirical element and scepticism in its outlook but does not escape relegation to the confutable realm of metaphysics.

Nonetheless, as with all belief systems, perception begets reality and there is a growing recognition (evinced in the publication of this volume and many related works) that 'super-empirical' understandings of the nature of universal relations are the heart of cultures and societies. This is because they configure human motivations, goals, actions, and behaviour that underlie environmental outcomes and must therefore be considered to be a primary object of adaptive change.

The relevance of a focus on underlying religious and other beliefs is also enhanced by the fact that metaphysical-based worldviews are still very prevalent across the global population. Estimates of affiliation with major religions range from 66 to 84 per cent globally (Veldman 2012, 2014). A broad-based, eclectic and often 'green' spiritualism has also emerged to influence the behaviour and choices of its adherents – especially in the higher income nations (Taylor 2010). This new age spiritualism is often informed by wisdoms such as that of Buddhism.

The translation from beliefs and values to action or behaviour will be imperfect, and has been a major source of debate and research in the social sciences for many decades (Bardi and Schwartz 2003). However, underlying individual and community beliefs about the actions that will reduce suffering and ultimately bring about better well-being are bound to have an influence on decisions and behaviour, if rationality prevails. While they may be based on incomplete or erroneous information about outcomes, or entail incorrect 'theories of happiness' (if happiness or well-being enhancement is the goal), beliefs should still play an important role in behavioural decisions and represent 'action shaping worldviews' affecting anthropogenic climate change (Jenkins and Chapple 2011:443).

All of the religious and related worldviews, such as Buddhism, inevitably have their own cosmologies and belief systems that represent, perhaps over the very long-term, theories of 'happiness' (or well-being or suffering reduction) that will motivate and guide people's daily and life aims and moral conducts. Despite assumed objectivity, most economic systems are also driven by an implicit theory of happiness. This is the materialistic utilitarianism of neoclassical economics with its profound belief in the welfare-enhancing impact of increased consumption of goods and services.

Of course, there is no singular view or interpretation of religious and other collective belief systems. In addition to the dissonance between belief and behaviour, historical divisions, and individual person and local contextual variations ensure this (Haluz-Delay 2014). In this chapter, the focus is on the climate change–relevant worldview, or *Weltanschauung*, of the general tradition of Buddhism rather than on its social, organisational, or institutional dimensions that might link to climate change.

In taking this philosophical position, this chapter does not propose that the diverse range of mythologies and rituals of Buddhism in practice is of general use for addressing climate change. Even an underlying premise that climate change is the result of *kamma* from excessive greed and insatiable desire (facilitated by free market consumerism) (Mori 2002) does not need to be accepted. However, the essence of Buddhism certainly has much to say that is relevant to climate change and how to respond to it. Adopting the market economic parlance, Buddhism is rich in its views on consumption, livelihood, and production activity in general. These are major human activities covering much human effort and time, our consequent impact upon the world, and its effects back upon us. In this way, Buddhism's underlying metaphysical base should translate directly into daily life and the choices that we make.

The rest of this chapter is structured as follows. Section 11.2 outlines the essential aspects of the Buddhist diagnosis of anthropogenic climate change – from the profound interconnectedness that forms Buddhism's cosmological view, to the nature of the human condition and the misguided beliefs that tend to thwart attempts at improving well-being. Then, building on these insights, the third section of the chapter outlines the logic for appropriate action that flows from this worldview. This sets the style and guidelines for decisions, and the particular social, economic, and physical changes that will be required to address climate change.

In turn, the fourth section presents some of the key practical strategies that could help operationalise the wisdom and logic proffered by Buddhism. The section covers a range of strategies, from changing knowledge, expectations and accuracy of human motives for well-being, to developing new technological and economic means to minimising intervention and disturbance to the world, to promoting mechanisms for enhancing people and nature connections. A Buddhist-inspired version of 'practical wisdom' is proposed as a useful way to deal with the extreme interconnectedness and complexity that confronts the evaluation of major social and economic strategies and actions. The conclusion includes an overall summary as well as a review of the limitations that potentially confront such changes.

While this chapter focuses upon understanding and addressing climate change from the theoretical and somewhat prescriptive worldview of Buddhism, empirical dimensions are included where possible. The need for empirical work to support discourse on the religion–environment interface has been stressed in previous studies (Veldman *et al.* 2014). Buddhism itself promotes a form of scientific scepticism, open-mindedness, and empirical verification in the acceptance of its wisdom.

11.2 The Buddhist View of Climate Change

The Buddhist worldview helps provide an understanding of the 'deep causal roots' of anthropogenic climate change. According to the philoso-religion, personal inner suffering and environmental damage are related and stem from the same source. Under Buddhism's life philosophy, the basic problem is the dysfunction associated with craving for material and personal gain, which leads to consumption and large-scale biophysical disturbance such as climate change and other major environmental, economic, and social problems. To elaborate, the key idea is that our search for enhanced well-being is largely a matter of striving to reduce a rather pervasive condition of suffering, or at least 'discontentment' or 'unsatisfactoriness', that is associated with existence (*dukkha*). In the quest to do this, people seek pleasures from social relations, material accumulation, and other aspects of the external world (Reat 1994). However, in Buddhism, achieving such pleasures does not realise better well-being. In fact, the expectation that such attachments and gains will make us 'happy' is considered the cause of our inner suffering. In a fossil fuel and carbon-based high-throughput world economy, burgeoning levels of consumption bring further costs to our well-being because of the accompanying disruption that people wreak upon the habitats that sustain them. Climate change is often viewed as the most cogent and powerful case of negative 'externalities' associated with the hegemony of the consumer market economy (Greenstone 2014).

In more Western-based scientific parlance, the problem stems from an incorrect 'theory of happiness' – one which persists despite failure to bring lasting well-being and now threatens to cause immense levels of future suffering. However, market economics do not subscribe to this view. Even the sustainability sciences tend to describe the situation in terms of externalities rather than a fundamental problem with underlying welfare theories. In ecological economics, for example, the focus for addressing major environmental problems has often targeted population (P) and affluence (A) or economic growth. This is per the classic IPAT equation, which states that environmental impact (I) equals population (P), multiplied by affluence or output per person (A), and multiplied by the environmental demands per unit of output or 'technology' (T) (Ehrlich and Ehrlich 1970). These perspectives hold little confidence in material and energy-saving technological change (the 'T' in IPAT) to bring about sustainability given the preference for achieving 'strong sustainability'. This is founded on the notion that achieving sustainability must be based on the reality that there are few substitutes for 'critical' natural capital (Daly and Farley 2004; Ayres 2007).

However, in spite of these challenges, there is much hope placed on the potentially important role of transformations in material and energy-saving technologies, and the nature of production and consumption, across the range of social science–based sustainability and well-being studies. There is also increasing interest in the link between affluence or economic growth and well-being (Daniels 2010a, 2010b). Buddhism, therefore, has many offerings to inform this transformation towards addressing climate change.

11.2.1 Profound Interconnectedness and Kamma-Vipaka

At the core of the Buddhist belief system are the notions of interconnectedness or 'dependent origination'. This foundational premise is not unique to Buddhism and can be clearly identified in much earlier Vedic thought (Reat 1994). It tends to be a defining theme across many worldviews rooted in the ancient Eastern religious traditions. Although there are alternative interpretations across the various strands of Buddhism, the Indra's net of existence remains as a shared immutable basis and provides the rationale for the doctrine of the law of karma or *kamma-vipaka*.

Feldman (1998:1) describes how *paticcasamuppada* (the Law of Dependent Origination) 'is a vision of life or an understanding in which we see the way everything is interconnected – that there is nothing separate, nothing standing alone. Everything affects everything else. We are part of this system. We are part of this process of dependent origination'. Similarly, Wagner (2007:333) notes that the 'Buddhist approach is based on a concept of universal interconnectedness, mutual conditioning and a radical interdependence of all phenomena, and in this respect quite close to modern system theory. In the classical scriptures, reality is compared to a sacred net of many mutually interwoven strings at countless levels'.

The idea of dependent origination, therefore, provides a strong connection between Buddhism and contemporary ecological worldviews, and in the acceptance of the value and means of achieving sustainability. Buddhist cosmology is centred upon the fundamentally interconnected nature of the three spheres or realms of human existence:

(1) the individual realm (covering existence, thought, and action);
(2) the collective interrelations or institutions that form society; and
(3) the rest of the natural world (Yamamoto 2003).

In this universal ecology, the 'ripples' from events or state changes in one realm directly spill over into the others – spatially, temporally, and transcendentally – and then bounce back upon the originator's own welfare. In accordance with a contemporary Zeitgeist that appears to embrace holism, this outlook fits well with the new wave of environmental sciences (for example, ecological economics, ecology, contemporary social ecology, and natural health studies). It reflects the recent, growing interest in the primacy of nature and the 'embedding' of humans, and their artefacts and built environments, within the rest of the natural world (Bookchin 1993; Daly and Farley 2004).

In Buddhism, the law of dependent origination explains how all outcomes, results or effects (*vipaka*) of speech, action or body arise from multiple causes or actions with intent (*kamma*). In turn, these causes arise from other *vipaka*, and this is the basis of the law of *kamma-vipaka*. The law identifies how ignorant action with 'unskilful' or bad intent will lead to adverse results across the three realms of existence (from individual, to society, to nature, and back on the self). 'Skilfulness' is gauged by the extent to which craving, greed, delusions, or aversion are embodied in the underlying motive and intent of the original action (Attwood 2003). One's mental orientation is critical here – deliberate, wilful action is the source of (un)skilfulness.

11.2.2 The Four Noble Truths

The Four Noble Truths are a central pillar of the Buddhist worldview and form a primary part of the Buddhist explanation of the roots of the major adversities confronting humanity, including climate change. They reveal much about the driving forces relating to aspirations for overall affluence and the nature of consumption. The first two Noble Truths, outlined below, provide much of the necessary foundation.

The First Noble Truth is that conscious experience tends to be dominated by *dukkha* – a difficult translation but most commonly equated to 'suffering', though perhaps better described as 'disquiet' or 'unease'. Although it has many positive experiences to offer, life is thought to be generally imperfect and infused with dissatisfaction and discontent. This condition is sometimes described as 'pervasive dissatisfaction' (Epstein 2005) and a pertinent observation from the Buddhist worldview would be that dissatisfaction, despite the apparent 'success' of consumer economies over the past 200 years, remains largely unabated.

The Second Noble Truth reveals the source of this persistent dissatisfaction or disappointment. It proposes that discontent is a result of persistent attachment to external, worldly phenomena in the belief that they will bring sustained and consummate satisfaction or happiness (French 2003). These objects of our desire include not just material goods or assets and the services that they provide but also people and other animate beings, as well as ideas, social and economic roles, art, success, and status (Webster 2005). Unfortunately, the satisfaction derived from external objects typically tends to disappoint with respect to expectations, or diminish with familiarity and saturation. This is because all worldly phenomena are impermanent and eventually change into a different form or state where they no longer comprise the source of benefit originally expected. Hence, their loss is inevitable, and dissatisfaction and disappointment inevitably ensue.

The Third and Fourth Noble Truths describe how there is a way out of the cycle of pervasive discontent. This is the 'Eightfold Path', which details the required changes in understanding mental processes, patterns and thoughts, and actions and behaviour that are required for progress towards reduced suffering (Sangharakshita 2007). The eight aspects have a natural flow from wisdom (right understanding and right aspiration) to moral commitment (right speech, action, and livelihood) to mental regulation (right effort, mindfulness, and concentration) but they are presented as mutually reinforcing goals rather than a necessary linear sequence.

A central theme throughout the Eightfold Path is the principle of moderation or 'the Middle Way'. The effective path to eventual release from suffering (*nirvana*) is seen to lie between the extremes of hedonistic self-indulgence and sensual pleasure, and excessive self-mortification or asceticism (Gunasekara 1982). Accordingly, the Eightfold Path contains a host of guidelines that directly shape the motives and nature of the human interface with the external world, and has great relevance for sustainable economic activity and management.

The principal features of the Eightfold Path lead to the need for thought and behaviour that are fully mindful of their effect on others and all facets of the 'three realms'. The Path's essence is characterised by moderation of desire and economic transformation, minimum intervention, and choices and behaviour that are non-harmful. This theme is the foundation for the rest of the chapter.

11.3 A Buddhist-Inspired Logic for Action

To summarise, in the Buddhist view of the universe and the way things work, the key concepts of relevance to acting on climate change are:

(1) pervasive, profound interconnectedness and extensive unintended consequences (or 'externalities' in the modern sustainability sciences),
(2) the law of karma, which suggests that disruptive action with selfish intent will push back on the instigators of the action, and
(3) the need for a Middle Way and the Eightfold Path to attain relief from suffering and lasting well-being.

Building on these concepts, this section describes two major conclusions that inform Buddhism's guiding logic as a useful means of addressing climate change and other major sustainability and well-being issues. The conclusions are: (1) 'practical wisdom', which is an effective concept and approach for individual and collective decision-making in a highly complex world; and (2) the criterion or maxim of non-harm and non-violence, embracing minimum disturbance or intervention.

These two concepts are necessary due to the fact that despite a human tendency towards reductionism, it is impossible to 'do just one thing', as there are no singular causes or effects. One major result of the extreme level of interconnectedness is extreme complexity. The primary result of this is difficulty in assessing the full well-being, sustainability, and virtue implications of any behaviour or intervention (Dalai Lama 2001). While it is important to try to understand, as much as possible, the sources and cause–effect relations associated with major problems, it is simply not possible to accurately forecast the full flow-on effects of a course of action. Contemporary natural and social sciences can try their best, and the endeavour is worthwhile up to a point. A common way of analysing the full range of impacts resulting from decision-making is the analysis of 'externalities' and the execution of detailed social benefit–cost analyses that reveal the complete, long-term and rich diversity of effects, and overall impact ('net benefit'), of major physical, technological and policy interventions (Mouter *et al.* 2015). However, while these procedures can help understand the general nature, range, and potential magnitude and distribution of impacts, complexity will ensure that it is next to impossible to assess outcomes with any degree of precision. The aim of the rest of this section is to explore the ideas of practical wisdom and non-harm in greater detail.

11.3.1 The Practical Wisdom of Buddhism

Buddhism provides an alternative and potentially much more efficient means of assessment and decision-making. Its approach builds upon, but also transcends, scientific cause–effect and subsequent benefit–cost analyses of climate change response options. Its approach is a unique form of 'practical wisdom' in the sense of Aristotle's 'phronesis'. Phronesis is often interpreted as prudence, intellectual virtue and 'the ability to determine and undertake the best action in a specific situation to serve the common good' (Nonaka and Toyama 2007:378; Flyvbjerg 2001).

Buddhism's practical wisdom basis for its decision-making and strategic problem-solving approach is worthy of some explanation. Although it can be a rather abstruse concept, phronesis is a type of wisdom or intelligence to guide action. To evaluate options, a modern view would see it as embracing the importance of scientific, evidence-based knowledge to understand cause–effect relations that can be applied with some degree of universal relevance. However, the value of applying this knowledge will vary depending upon the context, taking into account variables such as time, place, culture, and situational specifics.

Assessing the impact of contextual variation upon the value of action choices requires scientific understanding – deductive analysis and objective empirical endeavour – but must also draw upon experience, intuition, and context empathy. Scientific knowledge will help evaluate decisions but, alone, cannot ensure that the deliberation will lead to 'good' (or even intended) outcomes. To make decisions that serve the common good, there needs to be a balance of instrumental and value rationality where choices are contingent and appropriate to specific and dynamic situations, and guided by values and ethics (Flyvbjerg 2001; Nonaka and Toyama 2007). This is the essence of practical wisdom for action – it has general principles or wisdom but varies in context application or practice.

This practical wisdom approach is advantageous as it takes into account the inevitable and powerful interplay between effectiveness and ethics when making decisions. As such, it provides a neat or efficient path for Buddhism, ecological economics, and any of the sustainability or other world perspectives that recognise the complexity confronting choices from the personal and local to global domains. Practical wisdom involves (a) the skill or ability to combine sufficient knowledge (episteme) regarding relations in the physical universe, as well as (b) context-specific conditions and effects, together with (c) some principles for ethical guidance.

For Buddhism, the extended nature of cause–effect relations, from the physical to the moral realm, is a form of episteme. It extracts a practical wisdom maxim based on this understanding – but one that must also take context into account. Hence, practical wisdom provides the ability 'to identify salient features of complex and particular situations' (Roca 2008:610) and evaluate what is good, in totality, for oneself and society. This informs both judgement about a general course of action and more detailed strategic planning. The six skills it draws upon include experiential/empirical learning, scientific knowledge, knowledge of the specific relevant context, intuition, metaphysical understanding of the general nature of cause–effect relations across the physical and moral realms, and the nature of and influences upon social well-being. Buddhism has much to offer the last two sources and espouses the benefits of most of the others.

While underlying beliefs may guide the latter three wisdom sources, the Buddhist-inspired practical wisdom-based mode of evaluation of decision-making on climate change would need extensive research on, and accounting for, the full range of option impacts across time and space. Anticipating and effectively predicting the consequences and effects of developing situations or trends, and actions and responses, on interconnected well-being is critical. This is no easy task and one that highlights the need for practical wisdom 'shortcuts'. The extended web of biophysical flow-on effects can be informed by empirical research, and by externality and other societal 'metabolism' methods. However, according to the

Buddhist perspective, well-being is very much about knowing the full spill-over effects of actions spanning the physical and the moral realms. Biophysical environmental effects must be complemented by efforts at knowing long-term and extensive social, economic, and non-human life effects – with a refinement of *kamma-vipaka* analysis and the estimation of extended eco-footprints to 'karmic footprints' (Daniels 2015). This would comprise Buddhism's practical wisdom perspective for assessing the virtue and 'goodness' of societal interventions.

11.3.2 Non-harm and Minimum Intervention

The non-harm principle is based on an understanding that our actions send ripples through Indra's net of existence, producing numerous physical and moral effects and side effects. It follows that the correct path is to inform ourselves about the likely nature of these effects, and pre-empt and respond accordingly. Although Buddhism would suggest that there is a need to know what these effects are, the complexity associated with interconnectedness suggests that it is not possible to know all the consequences, yet alone control them. Practical wisdom within the Buddhist framework helps provide a viable primary maxim here. In this setting, the law of *kamma-vipaka* and the Four Noble Truths lead to a fundamental guide for action – that of non-harm or non-violence, and minimum intervention or disruption of the surrounding physical and social world. This principle is shared with Hindu, Jain, and Gandhian practices of non-violence (Jenkins and Chapple 2011).

Altruism, compassion, and self-interest all align with a context of profound interconnectedness and the law of *kamma-vipaka*. In such a universe, 'enlightened self-interest' prevails, as harm to others will harm the agent that caused that harm in the first place (Dalai Lama 2001:61). True and rational self-interest will recognise this and heed empathetic connections, meaning that compassion for others and non-harm are the only ways to achieve enhanced well-being or relief from suffering. Buddhism helps explain the objection by ecological economists to the assumption in traditional economics that generosity, non-maximum consumption, and uncompensated care are irrational (Heinberg 2011). Instead, enlightened self-interest leads naturally to compassion, loving kindness, and non-harm, because disruption and violence to the world is violence to oneself. This compassion extends logically to all people and nature. The Buddhist worldview therefore gives a logic to caring and compassion about longer-term repercussions to people and nature.

The essence of profound interconnectedness and karma operating as powerful forces in determining the outcomes of our actions potentially has intergenerational influence. For people without descendants, deep-seated humanitarian care, or belief in re-birth, it may be more difficult to consider cross-generational impacts and there may be less motivation to support substantive change for sustainability. However, appreciation of interconnectedness and even short-term *kamma-vipaka*-styled effects may well be enough to help foster a sense of perpetual connection to a universe and an understanding of well-being linked to non-harm – one that transcends and debunks individual short-term gains from harmful and disruptive actions.

Having said this, there is one quite simple response that flows from the integration of the Buddhist worldview with practical wisdom – do less. In the driving edge of a globalising consumer economy, this proposal is almost heresy. However, as Buddhist ideas concerning well-being, less disruption to the world from unmitigated craving, and material accumulation remind us, renouncing these desires will not lead to suffering. Rather, renouncement should lead to the opposite outcome, as extolled in the logic of the Four Noble Truths. Like drinking saltwater to quench one's thirst, the pursuit to fulfil selfish craving will only lead to greater desire to appease the disappointment and dissatisfaction that results. Conversely, as discussed, that welfare can be achieved through the maximisation of consumption is the fundamental assumption of neoclassical economics and the burgeoning global market economy.

The empirical evidence on the link between subjective well-being (SWB) or life satisfaction and economic growth is unclear (for example, see Deaton 2008). Despite this complexity and ambivalence, two general conclusions are drawn here about the nature of the link between SWB and income from economic growth. Firstly, the relationship appears inconsistent (Helliwell, 2003). This is because income growth is likely to produce many social costs and the evidence in support of its net long-term benefits is not unequivocal. Increasing income tends to bring diminishing returns at best (see Frey and Stutzer 2002; Gowdy 2003; Steinberger and Roberts 2010).

In addition, despite considerable progress in dematerialisation and eco-efficiency in some sectors, global and regional levels of environmental pressure are on the rise. This may well lead to a range of severe, adverse state changes and impacts, such as those produced by climate change, upon economic well-being and related life satisfaction. The result is a 'double whammy' that echoes the cosmological predictions of the Buddhist worldview. This double whammy is comprised of:

(1) An outcome where the craving for worldly gains that has driven economic expansion and massive natural resource depletion has had dubious well-being gains beyond meeting basic needs.
(2) Serious, if not potentially catastrophic, well-being losses from over-exploitation of the biophysical resources for growing consumption.

11.4 Operationalising Buddhist-Inspired Strategies for Addressing Climate Change

To address the causes of anthropogenic climate change, the task at hand appears simple – drastically reduce greenhouse gas emissions (GHGs). A suite of potentially effective policies and strategies have been devised towards this end. There are regulation and subsidy schemes, but using market incentives to encourage technological efficiency and substitution away from fossil carbon have generally been favoured as the 'least-cost' solutions (Swift 1998). Examples include carbon taxes and emission trading schemes. These strategies extend neoclassical economic market logic to include externality effects, which are considered to be market failures, and aim to correct prices to reflect true social costs and benefits. The case for them, based as it is on logic and potential effectiveness, is strong, and even the radical perspective of ecological economics recognises the benefits of this approach, as at least part of the solution.

However, in practice, any climate change mitigation strategy is faced with considerable challenges when dependent upon financial, regulatory, or legal mechanisms. This is partly due to climate change uncertainty and scepticism, but is also linked to the massive tragedy of the commons and free-rider problems associated with open access resources such as the atmosphere. When isolated self-interest rules, there is no primary motive to manage common property. The negative implications of this perspective, for effective climate change policy, are readily observed at international levels. There is also substantial resistance to increased energy prices, even within nations.

Without a clear and motivating appreciation of interconnectedness and intelligent self-interest, market and regulatory approaches lack the deep underlying basis required to achieve the behavioural changes required to effectively address climate change, and are unlikely to be effective. If there is broad acceptance of the practical wisdom and *kamma-vipaka* understanding from Buddhism, there will be a significant 'force' countering the environmental degradation associated with the tragedy of the commons.

So, according to a Buddhist perspective, what are the main social and economic strategies that could produce the new conditions required to address climate change? The underlying nature of Buddhism's worldview and belief system, and related guides for goals, wants, and choices, would configure the design and implementation of appropriate strategies. As discussed above, the third and fourth Noble Truths and the Eightfold Path create a decision-making logic and criteria for practical action.

In this respect, a good place to start, and to systematically review the potential contribution of Buddhist wisdom, is the classic environmental analysis concept of the IPAT equation. This describes the major, interactive sources of environmental pressure, including GHG emissions that lead to climate change. IPAT presents a structure for examining the observable sources or immediate social and economic drivers of the problem. Understanding the main drivers behind environmental and other major global challenges is increasingly recognised as an ideal way to systematically tackle climate change and other 'mega problems'.

Arguably, the preferred end states of consumer market economies (CSEs) and Buddhism are the same – to enhance well-being. As discussed, the CSEs still tend to adhere to traditional utilitarian welfare theories focused upon maximum goods and service consumption, while the Buddhist theory of welfare is based on reducing the craving, attachment, and well-being expectations related to impermanent worldly phenomena. In this sense, the corollary of Buddhism might seem anti-economic or at least anti-market. This is probably correct to some extent, especially in terms of questioning desire and attachment to external satisfaction sources. However, the Buddhist perspective can also be seen to have many positive views in terms of a healthy, vibrant, and effective economy, one that would largely remove the sources of anthropogenic climate change.

We begin with A (affluence). The notion of affluence, or total consumption levels overall, has little meaning in terms of environmental (or social) pressure or harm. It is typically measured as GDP per capita – a monetary value that has historically been correlated with material and energy use. This is commonly overlooked in emotive anti-growth movements (Kallis *et al.* 2012).

Although Buddhism may see craving and attachment associated with A as misguided, it does not reject the important role of economic activity as part of relief from suffering. For example, the path to liberation from suffering, the Middle Way (Gunasekara 1982), is a balanced approach in which basic needs and wants that genuinely enhance welfare can, and should be, satisfied for all people. A significant proportion of the world's population does not have food or water security. Low impact economic growth to meet these basic needs is entirely consistent with the Buddhist perspective. Basic needs should cover food, clothing, warmth, shelter, and most ecosystem services, as well as psychological security linked to social and community-based needs.

Affluence, or total output per capita, provides limited understanding of the well-being-based or *kamma-vipaka* outcomes of human actions. According to the latter, it is more important to analyse and assess both:

(1) the *nature or type of output*. This covers both production and consumption by function and form. Output can be comprised by, for example, feedlot meat production or organic vegetables, heli-skiing or surfing, and car or public

transport use. This involves detailing A into categories that reflect the mix of consumption, and nature of lifestyles and time use.

(2) the environmental impact associated with a unit of each category of A. This can vary for the same type of output depending on the technology (T) used.

'Technology' (T) is perhaps not the best descriptor of this latter aspect. The attribute is more accurately described as the 'impact-intensity' of a unit of each type of good or service (sometimes labelled 'N') so that environmental pressure per person equals $\Sigma(N_i * A_i * T_i)$ for the full range of consumption or time use. While this might seem a little technical, it provides an effective way to see how Buddhism can positively affect the underlying drivers of climate change.

Each unit of output can be quantified physically or in monetary terms. The impact measured has been targeted mainly at environmental demands in the past. However, in a broader, *kamma-vipaka* (k-v) assessment process, the impact would include the full range of social, economic and environmental flow-on effects of that unit of output. These effects are dynamic and evolve with technological change, policy and consumer demands, and will vary within specific types of output. For example, a car may be powered by different types of primary energy sources with very different environmental, economic, and social impacts.

The nature and impact of human economic activities and lifestyles (reflected in N and T) are closely linked and, together with the level or quantity of each component, mutually determine material and energy flows or the socioeconomic 'metabolism' of society. This is mainly a biophysical concept and is likely to be correlated with disturbance and disruption (or perhaps seen as 'violence') to the existing state of nature. However, the harm to *social* well-being and broader life systems must also be considered closely under the Buddhist worldview.

Another important aspect to note regarding Buddhist contributions to addressing climate change is that substantial well-being is drawn from untransformed and 'managed' nature, and from many forms of reflective, artistic, health-based, and recreational activities that have very low demands upon the ecosphere. The connections between increased material and energy use and vital economic activity may have applied in the past but are no longer necessary now. Minimised intervention and likely reduced harm can result from changes in both N and T, regardless of trends in A.

In recent years, a new wave of related conceptual and analytical 'green economy' approaches have been targeting the environmental impact of output (T) and, to a lesser extent, the nature of output or consumption (N). Such approaches include concepts, strategies, tools, and practices such as eco-efficiency, dematerialisation, and ecological modernisation, and all focus on saving material, energy, and waste over the full life cycle of economic output. This brings joint economic advantage and environmental sustainability benefits. Arguably, the focus is on decoupling environmental impacts (T) for given output levels and the need to ecologically restructure the relationship between production and consumption. However, the importance and recognition of these dimensions for achieving sustainability seems to be growing (OECD 2001).

Buddhism is certainly consistent with a focus on reducing T (i.e. the biophysical disruption per unit of output). However, it can also be seen to go deeper, with profound implications for all three major non-population sources of environmental pressure or harm, which is the level, type, and per unit impact of economic activity and time use.

To summarise, a Buddhist worldview should influence societies to (i) moderate consumption in recognition of its limits as a source of well-being; (ii) reduce levels of socioeconomic metabolism via technological change and shifts towards less harmful forms of economic activity and time use; (iii) develop and encourage policies and norms that minimise harm and disruption to the natural world; and (iv) promote full decision-making-based analysis of environmental, economic, and social well-being effects. Of course, all these facets are related.

11.4.1 Buddhism for Sustainable Production and Consumption

Appropriate climate change action aligned with the Buddhist worldview includes several major dimensions or target criteria. A good starting point for identifying these targets is the conventional economic realms of production and consumption and the quest to convert these into sustainable forms. Although these two spheres of economic action may not encompass all forms of human activity, they are helpful as a base to build more effective interdisciplinary and trans-disciplinary approaches. Given profound interconnectedness in the universe, it is not surprising that the nature of the targets described overlap considerably.

The Buddhism-compatible transformation criteria include:

A. Sustainable production. Akin to the Right Livelihood aspect of the Noble Eightfold Path, sustainable production is about engaging in work or livelihood activities that are not harmful. Under this approach, individuals and societies consider the disturbance and suffering or other adverse consequences of making their living. Of course, this assumes that people have surpassed a minimum living standard and that they are not constrained by structural forces.

B. Sustainable consumption and *kamma-vipaka* footprints. A similar responsibility would be attributed to people as 'consumers'. This is a rather derogatory term, almost a misnomer, in the way that it represents people as they seek well-being from both transformed and untransformed goods and their services (including ecosystem services). The aim of sustainable consumption to address climate change is to minimise the carbon 'footprint' of the goods and services used in meeting wants and needs. Buddhism guides us to make consumption choices to minimise our *kamma-vipaka* 'footprint' based upon the full flow-on effects of our consumption, lifestyle, and time use choices and their environmental, social, and economic sustainability impacts through the lens of the law of karma.

C. Positive effects of trade. Trade is very important when it comes to analysing and changing the sources of GHG emissions but is a difficult topic to resolve in terms of sustainability and disruption. Self-sufficiency is a common theme associated with sustainability and minimising material and energy demands. Schumacher (1973) thought that self-sufficiency and local community would help reduce the chances of conflict, competition, and violence, and also reduce transport costs. However, interdependence can breed harmony, innovation and ideas, and efficiencies that can lead to lower environmental pressures without loss of well-being.

D. Well-being without transformation. The notion that what is available for consumption (i.e. getting well-being) is restricted to production or economic output is viewed as incorrect according to the Buddhist worldview. In contrast, many major sources of enhanced well-being involve inner peace and non-action, or passive engagement with and appreciation of the natural world. Indeed, the positive effects of eco-efficiency gains (the 'T' in IPAT) need to be accompanied by changes in the nature of consumption (N) towards less harmful goods, services and activities. This entails engendering a feeling of sufficiency, rather than just efficiency, with regards to external objects, as well as the mindful enjoyment of what exists without the need for human intervention and transformation (Alexander 2014).

E. Align with the Earth's natural cycles. Another strategic dimension where Buddhism and sustainability align is the goal of minimising disruption on the external world by 'fitting in' with nature and its natural cycles. Again, the emphasis becomes one of reduced throughput and blending with natural cycle flows, which tend to be inherently renewable. Fossil carbon economies and other non-renewable resources are typically antithetical in this respect. Notions of bioregionalism, permaculture, and circular economies align with the practical wisdom of minimum intervention. Achieving these criteria will require substantial innovation and changes in social relations that are built upon an understanding and implementation of a revised theory of happiness – one that highlights that well-being is collective in nature and is likely to be adversely affected, in time, by consumption and activity that is high impact. This will help reconfigure the belief-behaviour chain of reasoning and remove dysfunctional structural constraints upon addressing climate change.

In the task of moving towards these target criteria, there are at least three main sets of proactive strategies that might help. There are many barriers to such change, but the path to an effective and deep response to climate change seems logical and feasible – even without any explicit adherence to the Buddhist perspective.

11.4.1.1 First Set of Strategies

The first major set of strategies involves a societal focus upon collecting and disseminating information about the nature of universal cause–effect relations across the physical and moral realms. This would promote greater understanding of how profound interconnectedness, in the physical and moral realms, affects human well-being. As discussed above, Buddhist cosmology helps produce a moral code and set of directives that prescribe action to help mitigate climate change sources.

Closely linked to this strategic emphasis is the need for responses that involve the extensive development and implementation of systems and methods for assessing and measuring the externalities, flow-on effects or 'kamma-vipaka impacts' of our choices. These include biophysical as well as social, economic, and lifeworld consequences. Assessing the intricate web of effects of our actions may be a very difficult task. However, concerted efforts and accumulated experience and data in this field, combined with the Buddhist practical wisdom of non-harm and minimum disruption as decision-making guides, will help create a more effective process in the longer term.

Overall, these strategies are about cultivating a universal responsibility or consciousness that enables people to understand the implications of their actions or choices in production and consumption on the welfare of others and ourselves. Key to this effort is empirical, scientific research. This includes research into the nature of well-being as well as extended externality analyses covering material and energy flow mapping techniques augmented by social benefit–cost analyses. These methods for measuring metabolism and *kamma-vipaka* footprints provide a basis for additional tools such as economic valuation techniques that assess the total economic value (TEV) of natural resources. They complement, rather than compete with, 'religious valuations' in designing and implementing effective climate change policy (Veldman *et al.* 2012:271). This is a core area for reaping the benefits of an interdisciplinary research approach that blends economics, ethics, religious studies, and physical sciences. These ideas already enjoy vital levels of activity independent of any explicit recognition of the Buddhist worldview. Their close alignment with Buddhism affirms the relevance and potential wisdom of this ancient tradition.

11.4.1.2 Second Set of Strategies

Once there is a substantive body of evidence for the nature of these linkages, communities could institute education programmes about the flow-on ripple effects of our actions on the biophysical and social world, and then back onto our codetermined individual and collective well-being. Being educated and cognizant about the likely well-being consequences of the goals, motives, choices, actions, and behaviour and means pursued is essential knowledge to be gained in our schooling. There is a strong case to have this content as a foundational part of schooling – the life skill base for broader community change to enhance the conventional emphasis on instrumental, technical knowledge, and learning.

11.4.1.3 Third Set of Strategies

A third set of strategies to move societies and their economies towards more sustainable outcomes is to apply the range of market 'internalisation' techniques, such as carbon taxes and emissions-trading schemes. Under the Buddhist worldview, internalisation should be designed to account for more complete assessment of the impacts on humans and other lifeforms. The changing nature of trade-offs given the new prices should direct people towards less harmful activity. While this may have beneficial effects, the change in understanding and knowledge resulting from the first two strategies would likely provide a deeper, more fundamental, and ultimately effective, basis for addressing climate change.

From the Buddhist perspective, the transformation in individual thinking, and broader economic and social systems, that is required to address climate change is the same as that required to attain the general reduction in 'suffering' of existence and, hence, the path to better human well-being. The aim is to create socioeconomic systems that are neither physically nor emotionally destructive, replacing competitive self-interest with 'intelligent' self-interest and providing the compassion and care necessary to promote interconnected well-being (Batchelor and Brown 1992).

11.5 Conclusion

Buddhism offers practical and effective ideas and means to help address climate change. Two of its most important contributions are closely linked. These are (1) its explanation of the sources of human suffering and, hence, well-being, and (2) the centrality attached to interconnectedness across the universal physical and moral realms. Together, these elements provide an analysis of the deep-rooted causes of climate change in terms of human desire and craving, and consequent disruption and disturbance to the natural environment. Climate change becomes a *kamma-*

vipaka outcome of a misguided but fundamental and very powerful belief – that people can gain lasting well-being via the selfish pursuit of pleasure based on material accumulation, greed, competition, power, and status, *without* concern for the highly interdependent nature of social and physical reality. Moreover, in explaining the deep underlying causes of the problem, Buddhism also provides a logic for efficacious action.

A real strength to be gleaned from the Buddhist worldview is in the guidelines that it provides for changing the productive and consumptive choices that drive the environmental pressures behind climate change. Its operational response rests upon moderated and mindful consumption, and draws upon scientific research on the biophysical and well-being impacts of economic, social, and lifestyle choices. The fundamental idea is that selfish external attachment leads to the disturbance of nature, and this is not in the best interests of people and societies. Understanding, awareness, and consideration of the full biophysical and socio-economic effects of actions are the basis for change. The next important step is to analyse these effects upon human well-being.

This provides a framework to assess the relationships between what humans do, and how they do it, and whether it will produce a positive result. This can be a very complex task and the chapter has described how Buddhism's practical wisdom can help resolve such challenges by providing some simple decision-making maxims. Hence, awareness and knowledge, and a sense of universal responsibility, provide the basis for mindfulness and behavioural change to create communities that have a minimal *kamma-vipaka* 'footprint' on nature and other people (and, in return, will bring about less harm upon themselves).

It would be idealistic to presume that the world might 'turn Buddhist' and engender the massive sociocultural, economic and technological transformations required to save us from the ravages of climate change. Embracing a single value system is neither realistic nor necessary. Most traditions will hold positive and negative aspects regarding sustainability (Jenkins and Chapple 2011). The Buddhist community (which is, in itself, a highly diffuse entity) would be incapable of galvanising the changes required even within its own. Rather, the value in a Buddhist discourse is to identify, analyse and deploy appropriate practical wisdom for sustainability – when the concept includes enhanced well-being in an internal, socio-psychological and/or spiritual sense, as well as in an external, environmental sense. This wisdom can be expressed in social and community strategies, personal choices, business and management ethics, and broader policies including everything from education, to social policy, to the nature of markets and the financial system, to infrastructure and technology support.

Moreover, it should be borne in mind that the wisdom Buddhism offers is not just for Buddhists. Many current trends in attitudes and lifestyles already suggest that there is neither a need to push the worldview, nor to fully subscribe to Buddhism as a religious tradition, to affect change. Kamble and Vaidhya (2012) note some of the influential intellectuals, such as Einstein and Bertrand Russell, who drew upon, and found useful, the wisdom offered by this ancient worldview for holistic scientific problem-solving. There are also many scientific and popular works that discuss the alignment between Buddhism and other closely related Eastern philosophies, and modern scientific theory and research (e.g. see Capra 2010; Lopez 2009).

One major, positive trend is the growing awareness of the profound interconnectedness between humans, and humans and nature, as well as the need to include and analyse these links in evaluating any action. This is evinced in the strength of an array of scientific and para-scientific perspectives that focus upon complex systems, inter- or trans-disciplinarity, and holism – from the halls of pure physics to applied sustainability sciences and beyond. There is also a widespread recognition of the universal cause–effect relations akin to the law of karma, and a shift towards 'new age' post-materialism and changes in preferences favouring environmental quality and other non-material aspects of life as incomes rise. These are now key features in popular culture of Western and other high-income nations, and it seems plausible that they will also emerge with material satiation in the Asian Tiger Economies.

A parallel trend can be found in the growing interdisciplinary field of the science of happiness (Frey and Stutzer 2002; Diener and Seligman 2004; Layard 2011). In contrast to narrower perspectives, such as that of economics, this new development entails a much more comprehensive and detailed analysis of a broad range of psychological, economic, social, community, geographic, situational, and other factors influencing subjective well-being and life satisfaction. The research field is revealing a more holistic theory of happiness than the 'more consumption equals increased well-being' mentality upholding the powerful worldview espoused in neoclassical economics. It seems that Buddhism can contribute much to this research endeavour by promoting positive goals, practices, lifestyle choices, and attitudes (Flanagan 2011).

Of course, there are many other geopolitical factors and development potentialities and constraints that affect the influence that Buddhist worldviews may have on the social, economic, and technological sources of GHG emissions. However, a fruitful cross-pollination of ancient Eastern wisdoms, and new global technologies and economic and welfare models, may offer a wellspring of positive change. Over the coming century, there are bound to be other global mega-problems with similar human belief roots to those underlying climate change. For this reason, an expansive search for positive solutions rooted in Buddhism is well-justified.

References

Alexander, S. 2014. Life in a 'Degrowth' Economy, and Why You Might Actually Enjoy It. The Conversation, 2014. http://theconversation.com/life-in-a-degrowth-economy-and-why-you-might-actually-enjoy-it-32224.

Attwood, M. 2003. Suicide as a response to suffering. *Western Buddhist Review*, 4.

Ayres, R. U. 2007. On the practical limits to substitution. *Ecological Economics*, 61(1), 115–128.

Bardi, A., and Schwartz, S. H. 2003. Values and behavior: Strength and structure of relations. *Personality and Social Psychology Bulletin*, 29(10), 1207–1220.

Batchelor, M., and Brown, K. 1992. *Buddhism and Ecology*. Delhi: Motilal Banarsidass.

Becker, G. S., Glaeser, E. L., and Murphy, K. M. 1999. Population and economic growth. *The American Economic Review*, 89(2), 145–149.

Bhaskar, R. 2010. *Interdisciplinarity and Climate Change: Transforming Knowledge and Practice for Our Global Future*. New York: Taylor & Francis.

Bookchin, M. 1993. What is Social Ecology? In: Zimmerman, M. (ed.) *Environmental Philosophy: From Animal Rights to Radical Ecology*. Englewood Cliffs: Prentice Hall.

Capra, F. 2010. *The Tao of Physics: An Exploration of the Parallels between Modern Physics and Eastern Mysticism*. Boulder: Shambhala Publications.

Lama, D. 2001. *Ethics for the New Millennium*. New York: Riverhead Books/ Penguin.

Daly, H. E., and Farley, J. 2004. *Ecological Economics: Principles and Applications*. Washington, DC: Island Press.

Daniels, P. 2015. Sustainable energy for sufficiency economies: Methodological insights from Buddhism. In: Vuddhikaro, P., Dhammahaso, P., Cittapalo, P., and Peoples, D. (eds.) *Buddhism and the World Crisis*. Ayutthaya, Thailand: Mahachulalongkornrajavidyalaya University Press. pp. 256–271.

Daniels, P. L. 2010a. Climate change, economics and Buddhism – Part 2: New views and practices for sustainable world economies. *Ecological Economics*, 69(5), 962–972.

Daniels, P. L. 2010b. Climate change, economics and Buddhism – Part I: An integrated environmental analysis framework. *Ecological Economics*, 69(5), 952–961.

Deaton, A. 2008. Income, health, and well-being around the world: Evidence from the Gallup world poll. *The Journal of Economic Perspectives*, 22(2), 53–72.

Diener, E., and Seligman, M. E. P. 2004. Beyond money: Toward an economy of well-being. *Psychological Science in the Public Interest*, 5(1), 1–31.

Ehrlich, P., and Ehrlich, A. 1970. *Population, Resources and Environment*. San Francisco: W.H. Freeman.

Epstein, M. 2005. *Open to Desire: The Truth about What the Buddha Taught*. New York: Gotham Books.

Feldman, C. 1998. Dependent Origination www.seattleinsight.org/Portals/0/Documents/Study%20Materials/Dependent-Origin-Feldman.pdf. Accessed October 14, 2017.

Flanagan, O. 2011. *The Bodhisattva's Brain: Buddhism Naturalized*. Cambridge: MIT Press.

Flyvbjerg, B. 2001. *Making Social Science Matter: Why Social Inquiry Fails and How It Can Succeed Again*. Cambridge: Cambridge University Press.

French, S. 2003. *The Code of the Warrior: Exploring Warrior Values Past and Present*. Lanham: Rowman & Littlefield.

Frey, B. S., and Stutzer, A. 2002. What can economists learn from happiness research? *Journal of Economic Literature*, 40(2), 402–435.

Funtowicz, S., and Ravetz, J. Post-Normal Science. *Online Encyclopedia of Ecological Economics, International Society for Ecological Economics*. At www.Ecoeco. Org/Publica/Encyc. Htm, 2003.

Gowdy, J. 2003. Contemporary welfare economics and ecological economics valuation and policy. *Internet Encyclopedia of Ecological Economics, International Society for Ecological Economics*.

Greenstone, M. 2015. 'Paying the Cost of Climate Change. Planet Policy. Sept 14. 2014'. *Brookings* (blog), 2015. www.brookings.edu/blog/planetpolicy/2014/09/19/paying-the-cost-of-climate-change/. Accessed May 5, 2018.

Gunasekara, V. 1982. *Basic Buddhism: An Outline of the Buddha's Teaching*. Brisbane: Buddhist Society of Queensland.

Haluza-DeLay, R. 2014. Religion and climate change: Varieties in viewpoints and practices. *Wiley Interdisciplinary Reviews: Climate Change*, 5(2), 261–279.

Heinberg, R. 2011. *The End of Growth: Adapting to Our New Economic Reality*. Gabriola Island, Canada: New Society Publishers.

Helliwell, J. F. 2003. How's life? Combining individual and national variables to explain subjective well-being. *Economic Modelling*, 20(2), 331–360.

Jenkins, W., and Chapple, C. K. 2011. Religion and environment. *Annual Review of Environment and Resources*, 36, 441–463.

Kallis, G., Kerschner, C., and Martinez-Alier, J. 2012. The economics of degrowth. *Ecological Economics*, 84, 172–180.

Kamble, R. K., and Vaidhya, P. P. 2012. Buddhism: A scientific philosophy. *SPM-Journal of Academic Research*, 1(1), 6–14.

Layard, R. 2011. *Happiness: Lessons from a New Science*. London: Penguin.

Ledford, H. 2015. How to solve the world's biggest problems. *Nature*, 525, 308–311.

Lopez Jr, Donald S. 2009. *Buddhism and Science: A Guide for the Perplexed*. Chicago: University of Chicago Press.

Mori, S. 2002. Science and Buddhism. *Journal of Oriental Studies*, 12, 65–79.

Mouter, N., Annema, J. A., and van Wee, B. 2015. Managing the insolvable limitations of cost-benefit analysis: Results of an interview based study. *Transportation*, 42(2), 277–302.

Nonaka, I., and Toyama, R. 2007. Strategic management as distributed practical wisdom (Phronesis). *Industrial and Corporate Change*, 16(3), 371–394.

Reat, N. R. 1994. *Buddhism: A History*. Fremont: Jain Publishing Company.

Roca, E. 2008. Introducing practical wisdom in business schools. *Journal of Business Ethics*, 82(3), 607–620.

Sangharakshita. 2007. *The Buddha's Noble Eightfold Path (Buddhist Wisdom for Today)*. Birmingham: Windhorse Publications.

Schumacher, E. F. 1973. *Small Is Beautiful: A Study of Economics as If People Mattered*. London: Abacus.

Steinberger, J. K., and Roberts, J. T. 2010. From constraint to sufficiency: The decoupling of energy and carbon from human needs, 1975–2005. *Ecological Economics*, 70(2), 425–433.

Swift, B. 1998. A low-cost way to control climate change. *Issues in Science and Technology*, 14(3), 75–81.

Taylor, B. R. 2010. *Dark Green Religion: Nature Spirituality and the Planetary Future*. Berkeley: University of California Press.

Thiele, L. P. 2011. *Indra's Net and the Midas Touch: Living Sustainably in a Connected World*. Cambridge: MIT Press.

Veldman, R. G., Szasz, A. and Haluza-DeLay, R. 2014. *How the World's Religions are Responding to Climate Change: Social Scientific Investigations*. New York: Routledge.

Veldman, R. G., Szasz, A. and Haluza-DeLay, R. 2012. Introduction: Climate change and religion - A review of existing research. *Journal for the Study of Religion, Nature and Culture*, 6(3), 255–275.

Wagner, H. G. 2007. Buddhist economics ancient teachings revisited. *International Journal of Green Economics*, 1(3), 326–340.

Webster, D. 2005. *The Philosophy of Desire in the Buddhist Pali Canon*. London: Routledge.

Yamamoto, S. 2003. Mahayana Buddhism and environmental ethics: From the perspective of the consciousness-only doctrine. In: Dockett, K., Dudley-Grant, G., and Bankart, P. (eds.) *Psychology and Buddhism: From Individual to Global Community*. New York: Kluwer Academic, pp. 239–255.

12

Creating a Culture for Transformation

KAREN O'BRIEN, GAIL HOCHACHKA AND IRMELIN GRAM-HANSSEN

12.1 Introduction

The Paris Agreement on climate change and the United Nations' Sustainable Development Goals can easily be interpreted as 'marching orders' for societal transformations. Yet how do we transform at the rate, scale, speed and depth called for by global change research and international agreements? It has been argued that such transformations are unrealistic, unlikely or impossible, especially within a limited timeframe (Friedlingstein *et al.* 2014; Rogelj *et al.* 2016; Raftery *et al.* 2017). What are the alternatives? Some point to adaptation as a solution. However, the IPCC (2014) concluded with high confidence that *even with adaptation*, the risks of temperature increases of anywhere from 3.6°C to over 4°C over the next century, together with sea level rise and more frequent and intense extreme weather events, will lead to severe, widespread and irreversible impacts globally. Some argue for the need to invest in geoengineering research and development, even though some technologies for solar radiation management and carbon dioxide removal have been widely criticised for introducing new risks while failing to address the root causes of climate change (Dalby 2015). These alternatives are consistent with what Milkoreit (2017) refers to as a failure of imagination to create a compelling, shared vision of an alternative future that catalyses social transformations.

Can we imagine a *cultural transformation* that catalyses policies and actions to meet climate and sustainability goals? How would such a transformation come about? Specifically, what factors and experiences might contribute to cultural tipping points for sustainability, i.e. a point in time where sustainability is prioritised, promoted and more importantly, embodied in everyday life? In other words, how can culture be harnessed for transformation, rather than serve as an impediment to change? To foster a culture that actively engages with transformations to sustainability calls for insights on cultural change.

Culture can be defined and interpreted in many ways. Benhabib (2002:3) distinguishes between the Romantic notion of *Kultur*, which represents distinct expressions of shared values, meanings, linguistic signs and symbols, and a more egalitarian understanding that views culture as 'the totality of social systems and practices of signification, representation and symbolism that have an autonomous logic of their own, a logic separated from and not reducible to the intentions of those through whose actions and doings it emerges and is reproduced'. She criticises a reductionist approach, whereby cultures are considered clearly delineable wholes that are congruent with population groups. Rather than essentialising it and associating it with a homogenous group identity, Benhabib (2002:8) views culture as 'constant creations, recreations and negotiations of imaginary boundaries between "we" and the "other(s)"' that are formed through binaries based on evaluative stances, such as what is 'good' and 'bad' or 'pure' and 'impure'.

In this chapter, we adopt Benhabib's (2002) broader definition of culture to explore the idea of cultural transformation. We start by discussing the notion of cultural tipping points and exploring some of the barriers and potentials for rapid cultural change. Drawing on insights from Integral Theory, Self-Determination Theory and Dialogical Action Theory, we examine the relationship between individual and collective change. We then describe an experiment focused on individual behavioural change and consider how it can be used to trigger awareness of the dynamics of cultural change. We present some preliminary results from a transformative learning experience, then link theoretical insights to some reflections from this experiment. We conclude by revisiting the potential for rapid cultural transformations to sustainability.

12.2 Cultural Tipping Points

Many climate change projections for the future are based on integrated assessments that focus on variables such as population growth, gross domestic product, the energy efficiency of technology and carbon intensity of energy (Swart *et al.* 2004; Kriegler *et al.* 2012; Friedlingstein *et al.* 2014). Integrated scenarios, defined as 'coherent and plausible stories, told in words and numbers, about the possible co-evolutionary pathways of combined human and environmental systems' are derived from qualitative analyses that include cultural, institutional and value aspects of sustainability (Swart *et al.* 2004:139).

Projections and scenarios can be considered valuable tools for analysis, but they seldom include the possibility for large-scale cultural transformations. If they do, they do not elaborate on the types of social changes driving such transformations.

This reflects a gap in knowledge on how such changes might occur. The IPCC Shared Socioeconomic Pathways (SSPs), for example, represent narratives of future socio-economic development generated by state-of-the-art models (van Vuuren *et al.* 2017). The SSP1 narrative of 'Taking the Green Road' alludes to gradual yet pervasive shifts in development priorities that are based on increasing environmental awareness and changing attitudes, yet says little about how these changes might come about (O'Neill *et al.* 2017).

Research on social tipping points has explored the factors that may contribute to large-scale social change, including how changes in social norms influence behaviour at larger scales (Nyborg *et al.* 2016). Some have explored how mathematical models from the natural sciences might be used for identifying 'Early Warning Signals' to anticipate non-linear societal responses to environmental changes (Bentley *et al.* 2014). These authors found that it is difficult to apply natural-systems models and conclude that probabilistic insights from research on collective social dynamics may be more promising. Factors such as heterogeneity, connectivity and individual-based thresholds may be used to predict qualitative changes that cascade through social networks or systems (Bentley *et al.* 2014). Others have examined the implications of tipping points for the science–policy interface. For example, Werners *et al.* (2013) explore social and political thresholds as a relevant focus for sustainability, with an emphasis on situations where current management strategies no longer suffice to meet policy objectives and societal preferences. Turning points occur when social and political thresholds are exceeded and priorities shift (Werners *et al.* 2013).

What remains unclear, however, is how quickly cultural shifts can come about as individuals change their beliefs and attitudes. A few studies have explored this, showing that cultural shifts can be initiated by a minority, establishing new norms, rules and standards that then draw in the majority, such that change happens more broadly and rapidly. For example, Xie *et al.* (2011:1) used agent-based modelling to identify how 'a prevailing majority opinion in a population can be rapidly reversed by a small fraction p of randomly committed agents who consistently proselytize the opposing opinion and are immune to influence'. They found that when the number of committed agents exceeded a threshold of about 10 per cent, there was a dramatic decrease in the amount of time it took for the entire population to adopt the opinion of the committed minority. Xie *et al.* (2011) refer to the suffragette movement in the early twentieth century and the civil rights movement in the United States as examples where committed and inflexible minority opinions influenced the majority. More recently, Castilla-Rho *et al.* (2017) used agent-based modelling to explore the contextual factors that drive human cooperation and collective action, including monitoring and enforcement powers, social norms and cultural values. They found that social norms about groundwater conservation shifted abruptly with small changes in cultural values, combined with monitoring and enforcement provisions. Specifically, a small number of rule followers were found to have a strong, positive and non-linear influence on group behaviour (Castilla-Rho *et al.* 2017). Centola *et al.* (2018) show that theoretically expected dynamics do emerge within an empirical system of social coordination. In their experiment they found that 25 per cent of the population represented a critical mass but acknowledged that it was not expected to be a universal value.

Taking this promising research as a starting point, below we look more closely at the role of culture as both a constraint and catalyst for transformative change. Recognising the tensions between these two aspects of culture is important and can be useful in identifying the types of climate and social policies that support transformations to sustainability.

12.3 Culture as a Constraint

Culture is often described as a conservative characteristic of society that supports and maintains the status quo, particularly when interpreted by what Benhabib (2002) distinguishes as *Kultur*. Cultural change is considered to be particularly slow relative to socio-technical change. Indeed, while culture can be considered a property of individuals, it is carried collectively and creates its own momentum, maintained by norms, traditions and institutions that can tolerate and dissipate the impacts of non-conforming or 'radical' views, even if they are held by many individuals (Wilber 2004). Although culture is recognised as constructed and fluid, structures such as norms, traditions, rules, laws, policies, judicial precedents, protocols, institutions and bureaucracies often perpetuate discrimination based on gender, class, ethnicity or physical features, and they legitimate particular attitudes towards nature, resources and non-human life. Culture thus tends to be robust to the wavering trends, currents or fashions of the day.

Cultures can be slow and 'heavy' when it comes to responding to new ideas, as social structures feed on individuals' desires to conform or fit in. This desire may foster the avoidance or editing of thoughts or understandings that threaten one's sense of self-identity and social self (Wilber 2004; Swim *et al.* 2009; Norgaard 2011). Researchers have discovered a neural overlap between physical and social pain, which makes social connection and inclusion particularly important to the surviving and thriving of humans (Lieberman *et al.* 2009). The perceived weight of the structures that hold people within a particular cultural discourse tends to reproduce and privilege the mindsets or consciousness that gave rise to the structures in the first place, even when the current political and social context might otherwise encourage individuals to adopt more progressive (or regressive) perspectives (Wilber 2004). For example, Kahan *et al.* (2012:734) found that '[f]or the ordinary individual, the most consequential effect of his beliefs about climate change is likely to be on his relations with his peers'.

To shift cultures requires engaging with the collective itself, including with meanings embedded in social representations (Wilber 2004). According to Waddock (2015:259), memes or 'cultural artefacts that pass from one person or group to others' are the foundations that shape behaviours and beliefs within a culture, including attitudes towards change initiatives. Social representations, like memes, are shared assumptions and understandings about the world that are relayed into forms, such as in media, built environments and technology, and are used to collectively make sense of the world. They include material expressions of culture such as infrastructure, images, texts, technology and information that capture and reflect a particular worldview (Swim 2009). For example, a landscape with wind and solar farms may represent a culture that values sustainability, whereas oil and gas pipelines signal the values associated with a fossil fuel–based economy; both types of representations can be externally imposed on a culture. Social representations also include the consensual understandings and operating constructs, classifications, thoughts and ideals shared by members of a group that are produced and reproduced through everyday conversation and transactions, and through shared contexts.

Memes and social representations are not random and politically neutral. Drawing on the work of Pierre Bourdieu (2002), Stokke and Selboe (2009) discuss symbolic distinctions and representations, situating them within the context of power relations and the larger political economy. They discuss how political representation grants power to define the 'official version of the social world' (Stokke and Selboe 2009:62). However, they also point to the power associated with 'the agency of popular forces in appropriating and contesting symbolic representation' (Stokke and Selboe 2009:76). Just as culture can be used deliberately to maintain the status quo, it can also serve as a powerful force for social change.

12.4 Harnessing the Power of Culture

To understand how culture can serve as a catalysing force for change, we first explore several theories that provide insights into the relationship between individual and collective change: Integral Theory, Self-Determination Theory and Dialogical Action Theory. Our presentation represents neither an attempt to comprehensively cover what are robust, extensive theories resting upon their own canons of research nor an attempt to limit the understanding of the complex relationship between individuals and groups to this particular selection of theories. Rather, our objective is to highlight some specific insights that we consider relevant to cultural transformations, which will later be explored through an empirical case study based on an experiment with change.

12.4.1 Integral Theory

Integral Theory grew out of the transpersonal psychology and human potential movements to encompass a contemporary East-West philosophy or metatheory (Wilber 2006). Certain works in Ken Wilber's writing focus on transformation dynamics in groups, including in relation to cultural and systemic change, which is what we focus on here. Overall, Integral Theory describes the foundational dynamics of evolutionary systems as they arise in four interrelated quadrants, which correspond to four irreducible perspectives available for generating valid knowledge. These include subjective/psychological, intersubjective/cultural, objective/behavioural and inter-objective/systemic (Wilber 1996, 2000, 2001, 2006a, 2006b, 2007). Central to this theory is the integration of developmental psychology to clarify what is meant by transformation and to better understand the evolution of human consciousness in individuals and groups.

Drawing on research within developmental psychology, Integral Theory describes human consciousness as unfolding through stages of greater complexity across a lifespan (Kegan 1998; Torbert *et al.* 2004; Cook-Greuter 2013). Kegan (1998) considers transformation as a developmental shift in orders of consciousness represented by the ability to take as 'object' something that was formerly 'subject' within one's awareness. In other words, rather than remaining subjectively immersed within a perspective (i.e. being unaware of 'the water one swims in'), it is rendered visible to the conscious mind from an objective point of view. When these subject–object dynamics are studied in psychology, they refer to a central axis upon which psychological growth seems to occur (Wilber 2000).

Individual psychological transformations are important, as societal transformations are often catalysed by individuals who develop new social practices, technologies, or wisdom from a worldview of greater complexity or depth, then communicate and share these insights (verbally and cognitively) with others (Riddell 2013:132). Social consciousness, a term that refers to 'the level of explicit awareness a person has of being part of a larger whole', develops over time to embrace larger circles of care (Schlitz *et al.* 2010:21). Being aware of how one is influenced by others and how one's actions affect others can be an important catalyst for worldview transformations. When a more holistic and inclusive worldview is held by enough individuals within a particular group, it becomes meshed with the wider social milieu and creates or advocates structural and systemic changes to support it. Eventually, this worldview becomes the new norm, or 'centre of gravity', for a society that expresses a wide range of values. Examples of such shifts include the emergence in the twentieth century of a universalistic worldview that afforded rights to all humans regardless of colour, caste or creed (United Nations General Assembly 1948), and more recently the movement to recognise a diversity of identity-based cultures representing a range of sexuality and gender categories.

However, individual transformations in consciousness are neither enough to shift cultures nor are changes in political structures alone sufficient (e.g. the introduction of liberal democracy). The relationship between agency and structure cannot be considered directly causal in either direction, as discussed in Giddens' (1986) structuration theory. Wilber's integral approach explains how the dominant level of consciousness in a society can undermine or oppress those individuals who challenge an existing paradigm, while also pulling younger members of a population up to the culture's 'centre of gravity'. On the one hand, this 'gravitational pull' encourages young people to develop and replicate the basic norms for how people live together socially. On the other hand, this pull can stifle individuals who are critical of the cultural mainstream (Wilber 2000). This relates with the notion of a 'social imaginary', defined as broadly shared ideas that are associated with social norms and produce a 'massive background consensus' against which social reality is patterned and enacted (Habermas 1996:22). Von Helund and Folke (2014:254) applied this to social–ecological resilience and arrived at the idea of a social–ancestral contract that 'serves as a moral attractor that assembles the social–ecological entities of the system'. Such social attractors can be powerful, as they are constantly weighing on members of a society. Yet such attractors are not always consistent with notions of sustainability. Schlitz et al. (2010:20) suggest that 'more constricted, fear-based, threat-oriented, intolerant, or narrow views of the world and a person's place in it' can also be present, arising from a different process as compared to transformations that are more inclusive and prosocial. However, whether progressive or regressive, cultures often change through shifts in the dominant social discourse.

In summarising Wilber's integral theory of social transformation, Riddell (2013:132–133) notes that it highlights 'the power of the techno-economic base in determining the average societal level of consciousness, the importance of enacting new social practices/paradigms, and the need to test and spread new political and institutional forms as consciousness develops'. These are among various other insights from Integral Theory that are important to consider in understanding how to create cultures of transformation, which we summarise here and then further explain in the experiment described below, namely:

(1) personal, cultural, behavioural and systems change all play an important role in cultural transformations, and understanding the dynamics and drivers of change in each of these is important for effectively catalysing social change;

(2) personal transformations orient around a shift in the subject–object perspective, whereas cultural transformations reside in shifts within the wider social discourse; and
(3) harnessing the power of culture as an impetus for transformation will require helping individuals espousing and enacting progressive ideas and practices to shift the discourse within their cultural milieu.

12.4.2 Self-Determination Theory

Self-Determination Theory is an approach to human motivation and personality that investigates people's inherent growth tendencies and innate psychological needs that become the very basis for self-motivation and personal well-being. Though it comes from the discipline of psychology and is used in mental health work, it is relevant to social transformations, particularly when it comes to facilitating constructive social development. Ryan and Deci (2000) describe how the social context influences motivation and personal growth, moving a person towards agency or apathy. What distinguishes the indolent, passive, non-motivated tendencies found in some, from the persistent, proactive, positive tendencies found in others? These researchers distil from inductive empirical research three core needs to be met via the social context for such agency and empowerment to stabilise. These include autonomy, competence and relatedness, which appear to be essential for optimal functioning of the natural propensities for living an active, constructive life.

Autonomy here refers to the feeling of volition or choice that can accompany any act and it points to an internally perceived locus of causality (which is quite distinct from being independent, detached or selfish) (Ryan and Deci 2000:70). Competence refers to the quality of being adequate, sufficient and in possession of the necessary knowledge or capacity. Relatedness – a term that comes from attachment theory – places great importance on the social context in which an action is carried out; although many actions occur individually, the relational base has been found to be central for agency and motivation (Ryan and Deci 2000).

Self-Determination Theory examines how these three needs intersect with intrinsic and extrinsic motivation, self-regulation, personality development and so forth. Most notable in reference to sustainability, this theory emphasises the role that social contexts play in supporting agency and intrinsic motivation for living in an active, responsible manner. Social environments facilitate or forestall, for example, intrinsic motivation by supporting versus thwarting people's innate psychological needs. Ryan and Deci (2000:73–74) describe how 'contexts can yield autonomous regulation only if they are autonomy supportive, thus allowing the person to feel competent, related, and autonomous ... In this sense, support for autonomy allows individuals to actively transform values into their own.' Social–contextual events that lead to feelings of competence about the action being undertaken (e.g. appreciative inquiry and positive feedback) can in turn enhance intrinsic motivation for that action. In contrast, a controlling environment that emphasises an external locus of causality (i.e. less autonomy) often leads to a loss of intrinsic motivation. A controlling teacher, for example, can produce less intrinsically motivated students if their needs for autonomy are not met, just as a controlling social environment can undermine an individual or group's sense of autonomy and agency.

Self-Determination Theory thus describes how intrinsic motivation is more likely to flourish in contexts characterised by a sense of security and relatedness. Insights we can draw upon from this towards harnessing the power of culture for transformation include:

(1) clarity into the ways in which agency, empowerment and social responsibility are underpinned by basic, universal human needs of autonomy, competence and relatedness;
(2) a nuanced understanding of how social contexts can support or thwart these needs; and
(3) a psychologically informed framework for considering the social context carefully in cultural change endeavours.

12.4.3 Dialogical Action Theory

Cultural change is not always positive, and there is a need to be critically aware of how power and politics influence the potential for transformations. In discussing the relationship between power and climate change, Manuel-Navarrete (2010) calls for critical perspectives that recognise the importance of emancipatory approaches that go beyond treating humans as objects. Being treated like an object – or in Foucauldian terms, a subject of control through governmentality – can be oppressive and alienating. Paolo Freire (1970:173) emphasises the importance of people becoming the subjects or authors of their own lives, capable of critical reflection to name the world and transform it, rather than the objects of domination, which 'maintains the oppressed in a position of "adhesion" to a reality which seems all powerful and overwhelming, and then alienates by presenting mysterious forces to explain this power'. The shift in cognitive awareness described in Integral Theory (i.e. that seeing something as 'object' rather than being 'subject' to it can lead to a shift in cognitive awareness) aligns with what Freire describes as naming the world in order to transform it.

Restricting or eliminating critical dialogue can lead to the replication of existing systems of oppression, such that the oppressed, when liberated, merely take on the role of oppressors. This points to the important role that dialogue and narrative play in social change, and how detrimental 'echo chambers' can be when it comes to cultural change (d'Ancona 2017). Freire deconstructs the 'banking model' of education where teachers deposit knowledge into students and lays the foundation for a critical pedagogy based on dialogical action. With this approach, 'men and women develop their power to perceive critically the way they exist in the world with which and in which they find themselves; they come to see the world not as a static reality but as a reality in the process of transformation' (Freire 1970:12). Dialogical Action Theory supports the co-production of knowledge, recognising that the more that people problematise the current situation and deepen their critical awareness of reality, the greater responsibility they take for that reality (Freire 1970).

Dialogic action is consistent with transformative learning, described by Mezirow (2000) as a process through which taken for granted frames of reference become more inclusive. Through active dialogue between two persons, within a group or between a reader and author or viewer and artist, individuals can become critically reflective of established cultural norms or viewpoints and freed from distortions by power and influence (Mezirow 2000). The dialogic action approach is also conducive to what Stirling (2015) refers to as emancipatory transformations, which he suggests can be achieved through a combination of diversity, creativity and democratic struggle. Stirling (2015:56) is critical of the way that 'the roots of environmental change are increasingly located in the "behaviour" of ordinary people, rather than in the powerful vested interests that so actively constrain and condition associated growing individualism, consumerism and materialism'. Stirling (2015:67) sees the need for both knowledge and action to yield 'more distributed *culturings* of radical change'. Insights from dialogical action theory that may inform this 'culturing' include:

(1) the importance of dialogue in awakening critical consciousness for transformative learning and change;
(2) the importance of making people subjects or authors of change, rather than treating them merely as objects to be changed; and
(3) more emancipatory learning pathways involving opportunities for dialogue on 'generative themes' that become the foundation of personal and social change.

12.5 Experimenting with Change

To explore the power of culture in transformation processes, we present some preliminary reflections from an informal study conducted with university students after they had voluntarily participated in a facilitated change experiment. The change experiment, referred to as the cCHALLENGE,[1] involved identifying one small change that could be beneficial to the environment and committing to it for thirty days. In this section, we explain the design of the change experiment and how it relates to the insights on cultural transformation described above, after which we describe and discuss an informal study of an experiment with change.

12.5.1 Design

The cCHALLENGE was designed as a reflexive and experiential process for transformative learning, with an emphasis on the relationships between individual change, collective change and systems change. Based on the heuristic of the Three Spheres of Transformation (O'Brien & Sygna 2013; O'Brien 2018), inquiries and reflection questions are used to develop insights about behavioural changes and practical actions and how they are influenced by larger systems and structures, as well as by individual and shared beliefs, values and worldviews. It originated as a class exercise for graduate students studying the human dimensions of global environmental change at the University of Oslo and has been expanded through a social enterprise to reach other audiences, including secondary schools, municipalities and the general public.[2] It includes a transformative program that highlights key features of the curriculum on transformation, including insights from the three theories discussed above (see Table 12.1). It also includes a digital platform for sharing experiences, insights and reflections on the process of change (cCHALLENGE 2018).

The focus on one self-identified behavioural change as a starting point allows participants to both experience and assess change as an object, rather than as something to which they are subjected. It should be noted that there are many books, pamphlets, carbon calculators and other sources of advice and tools about small personal changes vis-à-vis sustainability. Some of these emphasise the importance of behavioural changes in impacting systems, and some highlight how structural incentives (such as a tax on GHG emissions) can support changing personal habits (Réquillart 2015). However, few of them account for the interactions among the *practical, political* and *personal* spheres of transformation, nor the intriguing and complex interplay between individual change and cultural change.

The cCHALLENGE invites participants to consider the interactions between the various dimensions of change processes and provides a facilitated process for them

---

[1] cCHALLENGE can ambiguously refer to a climate challenge, a change challenge, a conscious challenge, a courageous challenge, a collaborative challenge or a creative challenge.

[2] The social enterprise cCHANGE focuses on the role of collaborative, creative and conscious change in responses to climate and sustainability challenges; the cCHALLENGE is used as a tool for experientially engaging with the 'change' dimension of the climate challenge (cCHANGE n.d.).

Table 12.1 *Weaving insights from three theories of social change into the iterative design of the cCHALLENGE program*

Key insights from theories of cultural change that are important for creating transformations to sustainability	Design components of cCHALLENGE relating with these theoretical insights
Integral Theory	• Personal, cultural, behavioural and systems change all play an important role in cultural transformations, and understanding the dynamics and drivers of change in each of these is important for effectively catalysing social change • Personal transformations orient around a shift in the subject–object perspective, while whereas cultural transformations reside in shifts within the wider social discourse • Harnessing the power of culture as an impetus for transformation will require helping individuals espousing and enacting progressive ideas and practices to shift the discourse within their cultural milieu.	• **Supporting dialogue and interpersonal sharing** to step out of echo chambers and contribute to shifts in social discourse. • **Creating an emergent process to support a subject–object shift** in individuals. • *Valuing the agency* and emancipatory potential of individuals to propose and advance novel ideas from more complex worldviews than those of the status quo. • Acknowledging the upstream cultural current individuals traverse to advance a novel practice or paradigm, by **providing small supportive subcultures** (such as, a shared website for blogs and weekly seminars) in which to experiment with new ways of being and to support the uptake of new social practices and new social norms.
Self-Determination Theory	• Agency, empowerment and social responsibility are underpinned by basic, universal human needs of autonomy, competence and relatedness. • Social contexts can support or thwart these basic, universal human needs (above). • The social contexts should be tended to carefully so to include these psychologically informed considerations of needs when engaging in cultural change endeavours.	• **Creating an 'autonomy supportive' process** to invite participants into their own agency and to meet their basic need for autonomy and competence. • **Ensuring the subculture/context meets the participants' basic need of relatedness.**

Table 12.1 (cont.)

Key insights from theories of cultural change that are important for creating transformations to sustainability	Design components of cCHALLENGE relating with these theoretical insights
Dialogical Action Theory	• Dialogue is important in awakening critical consciousness for transformative learning and change. • It is important to encourage people to be subjects or authors of change, rather than treating them as objects to be changed. • More emancipatory learning pathways entail creating opportunities for dialogue on 'generative themes' that become the foundation of personal and social change.	• **Facilitating a process of critical dialogue** to support greater critical consciousness. • **Identifying generative themes** for participants to explore individually and together.

to become more conscious about the ways that culture and social norms exert pressure on people, and about the ways that they themselves influence others. It encourages participants to examine their experiments with change in relation to larger systems (e.g. consumption, transportation, energy) and structures (e.g. social norms, institutions, rules, regulations, incentives) and to reflect on whether and how these factors facilitate or prevent individual and collective change. Participants are also encouraged to consider the role that their own beliefs and assumptions play in change processes.

12.5.2 Informal Case Study

The informal case study involved eighty-two undergraduate students taking the Environment and Society course in the spring of 2018 at the University of Oslo. Students who participated in the cCHALLENGE identified and committed to one change for a thirty-day period. Although many students chose explicitly environmental themes, such as to reduce consumption or take new modes of transport, they were not restricted to changes that could be beneficial to the environment; some opted instead for personal changes, such as taking up meditation, embodied dance or contemplating the modern self. Over the thirty days, they wrote in a journal and were provided with a digital blog where reflections, inspiration, resources and information could be shared with classmates in a closed group. The blog also helped to track progress and it provided a space for comments from other students. Many of them also attended a two-hour seminar each week, where they reflected together on aspects of the experiment that were easy or hard, problematising the challenges they had taken on, and discussing the systems in which the challenge was occurring. Discussions involved first noticing or recognising the systems that the participants found themselves embedded in, and then surfacing which habits or choices were easily accommodated and which were obstructed by these systems.

The majority of cCHALLENGE participants were committed to the program and carried it out over the thirty days. Many expressed being inspired by it and reported that even if they felt they could not sustain the full challenge they had taken on, they learned something important and useful about how change happens. The majority of participants expressed that they experienced the process of change in a new way, with some recognising their agency and ability to impact others around them. Some reported changes that others in their family or social network took on as a result of conversations about their cCHALLENGE. Many committed to a reduced version of their challenge at the conclusion of the thirty-day experiment, such as eating less meat or driving less, which would still carry an impact in society. The students also shared experiences of frustration and doubt, often in reference to the discomfort associated with challenging cultural norms. Considerations about the role of culture in supporting and hindering engagement with the challenge were a key reflection among students.

Some preliminary results are described below and then discussed within the context of cultural transformations. Our analysis builds on participant observation carried out by two of the authors who served as seminar leaders, as well as a reading of the students' online blogs, analysis of reflection papers that were submitted by students upon completion of the experiment and a survey conducted with a 16 per cent sample of the class ($n = 13$). Students in this course ranged in age from twenty to thirty-five years old and came from different disciplines (mainly social sciences). The class included both Norwegian and international students, primarily from Europe, North America, Latin America and Asia. Although the reflection paper was obligatory, the degree to which students engaged with the cCHALLENGE was voluntary. Participation in the seminars and blogging was also optional, and students self-selected to respond to the survey. As this was an informal, qualitative study, the following discussion is based primarily on our impressions as participant-observers of the experiment, the student reflection papers and the students' responses in classroom discussions and in a survey.

12.6 Results and Discussion

We have organised the results according to four key themes that emerged from the participants' experience, and we present some quotes to illustrate the findings. We also discuss how the themes correspond to the theoretical insights presented above.

12.6.1 To Challenge Oneself Is to Challenge One's Culture, and Vice Versa

Students reported that in challenging themselves, they also challenged their very culture. This was perceived as difficult, especially when confronted with differing norms and values from family and friends. However, many students found value in being able to step outside their own cultural milieu and view it in a new way:

At home with the family, who live in the countryside, I experienced a bigger barrier connected to the challenge and felt like I was breaking with several norms when I talked about the purpose of the challenge and ate my vegetarian food next to their meat dishes.... I had to step outside of my own food culture and look at it from the outside with new eyes ... this gave me more self-awareness. (Student M, vegetarian diet) (Translated from Norwegian)

For this particular student, looking at her and her family's food consumption habits as if she were an outside observer allowed her to view her culture from a different perspective. This corresponds with the subject–object shift that Integral Theory considers central to personal transformation. While on the one hand the student challenged her surrounding culture, on the other hand she too was reciprocally challenged to take a new perspective of herself within her culture.

While challenging cultural norms was perceived as uncomfortable by most, some found it thrilling to discover that a simple behavioural change could have such impact:

Through the 30-day challenge, I learned to stand up for myself. As I was called both naive, idealistic, boring and 'that lady' when I ... gave 11 single potatoes without a bag to the person behind the register. It is a social challenge to stand out, and at the same time show pride, no matter what norm you are taking a stand from. (Student N, no plastic bags)

This student reflected on how standing up for her challenge, even when ridiculed, is linked to standing up for herself as a human being. This relates to Self-Determination Theory, which recognises the needs of autonomy and competence as underpinnings for agency, empowerment and social responsibility. When changes are linked to feelings of identity and agency, it is often easier to maintain behavioural changes and more likely to influence other actions.

12.6.2 Agency Is Contagious, through Both Action and Dialogue

Students reported insights on how small individual actions can create social ripple effects. As one survey respondent said, 'I could see that people were thinking about their food-choices just because they saw what I was eating for lunch or just because they knew about my challenge' (Survey Respondent A, vegan/vegetarian 2018). Most students who made this observation were positively surprised by how their challenge had inspired others:

I told my family about the challenge, and they immediately wanted to participate in their own way. My father decided to meditate every day, my sister wanted to learn new gymnastics tricks, and one of my brothers decided to try getting up earlier in the morning to have more time to wake up before school ... We also started a shared challenge for them all; recycling. I found four different buckets, and placed them in the kitchen. Then I labeled them 'glass', 'food waste', 'paper', 'plastic' and 'other rubbish', and told them how to use them. Two weeks later I came back, and they were recycling their heads off! This proves my previous point about planting an idea, and watching it grow in someone else's head. The influence one can have on other people. (Student O, recycling)

This influence on others often occurred through dialogue. Students encountered that in talking about their new practices, as well as the ideas and values that such actions were rooted in, they opened the possibility space for others to consider changing as well.

Friends who earlier had a negative view on vegans or thought that it only meant eating lettuce, have become positively surprised when I shared my experiences with them. During my cChallenge, several friends have chosen beans instead of beef in their tacos and even dared to taste a vegan burger. I definitely think that the more you talk about it, the bigger it gets. (Student P, vegan diet) (Translated from Norwegian)

A key insight of my cChallenge [was that] I had to get into conversations with other people. We often tend to believe that ... as an individual [one] does not have a chance to change something. This might be true if we do not convince other people of what we are believing in. To convince other people of your own beliefs, one also has to give other people the space to come to their own conclusion; this I believe is the way of making a long-term impact. (Student Q, biking)

This interpersonal sharing is an important part of the cCHALLENGE, allowing participants to contribute to shifting the social discourse. In Integral Theory, changing the dominant mode of discourse of a social group is a key to cultural change. What is talked about – and the way that it is talked about – influences social perceptions regarding what is socially condoned, acceptable and even possible. It is through dialogue that new ideas are absorbed into the mainstream. Ideas that previously may have been regarded as 'outliers' with the current social discourse become normalised.

Consistent with Dialogical Action Theory, the students underscored the importance of fostering critical awareness of social challenges, and the significance of transformative learning:

I have come to realize how crucial the constant awareness of climate change is to create necessary change to reduce global warming. Creating awareness around the issue of climate change can be done through challenges such as the cChallenge, but simply through conversation as well. Throughout the thirty days that have passed, I have noticed how spreading the word to the people around me has affected them to act more environmental friendly. (Student R, walking)

In encountering [my] family [with my challenge], I also became aware of how my actions, to eat vegetarian for 30 days, influenced others. For instance, my climate challenge quickly became a discussion topic without my initiation, both during dinner with my parents and when I visited my brother and his family. (Student M, vegetarian diet)

Some students also stressed the importance of reflecting on one's own autonomy in the face of a seemingly immense task. Breaking down that immensity to one single, tangible behavioural change helped participants to feel that they had gained some insights on the process of change. They also spoke of the importance of relating with others, both to be seen and to be held to account for the challenge they had taken on. As one survey respondent explained:

Creating an environment that allows inspiration and motivation is the key. Group meetings and 'peer pressure' so to say [were helpful]. [It] would be harder to accomplish if one was doing it alone on its own accord. Also having something more tangible like the cChallenge to start the process of change is immensely helpful, a metaphorical kick in the butt. (Survey Respondent B 2018)

Self-Determination Theory emphasises the importance of meeting the basic human needs of autonomy and relatedness in sustaining changes in habits. The seminar leaders facilitated a supportive process within the class structure to help students meet their basic need for autonomy and competence. They did this by inviting participants to explore their own agency and by providing positive feedback. One survey respondent reflected: 'I was very inspired; particularly for [the seminar leaders] that gave me so much encouragement to continue. [I felt] a sense of togetherness in a changing time' (Survey respondent C 2018).

Other students also reflected on this sense of togetherness, and particularly on the value of hearing about the difficulties other students had met and overcome with regards to their challenge. This gave them increased confidence in meeting their own challenges, or even in taking on new ones. One survey respondent described: 'I found it inspiring to hear from other students about their challenges; it made me think about how I can improve my own behaviour towards their topics' (Survey respondent D 2018).

12.6.3 The Role of a Small Supportive Sub-culture in Which to Experiment

Many participants emphasised how important a safe, supportive subculture was to try out their new practice:

The culture outside of [the] classroom made me feel less empowered, obsessive maybe, and kind of crazy, to try and change my habits when the structure and system in society still is the way it is (I was trying to cut out all plastic). The culture in the classroom made me feel empowered and hopeful of the future, because a lot of young people were willing to change and feel the need for protecting the environment. (Survey Respondent E 2018)

Underscoring the importance of group support, one student noted, 'You need help; from the people in your surrounding[s] and by society'. (Student F). Several survey respondents described the inspiration derived from witnessing how others undertook their challenges: '[I realised I] could seek inspiration from reading [sic] others' blogs', (Student H) and '[i]t was very inspiring discussing with classmates; without doing that, cChallenge would be much harder' (Student E). Others reflected on how their ability to step outside their cultural comfort zones was aided by the educational setting where other students and educators provided encouragement and support.

The most important outcome of the challenge was nevertheless how a new topic was introduced in my life. I consider myself as environmentally friendly, with a role in social settings who might [disrupt things] as 'that lady' that has something to say about sustainability. [And yet] leaving [my] comfort zone through the cChallenge and ... taking a stand [to] some aspect of the actual reality we live in, was overwhelming. [I became aware of] how we are set in socially constructed trajectories, and the absence of willingness to [engage] transformation. (Student N, no plastic bags)

During the thirty-day experiment, many students encountered how 'locked-in' systems and worldviews influenced their own individual choices. Survey respondents reflected on, 'how difficult it is for an individual to change a system' (Survey respondent H 2018) and 'how it can be very hard to "move against the tide" of the given norm' (Survey respondent I 2018). Importantly, some students reported that getting other people to join their challenge was a supportive factor in sticking with it. In the quote below, one student reflects on how this is not only about finding other people with similar worldviews or mindsets; it also relates to the (cultural) systems governing certain behaviours:

Change is hard when you're doing it alone. All of us as individuals will have to do major and minor changes in our everyday lives, but perhaps it will be easier if individually making environmentally friendly changes isn't clashing with the bigger systems and structures. (Student S, spending an hour outside every day)

A survey respondent summed this up with the statement: 'Creating a sense of group affiliation that facilitates change is important' (Survey respondent F 2018). Integral Theory supports this finding, recognising how hard it can be to go against the cultural current to advance a novel practice or a new paradigm, and acknowledging the importance of small supportive sub-cultures in which to experiment with new ways of being. The shared website for blogging and the weekly seminars provided this subculture, and the students clearly noted the importance of such a social haven, where they would be sheltered from strange looks, overt challenges or possible ridicule. This allowed them to metaphorically flex their wings like newly fledging birds before flying into the wider culture.

12.6.4 Culture Has a Gravitational Pull

Culture can pull individuals in different directions. It can restrain and discourage the thought-leaders who attempt to promote new ideas, just as it can inspire and challenge those who are resistant to progressive change. Some students reflected that one of the most challenging parts of this experiment was managing other's reactions to their choices, especially when these 'others' were people close to them. Learning how to navigate the subtleties of cultural norms and relations in a skilful, compassionate way is part of the transformation process. The following quote reflects the push and pull of culture and social relations, pointing to the sensitivity with which one has to go about implementing change:

When it comes to [sic] others' reactions to my choice of becoming vegan, this has been the most challenging part ... The most uncomfortable part has been situations when I have turned down good food made with love by the people I care about the most. For instance, my mom made waffles for the whole family, that I didn't eat, and my father came to my room when I was studying and brought me a cup of hot cocoa, which I had to turn down. These situations show that each individual does not live in a vacuum, but that we all make up a part of a collective that we care about and have to relate to ... At the same time, it is exactly my relationship to the rest of the world that inspires me to become vegan. (Student V, vegan diet) (Translated from Norwegian)

New ideas may at first provoke resistance, but they can eventually introduce people to new ways of being and new ways of doing. This quote shows the possibility for meeting others halfway, navigating cultural norms and assumptions within a concrete social setting:

Telling my parents about the challenge gained a lot of questions about 'why?' I explained about the environment and how our diet has an impact on it and they disagreed. It became clear that they saw our world as [a] playground with separate enclosures and Norway [as] a small contributor to pollution. Clearing oneself of responsibility and handing out guilt to others is not typical of my family alone but is a part of how many Norwegians compare and measure instead of changing. They understood the experimental part of the challenge, but probably not the [reason] why. To my surprise they agreed to engage in a vegetarian meal every Monday for the four-week period and it felt like a small victory on behalf of the world. (Student T, vegetarian diet)

In other cases, students found that they downplayed or suspended their challenge out of regard for another person's background or out of respect for their worldview. The following quote speaks to the power of the social domain to inhibit experiments with changes that challenge 'social imaginaries', social-ancestral contracts or moral attractors:

During the cChallenge I caught myself not wanting to challenge my grandparents and instead chose not to tell them about the experiment. This led to me eating different dishes with dairy. I also did this when I had dinner with others, because I didn't want to feel like a burden. With my grandparents, however, there was an extra factor present because I didn't want to end up in a discussion with them, since I knew that they would think the whole experiment was stupid. This is because we have grown up with different traditions and opinions. My grandparents grew up on a farm and beef and dairy was a much more important part of their diet than it is in mine. They have more of a traditional worldview than me. (Student U, vegan diet) (Translated from Norwegian)

Overall, many students experienced a tension between expanding their circle of care to embrace more people and species in response to larger global issues and inadvertently offending or hurting people in their immediate social sphere of family and friends.

12.6.5 Discussion

Drawing on the key insights from the experiment, woven with the insights from the three theories we have raised here, we now summarise our findings on some potential ways that experiential approaches can catalyse change. First, the cCHALLENGE experiment provided an opportunity for participants to feel an enhanced sense of agency and empowerment. This is often accompanied by an awareness of the cultural and institutional systems and structures that support certain behaviours and limit others. Second, the cCHALLENGE introduced a process whereby people could recognise their immense potential as individuals to enact change through the various networks and relations formed with others, as well as their ability to impact the institutions of which they are a part. This relationship between the individual and the collective is forged through dialogue and interpersonal connection. Third, the cCHALLENGE provided a process in which participants became aware of their own assumptions and understandings of change. No matter how small and brief the experiment, participants gained experiential insights on the challenges and possibilities for social change, while at the same time remaining humble to the challenge ahead. Fourth, the collective reflection and sharing of both successes and failures was key for turning the experiment from a novel experience into new and operationalised knowledge about how social change can happen, including how being part of a small subculture and experimenting with new practices can be a helpful conduit for shifting individual behavioural changes into the wider social and cultural context.

Turning to further implications of this experiment for other change efforts, it becomes apparent that an understanding of *how culture actually changes* is often a missing piece in discussions of transformation, particularly in the development of future emissions scenarios. Some approaches to cultural change use the cultural substrate as a means of informing, inspiring and engaging people with climate change and sustainability issues. While this has some important aspects to it, such an approach tends to be top-down, considering people as 'objects' to be transformed, even in participatory approaches that include stakeholders and interest groups. As a result, it can only go so far in actually fostering cultural change (Stirling 2015). An alternative approach to cultural transformations, as described in this chapter, focuses on people as the *subjects of transformation*. This involves seeing people as creative agents who are capable of working collectively to shift systems to achieve shared goals, transforming not only themselves and their carbon footprints but also entrenched economic systems and power relations – not just individually, but collectively through social movements (Brand 2016).

The difference between 'changing people' versus recognising 'people as change-makers' represents two contrasting views of cultural change. It also leads to very different conclusions regarding cultural tipping points. In the former, cultural transformations emanate from an elite group that imposes its visions and solutions for sustainability onto 'the masses', whereas with the latter, every person is seen as a potential agent of change with the capacity to shift systems and cultures (Sharma 2017). Freire (1970) described the former approach as cultural invasion and the latter as cultural synthesis, recognising that all cultural actions serve either domination or liberation and create dialectical relations of permanence and change. Building on this distinction with theoretical and practical examples from a short-term change experiment carried out by university students, this section has discussed some important dynamics on how to reckon with and release the power of culture in transformations to sustainability.

12.7 Conclusion

In this chapter, we have shown how deliberately experimenting with change has the potential to support the emergence of cultural tipping points, particularly if it is carried out in a supportive setting that takes into consideration lessons from cognitive and developmental sciences. To fully understand and evaluate the potential and limits of experiments such as the cCHALLENGE, more in-depth follow-up studies will be necessary. Regardless of whether a change experiment materialises in sustained action over time or not, promoting a deeper understanding of and engagement with the cultural field can help support transformations to sustainability. Situating the individual as a change-maker is an important starting point for generating solutions to the climate change challenge.

There may be implications of this work for climate change engagement efforts, which we recommend as areas of further study. As one example, in the face of climate change, this informal study suggests that what may be more important is to find ways to help people manage new practices and a new worldview within their existing cultural frames. This help could come in the form of dialogue-supporting public spaces or political processes in which citizens can explore how the climate challenge relates to their own lives and those of others. Researchers and practitioners alike have a unique responsibility to not only inform but to facilitate such processes, paying attention to their own assumptions about what transformation involves. This calls for less 'control' and more guidance and support to allow new perspectives and emotions to find a home within a larger, culturally influenced landscape.

Climate change introduces a complex emotional terrain that many are struggling to manage, both individually and culturally (Head 2016). Managing dissonance, grief, fear and uncertainty takes energy that could be transmuted into agency and action that transforms the social discourse. Discussing the deeper cultural shifts needed for systemic transformations to sustainability, Gerst *et al.* (2013:131) remark, 'The social agency for fostering such a systemic shift seems not yet on the world stage; indeed, it is difficult to imagine a Great Transition without the emergence of a vast cultural and political citizens' movement for one.' As we have argued in this paper, activating individual and collective agency can be a powerful lever for social change, and a potent way to generate the cultural tipping points needed to realise transformations to sustainability.

Acknowledgements

We would like to thank the students in the Environment and Society course (Spring 2018) at the University of Oslo for their commitment to the cCHALLENGE and reflections on the change process. We are also grateful to the Research Council of Norway and the University of Oslo for supporting the AdaptationCONNECTS research project.

References

Benhabib, S. 2002. *The Claims of Culture: Equality and Diversity in the Global Era*. Princeton, NJ: Princeton University Press.
Bentley, R. A., Maddison, E. J., Ranner, P. H., Bissell, J., Caiado, C. C. S., Bhatanacharoen, P., Clark, T., Botha, M., Akinbami, F., Hollow, M., Michie, R., Huntley, B., Curtis, S. E., and Garnett, P. 2014. Social tipping points and earth systems dynamics. *Frontiers in Environmental Science*, 2 (35), 1–7.
Bourdieu, P. 2002. *Outline of a Theory of Practice*. Cambridge: Cambridge University Press.
Brand, U. 2016. How to get out of the multiple crisis? Contours of a critical theory of social-ecological transformation. *Environmental Values*, 25 (5), 503–525.
Castilla-Rho, J. C., Rojas, R., Andersen, M. S., Holley, C., and Mariethoz, G. 2017. Social tipping points in global groundwater management. *Nature Human Behaviour*, 1 (9), 640–649.
cCHALLENGE, 2018. *HOME* [online]. cCHALLENGE. Available from: www.cchallenge.no/ [Accessed 7 Aug 2018].
cChange, n.d. *cCHANGE* [online]. Available from: https://cchange.no/about/ [Accessed 10 Jul 2018].
Centola, D., Becker, J., Brackbill, D., and Baronchelli, A. 2018. Experimental evidence for tipping points in social convention. *Science*, 360 (6393), 1116–1119.
Cook-Greuter, S. R. 2013. Nine levels of increasing embrace in ego development: A full-spectrum theory of vertical growth and meaning making, Retrieved from: www.cook-greuter.com/Cook-Greuter%209%20levels%20paper%20new%201.1'14%2097p%5B1%5D.pdf
Cook-Greuter, S. R. 2004. Making the case for a developmental perspective. *Industrial and Commercial Training*, 36 (7), 275–281.
Dalby, S. 2015. Geoengineering: The next era of geopolitics? *Geography Compass*, 9 (4), 190–201.
d'Ancona, M. 2017. *Post-Truth: The New War on Truth and How to Fight Back*. 1st Edition. London: Ebury Press.
Freire, P. 1970. *Pedagogy of the Oppressed*. New York: Herder and Herder.
Friedlingstein, P., Andrew, R. M., Rogelj, J., Peters, G. P., Canadell, J. G., Knutti, R., Luderer, G., Raupach, M. R., Schaeffer, M., van Vuuren, D. P., and Le Quéré, C. 2014. Persistent growth of CO2 emissions and implications for reaching climate targets. *Nature Geoscience*, 7 (10), 709–715.
Gerst, M. D., Raskin, P. D., and Rockström, J. 2013. Contours of a resilient global future. *Sustainability*, 6 (1), 123–135.
Giddens, A. 1986. *The Constitution of Society: Outline of the Theory of Structuration*. Reprint edition. Berkeley, CA: University of California Press.
Habermas, J. 1996. *Between Facts and Norms: Contributions to a Discourse Theory of Law and Democracy*. Cambridge: Polity Press.
Head, L. 2016. *Hope and Grief in the Anthropocene: Re-conceptualising Human–nature Relations*. New York, NY: Routledge.
IPCC, 2014. Climate Change 2014: Synthesis Report of the Fifth Assessment Report of the Intergovernmental Panel on Climate Change. Cambridge, UK and New York, NY.
Kahan, D. M., Peters, E., Wittlin, M., Slovic, P., Ouellette, L. L., Braman, D., and Mandel, G., 2012. The polarizing impact of science literacy and numeracy on perceived climate change risks. *Nature Climate Change*, 2 (10), 732–735.
Kegan, R. 1998. *In Over Our Heads: The Mental Demands of Modern Life*. Boston, MA: Harvard University Press.

Kriegler, E., O'Neill, B. C., Hallegatte, S., Kram, T., Lempert, R. J., Moss, R. H., and Wilbanks, T. 2012. The need for and use of socio-economic scenarios for climate change analysis: A new approach based on shared socio-economic pathways. *Global Environmental Change*, 22 (4), 807–822.

Lieberman, M. D. and Eisenberger, N. I. 2009. Pains and pleasures of social life. *Science*, 323 (5851), 891.

Manuel-Navarrete, D. 2010. Power, realism, and the ideal of human emancipation in a climate of change. *Wiley Interdisciplinary Reviews: Climate Change*, 1 (6), 781–785.

Mezirow, J., 2000. Learning to think like an adult. Core concepts of transformation theory. In J. Mezirow, & Associates (eds.), *Learning as Transformation: Critical Perspectives on a Theory in Progress*, San Francisco, CA: Jossey-Bass, 3–33.

Milkoreit, M. 2017. Imaginary politics: Climate change and making the future. *Elementa Science of the Anthropocene* [online], 5 (62). Available from: www.elementascience.org/article/10.1525/elementa.249/ [Accessed 20 February 2018].

Norgaard, K. M. 2011. *Living in Denial: Climate Change, Emotions, and Everyday Life*. Cambridge, MA: The MIT Press.

Nyborg, K., Anderies, J. M., Dannenberg, A., Lindahl, T., Schill, C., Schlüter, M., Adger, W. N., Arrow, K. J., Barrett, S., Carpenter, S., Chapin III, F. S., Crépin, A. S., Daily, G., Ehrlich, P., Folke, C., Jager, W., Kautsky, N., Levin, S. A., Madsen, O. J., Polasky, S., Scheffer, M., Walker, B., Weber, E. U., Wilen, J., Xepapadeas, A., and de Zeeuw, A. 2016. Social norms as solutions. *Science*, 354 (6308), 42–43.

O'Brien, K. 2018. Is the 1.5°C target possible? Exploring the three spheres of transformation. *Current Opinion in Environmental Sustainability*, 31, 153–160.

O'Brien, K. and Sygna, L. 2013. Responding to climate change: The three spheres of transformation. In: *Proceedings of Transformation in a Changing Climate*. Oslo, Norway: University of Oslo, 16–23.

O'Neill, B. C., Kriegler, E., Ebi, K. L., Kemp-Benedict, E., Riahi, K., Rothman, D. S., van Ruijven, B. J., van Vuuren, D. P., Birkmann, J., Kok, K., Levy, M., and Solecki, W. 2017. The roads ahead: Narratives for shared socioeconomic pathways describing world futures in the 21st century. *Global Environmental Change*, 42, 169–180.

Raftery, A. E., Zimmer, A., Frierson, D. M. W., Startz, R., and Liu, P. 2017. Less than 2°C warming by 2100 unlikely. *Nature Climate Change* [online], advance online publication. Available from: www.nature.com/nclimate/journal/vaop/ncurrent/full/nclimate3352.html?foxtrotcallback=true [Accessed 16 August 2017].

Réquillart, V. 2015. Small changes in diet can make a big difference to greenhouse gas emissions. *The Economist* [online]. Available from: www.economist.com/free-exchange/2015/11/26/small-changes-in-diet-can-make-a-big-difference-to-greenhouse-gas-emissions [Accessed 30 May 2018].

Riddell, D. 2013. Bring on the re/evolution: Integral theory and the challenges of social transformation and sustainability. *Journal of Integral Theory and Practice*, 8 (3/4), 126–145.

Rogelj, J., den Elzen, M., Höhne, N., Fransen, T., Fekete, H., Winkler, H., Schaeffer, R., Sha, F., Riahi, K., and Meinshausen, M. 2016. Paris agreement climate proposals need a boost to keep warming well below 2°C. *Nature*, 534 (7609), 631–639.

Ryan, R. M. and Deci, E. L. 2000. Self-determination theory and the facilitation of intrinsic motivation, social development, and well-being. *American Psychologist*, 55 (1), 68.

Schlitz, M. M., Vieten, C., and Miller, E. M. 2010. Worldview transformation and the development of social consciousness. *Journal of Consciousness Studies*, 17 (7–8), 18–36.

Sharma, M. 2017. *Radical Transformational Leadership: Strategic Action for Change Agents*. Berkeley, CA: North Atlantic Books.

Stirling, A. 2015. Emancipating transformations: From controlling 'the transition' to culturing plural radical progress. In: *The Politics of Green Transformations*. London: Earthscan, 54–67.

Stokke, K. and Selboe, E. 2009. Symbolic representation as political practice. In: Törnquist, O., Webster, N., and Stokke, K. (eds.) *Rethinking Popular Representation*. London: Palgrave Macmillan UK, 20.

Swart, R. J., Raskin, P., and Robinson, J. 2004. The problem of the future: Sustainability science and scenario analysis. *Global Environmental Change*, 14 (2), 137–146.

Swim, J., Clayton, S., Doherty, T., Gifford, R., Howard, G., Reser, J., Stern, P., and Weber, E. 2009. Psychology and global climate change: Addressing a multi-faceted phenomenon and set of challenges. A report by the American Psychological Association's task force on the interface between psychology and global climate change. *American Psychological Association, Washington* [online]. Available from: www.apa.org/science/about/publications/climate-change.pdf [Accessed 16 March 2017].

Torbert, B., Fisher, D., and Rooke, D. 2004. *Action Inquiry: The Secret of Timely and Transforming Leadership*. San Francisco, CA: Berrett-Koehler Publishers.

United Nations General Assembly, 1948. Universal Declaration of Human Rights.

van Vuuren, D. P., Riahi, K., Calvin, K., Dellink, R., Emmerling, J., Fujimori, S., Kc, S., Kriegler, E., and O'Neill, B. 2017. The shared socio-economic pathways: Trajectories for human development and global environmental change. *Global Environmental Change*, 42, 148–152.

von Heland, J. and Folke, C. 2014. A social contract with the ancestors: Culture and ecosystem services in southern Madagascar. *Global Environmental Change*, 24, 251–264.

Waddock, S. 2015. Reflections: Intellectual shamans, sensemaking, and memes in large system change. *Journal of Change Management*, 15 (4), 259–273.

Werners, S. E., Pfenninger, S., van Slobbe, E., Haasnoot, M., Kwakkel, J. H., and Swart, R. J. 2013. Thresholds, tipping and turning points for sustainability under climate change. *Current Opinion in Environmental Sustainability*, 5 (3–4), 334–340.

Wilber, K., 1996. A Brief History of Everything. Boston: Shambhala Publications.

Wilber, K., 2000. Integral Psychology: Consciousness, Spirit, Psychology, Therapy. Boston: Shambhala.

Wilber, K., 2001. A Theory of Everything: An Integral Vision for Business, Politics, Science and Spirituality. Boston: Shambhala Publications.

Wilber, K., 2006a. Introduction to the Integral Approach (and the AQAL Map). [online]. Available from: www.kenwilber.com/Writings/PDF/IntroductiontotheIntegralApproach_GENERAL_2005_NN.pdf [Accessed 30 October 2016].

Wilber, K., 2006b. Excerpt D: The Look of a Feeling–The Importance of Post/Structuralism, Retrieved from www.kenwilber.com/Writings/PDF/excerptD_KOSMOS_2004.pdf.

Wilber, K., 2007. Integral Spirituality: A Startling New Role for Religion in the Modern and Postmodern World. Reprint edition. Boston: Shambhala.

Xie, J., Sreenivasan, S., Korniss, G., Zhang, W., Lim, C., and Szymanski, B. K. 2011. Social consensus through the influence of committed minorities. *Physical Review E*, 84 (1), 011130.

13

Back to the Future? *Satoyama* and Cultures of Transition and Sustainability

JOHN CLAMMER

13.1 Introduction

Climate change is now a fact of contemporary life. It has become widely accepted that not only is the reality of change significant, and probably irreversible, but also that it is largely human-induced practices that are its basic cause. While climatic and geological events are certainly 'natural', there is massive evidence that the intensity and acceleration of climate change are triggered and sustained by our cultural practices and by the economic structures, e.g. non-renewable energy consumption and extractive pollution generation, that fuel these practices. We experience the biological and meteorological facts. It behooves us then to analyse the cultural and civilisational factors that have led us, wittingly or unwittingly, into crisis, and then, on the basis of this analysis, to search for alternative patterns of life that are sustainable and which contribute least to exacerbate the already serious situation. If such models can be found, it may well be that they can furnish us with blueprints for future life in the inevitable post-oil and post-extractive economy that cannot be too far in the future. This analysis is not attempted here – a well-documented literature (summarised in Clammer 2016) – has already done that. Rather, this chapter examines an example of a fairly widespread, and largely successful, attempt in Japan (a highly industrialised and extensively urbanised society) to create a sustainable pattern of agriculture and culture using an empirical case study. By reviving and modernising tried and tested agroecological and social patterns of the past, ones which were subsequently side-lined in Japan's postwar rush to economic growth, a model evolves to work with nature, reestablish patterns of community and revitalise crafts, culinary practices and forms of indigenous architecture. These collectively illustrate that a transition to a low-carbon and non-socially and non-ecologically exploitative form of life is indeed possible.

The notion of transition is crucial here since many who commentate on the inevitable future of a post-oil and post-affluent society anticipate a jarring, and even violent, transition in which interests vested in the old forms of economy, extraction, transport, consumption and livelihoods will struggle fiercely. Attempts to turn social, political, economic life and culture in new directions, ones compatible with the realities that environmental limits will necessarily impose on us, will fuel the struggle. The practice considered here, known in Japanese as *satoyama,* is the subject of exploration in this chapter. From this case, a number of important practical and theoretical conclusions are drawn about the elements desirable, and perhaps necessary, to create 'cultures of transition' that might well be the arks that carry us into a sustainable future. These elements are not simply examples of mitigation and adaptation to climate change; they are examples of societies that are not simply enduring the new future but actually desirable, enjoyable and fulfilling places in which to live. While not originally conceived as a response to climate change, the chapter attempts to show the possibility of cataloguing the elements of what might be called 'sustainable culture' through a specific ethnographically based case study of an actual practice. The key to this chapter is indeed the notion of culture and, while acknowledging that a case drawn from one particular geographical and historical experience may not be universal in all its implications, it demonstrates the significance of cultural practices and orientations, as well as more technical agricultural and economic ones, which might play a role in the conception of such a sustainable culture. It also suggests more general lessons that might be drawn from this and that have wide theoretical and practical applications.

13.2 *Satoyama*: The Concept and the Practice

In a study undertaken by the United Nations University (UNU) in Tokyo, *satoyama* is described in the following terms: '*Satoyama* is a Japanese term for a mosaic of different ecosystem types – secondary forest, farm lands, irrigation ponds and grasslands – along with human settlements, which has been managed to produce bundles of ecosystem services for human well-being. *Satoyama* found largely in rural or peri-urban areas of Japan is a way of life; in other words, a classical illustration of the symbiotic interaction between ecosystems and humans' (Japan *Satoyama* and *Satoumi* Assessment 2010:4). While the term has been extended to include coastal and marine ecosystems (*satoumi*), the focus of this chapter is on land-based examples. The study is based on ethnographic work undertaken by the author, then himself based at UNU, between 1998 and 2015 in a number of regions of northern, central and western Japan. Primarily, interviews and participant observation are utilised while drawing on the collective work of the larger research team from the UNU working on specific themes within the world of *satoyama*, such as the revival of traditional industries.

These ecosystems have been under pressure from a variety of directions, the main ones being Japan's ageing society, and with it the abandonment of many forms of traditional agriculture; rural–urban migration; industrialisation and land-use conversion. However, they also show significant signs of revival from a variety of directions. More younger people are attracted back to rural lifestyles, for example, and frequent food scandals have stimulated interest in the consumption and cultivation of organic produce. Additionally, agriculturally based communes, such as the now nation-wide *Yamagishi* movement, have proved resilient. Networks for the marketing of fresh produce to urban areas, perhaps the oldest and most extensive being the *Seikatsu Club*, have become a feature of the social landscape, especially in suburban areas, as have farmer's markets, such as the monthly Earth Day market held in Yoyogi Park in the very centre of Tokyo. Some areas, particularly in Western Japan, have successfully built on these trends to develop what is in effect a form of domestic eco-tourism – simple holidays in rural areas with fresh air, organic food, *onsen* (the very popular hot springs that dot volcanic Japan) and exercise – the latter taken not only in the form of trekking or hill climbing but also by helping out on the farm where one is staying. With the notion of community-supported agriculture slowly spreading in Japan, more and more people are not only now concerned about the source of their food, but they are also happy to go and help cultivate it.

The concept and practice of *satoyama* has ancient roots that can be traced at least as far back as the early Tokugawa period (1603–1868), and no doubt, in many respects, even longer. As a form of socio-agricultural organisation it was never, of course, perceived as a solution to climate change, although it was very much thought of as a system of conservation and ecological protection. From early in the period of Shogunal rule (1603–1868), forests were protected and logging without specific permission strictly prohibited, irrigation systems, ponds and tanks were communally managed, and an integrated system of recycling routinely practiced, not only of agricultural by-products, but even of architectural features. The traditional rural Japanese house is built of wood, stone (for foundations), straw and paper, all of which can either be incorporated into a new structure or recycled as compost, fuel or bedding for animals. Heating in such houses was, and still is, minimal, the only sources in most cases being either cooking heat from the kitchen, the small fire-pit over which a tea-kettle constantly hangs, and the *kotatsu*. The latter is a depression in the floor which contains a heat source, over which is placed a low, quilt-covered table at which people sit, to eat or to write, with their legs and feet snug in the warm cavity beneath the table and from which very little heat escapes. The result has been an integrated agricultural system with little waste, very low energy usage and largely local consumption. Additionally, rootedness in place is reinforced by community structures, the role of local Buddhist temples and Shinto shrines and kinship networks. The outcome is, unintentionally no doubt, not only a highly integrated agri-social system, and a classic example of localism and bio-regionalism, but also one that is climate-friendly to a high degree. These characteristics have attracted interest from a number of circles, including the Japanese government, which was host to the COP 10 (Conference to the Parties to the Convention on Biological Diversity) and which was eager to showcase local *satoyama* systems as prime examples of Japan's commitment to the Conference's goals.

While *satoyama* systems can be found all over Japan, the examples in this chapter are drawn principally from fieldwork in Ishikawa Prefecture, a largely rural district on the Japan Sea side of the country (facing towards Korea and China). This area has a more extreme climate than Tokyo to the east, or Kyoto Prefecture to the south, and often experiences heavy snowfall more commonly associated with mountainous areas, or Japan's northerly prefectures. The region's principal city, Kanazawa, is of considerable historical and cultural significance; it still contains districts of traditional housing and is famous for its crafts. While clearly local ecological conditions vary along the considerable length of Japan, in Ishikawa the warm summers and harsh winters mirror conditions to both the north and south. It has hilly and mountainous areas and substantial areas of forest and faces the sea; thus it possesses both coastal and inland environments. The term *satoyama* itself, as alluded to earlier, seems to have come into common usage in the mid-1700s and has since expanded its meaning to encompass the kinds of integrated ecological-social systems that are under discussion here. Again, while it is natural that individual farms vary in size, crops, population and other characteristics throughout the prefecture, an ideal type of *satoyama* can certainly be constructed based on the empirical nature of a range of such production/consumption/living units. Such an ideal would have the qualities to provide a range of ecosystem services, which would include, on the basis of primary agricultural and horticultural production: nutrient cycling; soil formation; flood, climate and crop disease regulation; and, provision of food, water, wood and other biomass products. On the basis of these primary qualities, it would contribute to the promotion of human well-being through the creation of zones of health (through food, water and air quality and, of course, exercise), aesthetic and recreational needs and the establishment of communities of mutual help and of considerable cohesion when they persist over long periods of time, as many have done.

In practical terms, such communities are made up of mixed agricultural zones and techniques, typically rice paddies; vegetable gardens; woodlands; grazing grounds for sheep and cattle; irrigation ponds; bamboo groves; common areas exploited for leaf collection for use in manure production; wild plants and herbs; mature fallen logs which provide the base for mushroom cultivation and wood for charcoal manufacture. Cooperative work includes the joint maintenance of irrigation canals, streams and pathways. In the pottery villages of Kyushu to the west, it also includes the digging of clay and provision of mutual aid for harvesting, thatching of traditional farm houses and performing shrine and temple festivals, which often provide an important focal point for social life in both rural and urban Japan. This rather ideal picture is of course challenged, as mentioned above: the ageing population in Japan has had many consequences, including the shrinking of the rural population to well under half its size since 1960 (Coulmas *et al.* 2008); urbanisation, including the encroachment of residential and commercial use onto former agricultural land (certainly not all of it *satoyama*); and changes in lifestyle and diet. The popularity of fast foods and more Westernised diets, for example, has led to a decline in rice consumption throughout the country, and hence the abandonment of former paddy land. Similarly, commercial (usually coniferous) plantations have, in many cases, replaced old growth broad-leaved forests; so while the total forested land in Japan has long remained constant, the old multi-use forests, a common characteristic of *satoyama* landscapes, have declined. The good news, however, is the great interest in the preservation and advancement of *satoyama* ecosystems on the part of the Ministry of Agriculture, Forestry and Fisheries and their increasing visibility internationally as sustainable eco-social systems. Added to this is the rising demand for organic and/or locally grown foodstuffs and the increasing number of younger people interested in returning to this kind of agriculture and lifestyle (as opposed to interest in taking up larger-scale agribusiness or commercial farming).

This is perhaps not so surprising given that *satoyama* systems operate as indigenous multitasking units which provide both desired and regulatory and provisioning services. The former includes: rice, mushrooms and a selection of fruits and vegetables; responsibly harvested timber; some aquaculture services (fresh water fish); charcoal, no longer used for heating and rarely for cooking, but which is still used in such traditional arts as the tea ceremony; and sericulture services for the still extant, although considerably diminished, Japanese silk weaving industry. Meanwhile, the latter services include: climate and air quality regulation; soil formation and prevention of soil erosion; flood control and natural water purification and natural pest regulation and pollination services. In turn, these qualities attract customers to the farm products and attract hikers and eco-tourists; thus, they help to stimulate the revival of local crafts, dishes and products. A remarkable instance of this latter process can be seen south of Ishikawa in the Seto Inland Sea, the body of water that separates the main island of Honshu from its smaller neighbor Shikoku and from Kyushu to the west. This sea is dotted with islands, most of which have experienced heavy depopulation and ageing of the remaining individuals, with a corresponding decline in agriculture. The island of Inujima for instance, once the site of a copper refinery, quarries and farming which supported a population of between 3,000 and 5,000, now has a resident population of about 40 with an average age of 75. However, beginning in the 1980s, the Benesse Corporation, a very successful educational publishing company led by an art-loving president, began to establish a chain of art galleries and art sites on three of the islands – Naoshima, Teshima and Inujima – and is now moving onto a fourth and larger one (Shodoshima). These galleries, the architecture of which blends with the hilly and sea-girt nature of the sites, have proved enormously popular and attract large numbers of visitors from the Japanese mainland and from abroad. This has provided the means to move beyond the art sites themselves, which provide relatively little employment to the small local population and to begin the revitalisation of local farming. The traditional rice terraces are being brought back into production, the local cuisines and traditional manufacturing methods for the production of such essential Japanese products as sake and soy sauce are being revived, and there are many down-stream effects, such as the emergence of restaurants which serve dishes made with the residues of soy and sake production. The result has been a revitalisation of *satoyama*, fed in this case by the establishment of art galleries (themselves non-polluting entities, and blending with the local landscape – in one major case the gallery itself, designed by the leading Japanese architect Ando Tadao, is almost entirely underground) and quasi-eco-tourism – bicycles are available to rent, cars are few and walking is encouraged and the little local buses are very cheap to use. In this case, one non-polluting 'industry' is supporting another to the mutual benefit of both, while showcasing a means of reviving a *satoyama* (and a *satoumi*) landscape (Muller and Miki 2011/12; Kaneshiro et al. 2015).

The key word often used in characterising *satoyama* landscapes is 'interlinkage', at one level the interlinkage of ecosystem services themselves – water, forests, mixed agriculture (usually plant-based and involving few animals), soil retention, preservation of biodiversity, low waste and low carbon-footprint and, in turn, their interlinkage with human systems. These latter include high levels of local cooperation, willingness to protect the common areas, preservation of traditional knowledge, continuity or reestablishment of rooted communities, reversal of the decline in rural populations to lower the average age thereof and the establishment of positive links between rural and urban areas. These can be achieved through such means as eco-tourism, community-supported agriculture, direct marketing of local farm products to urban areas and socialisation of small children. It is, for example, becoming increasingly common for Japanese kindergartens and elementary schools to take their charges to the countryside to plant (usually rice), to harvest the resulting crop and to take it back to school to cook and eat, thereby establishing a natural link between young people and the sources of their food. Studies of *satoyama* communities in Ishikawa have illustrated all these processes, both agricultural/ecological and cultural/social. These include: integration of a farm economy around both rice farming and oyster farming in a coastal district on an island off the Noto Peninsula in the northern part of the prefecture; revival of local varieties of soybeans (a major element in the Japanese diet); conservation of biodiversity through the use of totally 'natural' cultivation methods (no chemical fertilisers or pesticides); re-establishment of traditional sustainable charcoal making; and the preservation of traditional vegetable varieties. Other activities include: stimulation of rural tourism through farm-based holidays; maintenance of local water bodies to attract populations of sedentary and migratory birds; encouragement of local cultural expressions, such as dance forms; and conservation of the wonderful terraced paddy fields historically characteristic of much of the area, not only with local labor, but by inviting volunteers to participate in the weeding, reconstruction of bunds and other labour-intensive activities necessary for the reconstruction efforts. In an ageing society, the involvement of elderly people in community development projects, such as the conversion of a deserted elementary school into a tourist lodge in Noto Town, has proved to have many positive effects.

An effective synthesis of many of these activities is found in the *Maruyama-gumi* or 'team Maruyama', a project set up in a small village and pioneered by an organic farmer and a former housewife from Tokyo. The wife felt the need to return to her roots and, now settled in Ishikawa, she manages the team which works together to sustain traditional knowledge and the biodiversity of *satoyama* through art, foods, agriculture, welfare activities with the elderly and other means (UNU 2013). While *satoyama* landscapes face threats of many kinds, not least the movement in the 1960s, and after, to adopt the use of chemical fertilisers and pesticides, they also meet with huge encouragement. Indeed, many Japanese now report that they want spiritual richness, rather than mere material richness in their lives, a large contrast with the 1970s when exactly the opposite was the case. Now that climate change has moved to the forefront of global attention, *satoyama* has also been discovered as a real example of mitigation in practice and has been for several centuries.

13.3 Theorising *Satoyama:* Towards a Culture of Transition?

The *satoyama* integrated socio-ecosystem then might be seen as an example of climate responsibility *avant la lettre*. While its empirical interest lies in its characteristics as a mixed agro-ecological *and* sociocultural system, it also throws up many more theoretical challenges, and it is to these that I now turn. The notion of 'transition' was largely popularised by the British-origin Transition Movement, based on the idea that the world has now reached the point of 'peak oil' after which easily recoverable reserves will decrease, and so planning for a post-oil society must begin now, long before it runs out or becomes a rare commodity. The main thrust of the movement is, therefore, to prepare towns, such as the small West of England town of Totnes where the movement originated, to become 'transition towns' – ones actively preparing for a post-oil future and all that will entail (Hopkins 2008; see also Urry 2013). The range of adaptations and restructuring is naturally large. It involves addressing not only energy concerns but also such issues as food security based on the sustainability and resilience of local agricultural systems (Feola and Nunes 2014) and associated *cultures* of transition, such as the vast range of attitudes, shifts in expectations, cultural expressions and just and equable communities, spiritual or religious orientations (Baker 2009), legal structures (Cullinan 2011), the economics of the process or indeed of culture in its expressive sense – the arts in particular (Clammer 2016). The major Indian novelist and writer, Amitav Ghosh, has persuasively argued that climate change is as much a cultural issue as it is a technical one. In a recent book he reflects at some length on why fiction writers and other producers of culture have not addressed it with anything like the energy that it deserves. Rather, they have remained, for the most part, at a distance, assuming perhaps that such a subject is the proper province of the scientist, not the novelist (Ghosh 2016). This, he suggests, means that despite the severity of the situation, cultural workers (writers, artists and dramatists) have been slow to devote their skills to the creation of new and appropriate cultural forms. Given that stories or narratives provide the frames for many of our worldviews, the neglected role of culture in the debate needs rethinking (Russi 2016). Indeed, transition to sustainability and a climate-friendly society and economy will not take place without significant cultural shifts. A *transition culture,* and how to achieve it, then becomes a central concern.

Of course, if someone puts forward *satoyama* systems to demonstrate a successful example of this shift actually happening, someone else will inevitably bring up the question of 'scaling up', which does indeed suggest a number of theoretical possibilities and controversies. Is the 'great transition' – the large-scale structural shift away from polluting, green-house gas emitting, extractive industries – in any way advanced by these small-scale experiments? Is the big transition to be caused by the cumulative effects of many 'little transitions', or are such local examples always in danger of being wiped out by negative (for them) shifts in the larger economy? This vulnerability is certainly a fact, and the pressures to which the *satoyama* system has been subject attests to this. Many of the scenarios for the demise of the dominant economy (for which read 'neoliberal capitalism') assume collapse and a very painful process of transition as the supporters of the old version struggle bitterly to keep it alive, denying the reality of climate change or holding out for the promise of some technological solution that will allow us to essentially retain our high-energy, high-consumption lifestyles. Nonetheless, supporters of the idea that small is beautiful might well argue that community-based agro-ecological systems, despite their fragility (and what is not fragile, including an oil-based economy), offer a viable model of the future: ecologically sound, producing healthy food, stimulating community and cooperation, being virtually carbon neutral and, in some cases, requiring little input from the outside, adopting cradle-to-cradle design intuitively for many tools, implements and even houses. As such, they would be among the best candidates to survive both a collapse of the fossil fuel –based economy and the social patterns and cultural practices with which it both generates and sustains itself. In other words, it represents a *culture* of transition as well as a (highly sustainable) set of agricultural and land-management practices that have stood the test of considerable time and external pressures (on culture in agriculture, see Pretty 2002).

Why then might *satoyama* constitute such a model of a sustainable future and of one way of *getting* there? For those who have argued for localism and bioregionalism (e.g. Shuman 2000), it provides a living example and one with deep historical roots. Given the considerable length of Japan and geographical variation within the country, naturally each example of *satoyama* is adapted to its local environmental conditions and the specific qualities of local cultures. Of course, this is a necessity since *satoyama* represents both an agro-ecological system and a sociocultural one, the two being intimately integrated. An important and often overlooked aspect of this in the literature on ecological restoration is that such restoration does not only involve biological processes but also social ones, and that a vital aspect of those social ones is the creation of a *restoration economy* – one based on the sustainable management of local resources and the establishment of the social relations necessary to run such an economy (Jordan 2003). A cooperative economy requires cooperative social relationships for it to last, and as any economic system plays a large role in shaping human subjectivities (attitudes to consumption and 'needs', for example), the form of economy will shape the culture within which it is embedded. An interesting, but very under explored, aspect of culture in this respect is that of aesthetics, in particular the aesthetics of landscapes. The relatively new field of environmental aesthetics is now beginning to explore this area and shows how the attitudes to land and its use and preservation are greatly shaped by cultural concepts of beauty, order, harmony and grandeur, expressed not only in such media as landscape painting but in the everyday relationship to nature and its use (Kemal and Gaskell 1995; Brady 2003; Budd 2005; Carlson 2005; Parsons 2008). Each society, of course, has its own attitudes to landscape, influenced no doubt to a great extent by its physical qualities – deserts, mountains, plains and so forth. The Japanese self-image is very much one of being intimately linked to nature (seasonal variation in the colours of clothing, flower arranging as an art form, cherry blossom and autumn leaf viewing as national obsessions), and while I think that this is often overstated (Japan has plenty of ugly industrial landscapes), it is true that Japanese aesthetics are very much tied up with ideas of natural beauty. When talking with *satoyama* farmers and their families, this is something that is constantly and spontaneously referred to – both drawing attention to the beauty of the managed, but still 'natural' landscapes, and the expression of the sentiment that 'I would not want to live anywhere else.' Here there is a large, and certainly in the Japanese case unexplored, area of interest – the linking of such 'practical' landscapes as those of *satoyama* with the emerging field of environmental aesthetics and the burgeoning field of environmental psychology (for a representative example from among a rapidly growing literature, see Roszak *et al.* 1995).

The question of the 'exportability' of any given model of transition or sustainability then rests on two considerations: its ecological viability in another geographical context and the possibility of its cultural adaptation to a new social context. Japan certainly has its own cultural and social distinctiveness: a strong aesthetic orientation to nature; a long tradition of localism often engendered by geography, such as the narrow farmed valleys separated by mountain chains characteristic of much of the country; an equally long tradition of mutual help and cooperation in rural communities, whether agriculturally based or of mixed economy, e.g. pottery and farming; a history of communally based religious organisations located in the countryside, with an agricultural base, while marketing fresh and/or organic produce to the towns; and self-conscious attempts to revive a sense of the sacredness of nature, and in particular the spiritual power of mountains (Yamagata 2006). Nevertheless, the awareness of possible models is helpful in the creation of a repertoire of possibilities and a vocabulary of alternatives and, in this case, links the study of *satoyama* systems to much broader debates in climate change study. This point also links the analysis and appreciation of such systems once again back to the key question of evolving and strengthening a *culture of transition and sustainability,* not just a set of techniques for agro-ecological practices, vitally important as these of course are. (On the question of 'scaling up' localised movements, see North and Longhurst (2013), and in relation to the transition movement see: Bailey *et al.* 2010.)

In her passionate and deeply researched book on climate change (a book that contains many examples of local initiatives and places great faith in the possibility of social movements to transform the future), Naomi Klein essentially argues for a model that reflects the ideal-type characteristics of *satoyama* regimes (Klein 2015). She suggests, for example, that the *social context* of any transition movement is as important as its technical aspects: coming out of denial; heeding the 'wake-up calls' that nature is clearly sending us; and moving away from materialism understood as an addiction to the extractive- and consumption-based economy that is the basic cause of climate change (Klein 2015). In pursuing her agenda of looking for radical solutions to the cultural and economic conditions that create climate change, a number of suggestions emerge that are remarkably compatible with the *satoyama* model. These include an argument against the 'globalisation of agriculture' and the destruction of local food systems, the strengthening of local values and culture, managed de-growth towards a much more steady-state economy, less consumption and the promotion of agro-ecological systems as alternatives to the agri-businesses that currently dominate the world food-chain. Klein understands agro-ecology to be a 'less understood practice in which small-scale farmers use sustainable methods based on a combination of modern science and local knowledge' (Klein 2015:134). She unpacks this interpretation to identify agro-ecology as the integration of trees and shrubs with crop fields; inter-cropping; the use of green manures; soil conservation and the other characteristics of *satoyama*. Thus, agro-ecology is a system that also sequesters large amounts of carbon, increases food security, repairs and preserves soils and includes built-in recycling and waste management (there being essentially little or no 'waste'). Such a system is not simply 'resilient' but is actually regenerative or restorative; it brings fertility back to both nature and community. Any future economy must, of necessity, be an ecological one, and here, on a surprisingly large scale (spread throughout the length and breadth of Japan), we find an example of what this might look like. It is echoed too in the kinds of solutions being suggested around the globe. In India there are many examples, such as the 'eco-democracy' promoted by the writers and environmental activists Aseem Shrivastava and Ashish Kothari (Shrivastava and Kothari 2012) and by the well-known activist scholar Vandana Shiva (Shiva 2005) and the many actual experiments that dot the Indian countryside. These experiments simultaneously regenerate rural economies; combat climate change; harvest water; promote climate resilient crops, local handicrafts and culture; preserve wildlife; and resist the caste and gender inequalities with which the country is still infested (Kalpavriksh 2017).

While Japan is quite rightly thought of as a highly developed country, this status has not meant the total disappearance of traditional knowledge – whether of crafts, religious rituals, medical practices or agriculture. As the population ages and urbanisation of the country continues, there is, naturally, and as in many other situations globally, the grave danger of the erosion and eventual disappearance of such knowledge and bodies of custom. Interestingly, in almost every interview with a *satoyama* farmer, whether a long-established one or a new entrant, the preservation of traditional knowledge was cited as a major reason for maintaining or joining the system. In fact, it was particularly among new entrants that this was mentioned – the simultaneous reporting of the shallowness of the 'modern' culture with which they were surrounded or from which they were emancipating themselves, and the strong desire to maintain, and indeed revitalise, traditional forms, not only of agricultural techniques, but also in the expressive arts, crafts and traditional food products and means of production and preservation, as in the soy sauce and sake examples mentioned as part of the Shodoshima revival plan. The 'liberation' aspects of *satoyama* are not insignificant as they provide a powerful alternative imaginary to that of the urban, technology-driven consumption-based lifestyle of the majority. These activities do not, in themselves, necessarily contribute to climate change mitigation, but as scholars of traditional knowledge in a number of localities have argued, traditional land management techniques, the pursuit of low-energy (in terms of mechanical inputs) agriculture and the maintenance of biodiversity, forest cover and localisation in terms of both inputs and outputs (i.e. agricultural produce), all contribute substantially, in an unsung kind of way, to climate change mitigation (Oladele and Braimoh 2010). Seen in this light, *satoyama* is a perfect example of 'endogenous development' (Haverkort et al. 2002; Haverkort and Reijntjes 2010) which struggles to maintain and advance not only sustainable agricultural practices but also cultural ones and, in doing both, reestablishes a sense of community in an increasingly atomised and anonymous urban-industrial society. What makes this particularly interesting is that it is not happening in some remote 'developing' society but in a highly sophisticated and technologically driven one. Nor is this a case of an ancient and virtually 'third world' enclave somehow surviving as a social fossil in the midst of that complex larger society, but one well known to many urban dwellers who actively choose to vacation there, consume its healthy products, and to a great extent admire and idealise its philosophy, even when not practicing it personally in the middle of the large cities like Tokyo and Osaka. The link, in other words, is clear between the practice of agro-ecological styles of farming, community-construction and climate change mitigation.

13.4 Cultures of, and in, Transition

Transition to any form of sustainability, and with it the creation of climate-friendly regimes, cannot take place without major cultural shifts, and this is clearly one of the major reasons why, despite all the evidence so readily available, serious action at individual, societal and structural levels is so slow in coming. So slow in fact that, tragically, adaptation to potentially radical new forms of climate shift, rather than a cessation of the trends that have caused climate change in the first place, is becoming the most likely scenario. However, having said this, there are both signs of hope (the increasing attractiveness of agro-ecological systems and the desire for their products to be produced within such a system) and the possibility that when faced with real and inescapable crisis, human beings will have the energy and imagination to change. Revolutions have happened in the past, so possibly in the future as well? The good news is the fact that we are already seriously discussing alternatives to the run-away extraction and consumption-obsessed economy that we have created. To discover how to fast-forward such an effort is, consequently, perhaps the most pressing need of our generation. This involves a number of elements, such as exploring experiments that are already taking place (eco-villages, new social movements oriented towards environmental issues, permaculture and experiments in soil conservation, ideas emerging from solidarity economy thinking, among others) and understanding research, in the social sciences in particular, as to what the social movement theorists Max Haiven and Alex Khasnabich call 'pre-figurative research'. This is research directed not only at the past, or even the present, but which is conceived as 'a form of research borrowed from a post-revolutionary future. We wanted to imagine a form of common research, beyond enclosure' (Haiven and Khasnabich 2014:17). This they link to the central concept of their book – the 'radical imagination' – social imagination inspiring action, new forms of solidarity and remembering the past in such a way as to tell different stories about how the world has come to be as it is (for example, by remembering not only mainstream historical narratives, but also the underground ones of struggles, initiatives and movements). Going 'back to the future', as in the case of *satoyama*, implies just such a form of pre-figurative research. It highlights the necessity of mining the past for examples of climatically and ecologically friendly practices that contain lessons for the present and the future and which, in this case, suggest low-cost, highly productive, long-lasting, sustainable models of both climate change mitigation and satisfactory life-styles and modes of community.

Returning to the issue of culture, it is also the case that we need to emphasise two trajectories – the inward and outward, as it were. In the specific case of *satoyama* (and each locality would have its own specificities), we have already noted its affinity both to Japanese ideals of community and to Japanese conceptions of landscape and aesthetics. In other words, far from being an artificial or alien form of practice, it is deeply rooted in Japanese culture, history and ecology. This can be seen as yet another level of what might be called 'deep culture', notably the religious foundations of the Japanese relationship to nature, most notably seen in the indigenous religious tradition of Shinto. While most Japanese, if forced, would cite Buddhism as a religious label for themselves, underlying this remains a strong attachment to Shinto, especially as it relates to the more affirmative aspects of the life-cycle, such as birth, coming of age and marriage – Buddhism has a virtual monopoly on death rituals. Buddhism itself, as a collection of Mahayana schools including, most prominently, *Jodo Shinshu* or 'Pure Land' Buddhism, is strongly related to a vitalist conception of nature expressed as the potential Buddha-hood of all beings, and even the supposedly inanimate aspects of nature, such as plants and rocks. Meanwhile, Shinto, in the form of so-called 'State Shinto', had a considerable reputational crisis as it became associated with emperor-worship and pre-war and wartime fascism, and has since been actively trying to recover its status by reinventing itself (or perhaps rather returning to its original and authentic roots) as an ecological religion. Indeed, in many respects, this is what Shinto is with its emphasis on nature, the preservation of sacred groves, and the simplicity of its shrines and their integration with the landscape (International Shinto Association 1995; Rots 2017). I would certainly argue that underlying Japanese attitudes to nature and landscape are an essentially Shinto ideology. It is a kind of sophisticated animism (Clammer 2004), one which is now providing the basis for a 'return to nature' on the part of many Japanese. Such return is signaled not only by the resumption of agro-ecological farming but also in art, fashion, food preferences, architectural choices, in *anime* (the now internationally well-known form of animated film, especially as reflected in the movies of the celebrated producer Miyazaki Hayao). In this way Shinto has moved from a 'passive' status to become an active religious movement, no longer directed towards nationalistic ends, but now towards the recovery of nature (Clammer 2010) and, as such, it provides a religious/ideological basis for rural re-vitalisation. Here we see not only another 'return to the future', one which draws on an ancient tradition as a resource for the emergent situation, but in comparative terms the significance of one of the most significant aspects of any culture – its religions – for a shift towards transition, understood not simply as a 'secular' process, but as a kind of spiritual one. Indeed, the burgeoning literature on religion and ecology world-wide seems to suggest that this is no mere fad, but a significant shift in thinking across a whole range of societies and cultures (e.g. Gottlieb 2004; Kaza and Kraft 2000).

The other trajectory – what might be called the outer movement – is the linking of both specific examples of sustainable communities and emerging debates about alternatives to extractive/consumer capitalism with wider debates of ethical, philosophical and even theological, interest. The last decade has seen the emergence of the previously minor field of environmental ethics into considerable prominence (e.g. Curry 2012). This has widened debate among philosophers about the meaning of nature (Sopher 1995); it has led to the emergence of such literary fields as eco-criticism, and extended debate among theologians about the place of nature in their various schemes and scriptures. Many of these themes are brought together in a recent book by Michael Northcott, where he synthesises debates about the 'countryside' as it is understood in the United Kingdom. There is a sense that the countryside somehow enshrines something deeply English; he identifies pressures on the notion of countryside from developers through to frackers, and the perception that countryside somehow embodies the sacred. He locates the countryside as a subject of ethical discourse in other words, not simply as the object of planning (Northcott 2015). *Satoyama* itself, I would argue, can be read as a moral discourse as much as an agricultural one. This is not only because of the ways it situated itself in relation to other aspects of Japanese thinking and practice but also because it reminds us that climate change is itself an ethical issue and clearly one of the most pressing moral (and hence cultural) issues of the day. It affects the lives of millions, human and non-human, and may undo in decades what evolution has taken millennia to create. So many human, political and social conflicts become trivial in the face of this macro-issue, and all attempts to address it then take on a moral status, whether or not this is expressed in religious or secular terms.

The study of *satoyama* – one localised version of the search for long-term sustainability – proves to have many implications, even global ones. One of the most important of these is the dimension of culture which includes religion, aesthetics, the maintenance and expression of local traditions through the preservation of dance, crafts and architecture and the creation of a culture of cooperation and conviviality. A culture of transition – a willingness to embrace the lifestyle and consumption decisions that living sustainably on a finite planet entail – must, at this point in time, have two dimensions. One is a culture of *reconstruction,* not simply a critique of globalisation, neo-liberal capitalism and extractive industries, important as that critique is for our future. In other words, it must demonstrate workable alternatives, and I argue here that *satoyama* is such a workable and tested alternative. The other is the construction, no doubt in solidarity with other movements globally, of a culture of *resistance* to the negative aspects of 'development'. This is no easy task as it involves the active rejection of the blandishments of the mainstream system, the homogenisation of education, the seductions of the messages spread by Hollywood and Bollywood, the renunciation of violence and the equally firm resistance to those who practice such violence through war, terror or the quiet violence of enclosures, demeaning of local languages and practices, the law and racism (Meyer and Alvarado 2011; Gomez-Barris 2017). While cultures of transition and sustainability are by no means to be identified only with the countryside, given that more than half the world's population now live in cities (but that is a matter for another essay), in the context of the case study presented here, a number of final conclusions can be drawn.

There are four major ones. The first is the need for a new 'ruralism' far from Marx's contemptuous rejection of the 'idiocy of rural life', something exemplified in the *satoyama lifestyle.* Then, there is the need to discover and create 'narratives of transition' – the stories that shape cultural practices and self-perceptions of the very kind that Amitav Ghosh recommends (and laments that so few writers have yet attempted). The third is to recognise that *cultural sustainability* is as important, in its way, as ecological sustainability. It is enshrined in culture and language and is exactly that indigenous knowledge that has enabled many societies, as the anthropological record attests, to thrive and persist in climate-friendly ways over many generations (indeed, usually until assailed by the forces of 'modernisation' and 'development'). The fourth is that *cultural planning for transition* is vital, without which purely technical 'solutions' inevitably flounder. Gandhi famously argued that the salvation of India lay in its villages, and in the long run he may well be proved right. But, since he wrote and spoke more than a half-century ago, we have learnt a great deal about ecology, climate, community and processes of social change. Other than its intrinsic interest, the *satoyama* case shows one significant mode of negotiation with these forces, and it also shows that the 'indigenous' can flourish in the midst of the 'advanced', and may well prove to be the latter's future. It also shows that local pursuit of solutions to climate change can occur as seedbeds of change in one of the world's most industrialised and urbanised societies: that 'tradition' and 'modernisation' can indeed coexist in dialogue with one another. The realisation of that dialogue may, in a larger picture, be one of Japan's most significant bequests to our troubled world.

References

Bailey, I, Hopkins, R., and Wilson, G. 2010. Some things old, some things new: The spatial representations and politics of change in the peak oil relocalisation movement. *Geoforum*, 41, 595–605.
Baker, C. 2009. *Sacred Demise: Walking the Spiritual Path of Industrial Civilization's Collapse*. New York: iUniverse Inc.
Brady, E. 2003. *Aesthetics of the Natural Environment*. Edinburgh: Edinburgh University Press.
Budd, M. 2005. *The Aesthetic Appreciation of Nature*. Oxford: Oxford University Press.
Carlson, A. 2005. Environmental aesthetics. In: Berys, G., and McIver Lopes, D. (eds.) *The Routledge Companion to Aesthetics*. Abingdon and New York: Routledge, pp. 541–555.
Clammer, J. 2004. The politics of animism. In: Clammer, J., Poirier, S., and Schwimmer E. (eds.) *Figured Worlds: Ontological Obstacles in Intercultural Relations*. Toronto and London: Toronto University Press, pp. 83–109.
Clammer, J. 2010. Engaged Shinto? Ecology, peace and spiritualties of nature in indigenous and new Japanese religions. In: Clammer, J. (ed.) *Socially Engaged Religions*. Bangalore: Books for Change, pp. 50–61.
Clammer, J. 2016. *Cultures of Transition and Sustainability*. New York and London: Palgrave Macmillan.
Coulmas, F., Conrad, H., Schad-Seifert, A., and Vogt, G. (eds.) 2008. *The Demographic Challenge: A Handbook about Japan*. Leiden and Boston: Brill.

Cullinan, C. 2011. *Wild Law: A Manifesto for Earth Justice*. White River Junction, VT: Chelsea Green Publishing.
Curry, P. 2012. *Ecological Ethics*. Cambridge: Polity Press.
Feola, G. and Nunes, R. 2014. Success and failure of grassroots innovations for addressing climate change: The case of the transition movement. *Global Environmental Change*, 24, 232–250.
Ghosh, A. 2016. *The Great Derangement: Climate Change and the Unthinkable*. Gurgaon: Allen Lane.
Gomez-Barris, M. 2017. *The Extractive Zone: Social Ecologies and Decolonial Perspectives*. Durham: Duke University Press.
Gottlieb, R. S. (ed.) 2004. *This Sacred Earth: Religion, Nature, Environment*. New York and London: Routledge.
Haverkort, B., van 't Hooft, K., and Hiemstra, W. (eds.) 2002. *Ancient Roots, New Shoots: Endogenous Development in Practice*. Leusden: ETC/Compas and London: Zed Books.
Haverkort, B. and Reijntjes, C. 2010. Diversities of knowledge communities, their worldviews and sciences: On the challenges of their co-evolution. In: Suneetha M. S., and Balakrishna, P. (eds.) *Traditional Knowledge in Policy and Practice: Approaches to Development and Human Well-Being*. Tokyo, New York and Paris: United Nations University Press, pp. 12–30.
Haiven, M. and Khasnabich, A. 2014. *The Radical Imagination: Social Movement Research in the Age of Austerity*. London and New York: Zed Books.
Hopkins, R. 2008. *The Transition Handbook: From Oil Dependency to Local Resilience*. White River Junction, VT: Chelsea Green Publishing.
International Shinto Association. 1995. Shinto to Nihon Bunka [Shinto and Japanese Culture]. Tokyo: International Shinto Association.
Japan *Satoyama* and *Satoumi* Assessment. 2010. *Satoyama-Satoumi Ecosystems and Human Well-Being: Socio-ecological Production Landscapes in Japan*. Tokyo: United Nations University.
Jordan, W. R. 2012 *The Sunflower Forest: Ecological Restoration and the New Communion with Nature*. Berkeley: University of California Press.
Kalpravriksh. 2017. *The Search for Radical Alternatives: Key Elements and Principles*. Pune: Kalpavriksh.
Kaneshiro, K., Waki, K., and Hemmi, Y. (eds.) 2015. *Becoming: Bennesse Art Site Naoshima*. Naoshima: Fukutake Foundation.
Kemal, S., and Gaskell, I. (eds.) 1995. *Landscape, Natural Beauty and the Arts*. Cambridge: Cambridge University Press.
Kisala, R. 1999. *Prophets of Peace: Pacifism and Cultural Identity in Japan's New Religions*. Honolulu: University of Hawai'i Press.
Klein, N. 2015. *This Changes Everything*. London: Penguin Books.
Kaza, S. and Kraft, K. 2000. *Dharma Rain: Sources of Buddhist Environmentalism*. Boston and London: Shambhala.
Meyer, L. and Maldonado Alvarado, B. (eds.) 2011. *New World of Indigenous Resistance: Noam Chomsky and Voices from North, South, and Central America*. San Francisco: City Lights Books.
Muller, L. and Miki, A. (eds.) 2011/12. *Insular Insight: Where Art and Architecture Conspire with Nature*. Naoshima, Teshima, Inujima. Zurich: Lars Muller Publishers, Naoshima: Fukutake Foundation.
North, P. and Longhurst, N. 2013. Grassroots localization? The scalar potential and limits of the 'Transition' approach to climate change and resource constraint. *Urban Studies*, 50, 1423–1438.

Northcott, M. S. 2015. *Place, Ecology and the Sacred: The Moral Geography of Sustainable Communities*. London: Bloomsbury.

Oladele, O. I. and Braimoh, A. K. 2010. Traditional land management techniques for climate change mitigation. In: Suneetha M. S. and Balakrishna P. (eds.) *Traditional Knowledge in Policy and Practice: Approaches to Development and Human Well-Being*. Tokyo, New York and Paris: United Nations University Press, pp. 171–180.

Parsons, G. 2008. *Aesthetics and Nature*. London and New York: Continuum.

Pretty, J. 2002. *Agri-Culture: Reconnecting People, Land and Nature*. London and Washington: Earthscan.

Roszak, T., Gomes, M. E., and Kanner, A. D. 1995. *Ecopsychology: Restoring the Earth, Healing the Mind*. San Francisco: Sierra Club Books.

Rots, A. P. 2017. *Shinto, Nature and Ideology in Japan*. London: Bloomsbury.

Russi, L. 2016. Wild things: Stories, transition and the sacred in ecological social movements. *World Futures*, 72, 379–389.

Shrivastava, A. and Kothari, A. 2012. *Churning the Earth: The Making of Global India*. New Delhi: Penguin Books India.

Shuman, M. H. 2000. *Going Local: Creating Self-Reliant Communities in a Global Age*. New York: Routledge.

Sopher, K. 1995. *What Is Nature? Culture, Politics and the Non-Human*. Oxford: Blackwell.

United Nations University. 2013. *Satoyama and Satoumi of Ichikawa*. Kanagawa: UNU-IAS Operating Unit Ishikawa/Kanagawa.

Urry, J. 2013. *Societies beyond Oil: Oil Dregs and Social Futures*. London and New York: Zed Books.

Yamagata. 2006. *Yamagata: Seeking the Japanese Soul*. Yamagata: Tohoku Cultural Research Centre, Tohoku University of Art and Design (in Japanese and English).

14

Culture and Climate Change
Experiments and Improvisations – An Afterword

RENATA TYSZCZUK AND JOE SMITH

14.1 Introduction

For this afterword we have been invited to look back on our experience of working at the culture–climate change join across 25 years. It is an opportunity to try to identify useful discoveries and unwritten rules, and acknowledge some blind alleys, as we look back on a variety of design, media, arts and other creative collaborations. To begin, we offer some thoughts about the nature and role of this kind of work. We then describe some of our projects and reflect on what we have learnt along the way as we have sought to support, convene, catalyse and understand cultural work on climate change.

The chapters collected in this book together emphasise the importance of cultural work on climate change. This respects Mike Hulme's observation that 'however our contemporary climatic fears have emerged . . . they will in the end be dissipated, reconfigured or transformed as a function of cultural change' (2009:5). However, there are no blueprints for cultural work on climate change. Work in this area does not offer an instant remedy for public detachment or policy failures. But it can open up more expansive understandings of the many ways in which the world is being altered, or might be in the future, not simply physically but also imaginatively. Moreover, climate change calls for new strategies of deliberate transformation (O'Brien 2012) that recognise not only different understandings of agency and human–environment relationships but are an adaptive challenge in themselves (O'Brien 2016; O'Brien and Selboe 2015). These deliberate transformations are often latent with a political charge that requires or invites exploration and dispute. Cultural work can help to surface or support this.

Most climate research is rooted in the 'cultures of prediction' (Mahony, this volume; Heymann *et al.* 2017), which pervade the science and cultural politics of global environmental change. Other forms of knowledge (such as indigenous understandings) and meaning-making (for example, generated by the arts and humanities) struggle to achieve anything more than marginal status. Yet such contributions are within reach, as this volume demonstrates. For example, Ulloa, working in the Colombian context, argues for the importance of located, indigenous climate knowledges and yet also demonstrates the need for 'strategies of dialogue' if these voices and experiences are to be given appropriate recognition rather than a token presence. Bringing historical rather than geographic range, Endfield and Veale show how varied (though, as they note, not necessarily inclusive) archival accounts of extreme weather in Britain can contribute to a cultural imagination of climate change in the present. Postigo's presentation of an Andean case holds together both geographic and temporal dimensions of the experience of weather and climate with an eye on practices of adaptation. A further geographically rooted case of transition as opposed to adaptation is offered in Clammer's account of the re-emergence, or reinvention, of traditional food systems, *satoyama*, in Japan.

These insights support our argument that it is a profound mistake to view cultural work as a kind of communications 'finishing school' for the prior work of the natural science and policy communities or as part of the psychological 'rewiring' that some suggest is required in response to climate change (Marshall 2015). Furthermore, while we are in step with O'Brien *et al.*'s promotion of research practice that actively supports positive transformations, we find they pose a problematic question. They ask 'how culture can be harnessed for transformation rather than being an impediment to change' (this volume). We argue that cultural work is not available to be 'harnessed', let alone contained. It is an unruly field of practice that is energetic precisely because it can generate unanticipated outcomes.

Nevertheless, there has been growing recognition that cross-disciplinary, more culturally rooted, work will need to play a much more prominent role in shaping humanity's responses to the risks associated with climate change. This pressure will surely grow in the context of what Grevsmühl (this volume) terms increasingly 'mobile' climates. This has led Hulme, O'Brien and others to argue for more prominence for social sciences, arts and humanities contributions to climate change research. In our own work we have often argued that this should not be understood as some kind of resolution of communications challenges, or as a form of 'completion' of environmental research, but rather as an 'opening out'.

A shift in the 'intellectual climate' would involve incorporating a range of environmental humanities writing on, for example, values, responsibilities, rights, perceptions, faith and care pertaining to the 'human dimensions' of global environmental change (Castree 2016). An example is given in this volume discussing the pertinence of Buddhist and other world views for deliberation of climate change. The chapter nicely illustrates the potential of cultural perspectives drawn from a range of world views, from its attention to notions of interconnectedness to its provocative reading of the 'precarious' nature of the dominant model of happiness. There is plenty of work going on in the arts, in the media and in academia in this territory, and we don't intend to claim exclusivity or any exceptional status for our projects – on the contrary. Indeed, in the first in our series of Culture and Climate Change books (Butler *et al.* 2011), we attempted to place the range of cultural work on a timeline. Keeping up with emerging work was impossible and omissions embarrassing: we quickly gave up on this ambition. As the varied contributions to this volume demonstrate, this is now a vibrant field, both creatively and academically.

We come from two fields of study and practice, geography and architecture, which share much in common in relation to climate change and wider environmental research. They are deeply inscribed with multi- and inter-disciplinary working and are distinctive within universities in drawing together in one place insights and practices from across the humanities and the natural and social sciences. Both are concerned with space, place and processes of change – both social and natural. More recently, these disciplines are also among the most prominent centres of research and practice related to global environmental change and economic and cultural globalisation. Both architecture and geography are expected to respond to and, to some extent, to be responsible for these issues. We have explored what this means in our own disciplines; in architecture, in terms of agency (Kossak *et al.* 2009) and provisionality (Tyszczuk 2018); and in geography we also draw on the experience of working at the join between global environmental change issues and broadcast media (e.g. Smith 2000, 2005, 2011, 2013a, 2013b, 2014, 2017; Smith *et al.* 2018).

Our culture and climate change projects are rooted in collaborative and interdisciplinary approaches. They have also tended to be experimental and hence often risk taking. They have generally sought to support more plural and dynamic representations of global environmental issues rather than 'communicate the facts'. The work has often been driven by the objective of bringing together different communities of interest and experience. Related to this, the work has tended to take a less settled view of the underlying issues surrounding climate change than many would. For example we are wary, and on occasions directly critical, of attempts to drive society towards specific objectives that might be derived from the natural sciences but are packaged into particular conclusions or directions by NGOs or the wider policy community. Instead, we are learning by doing: improvising.

If thought of in terms of 'improvisation', discussions around climate change might serve as a context for exploring the future, by opening up different possibilities and potentialities for living on a fragile – for humans – and dynamic Earth. We could refer to this as 'constructing for the unforeseen' – acknowledging the root of the word improvise in the Latin *improvisus*, 'unforeseen' (Tyszczuk 2011). The centrality of experiment and improvisation in our work is informed by how we understand the cultural politics of climate change. We argue that climate change has six distinctive yet often interacting elements. These comprise: its global pervasiveness, its inherent uncertainties, its interdependencies (both social and ecological), the reverberations of history (particularly colonial and postcolonial), the centrality of interdisciplinary approaches in research, and a constantly shifting distribution of human vulnerabilities and responsibilities across time and space (see Smith 2011, 2014, 2017). These distinctive features of climate change mean that it is present in every aspect of human lives, politics and culture. Indeed:

climate change is too here, too there, too everywhere, too weird, too much, too big, too everything. Climate change is not a story that can be told in itself, but rather, it is now the condition for any story that might be told about human inhabitation of this fractious planet. (Tyszczuk 2014:47)

All six dimensions are relevant in diagnosing why climate change is a difficult story to tell. These are not properties that are unique to climate change, but they are unique in combination and are constantly being reconfigured by the generation of new knowledge, representations and events. The work we reflect upon here is all rooted in the fact that climate change is interesting as well as urgent and important. We have also developed our work with a clear understanding that our role as academics working from within arts, humanities and social science traditions is not to serve as adjuncts to policy or in the service of campaigners. Rather, we feel it is our responsibility to experiment, learn and share what we find in prototyping shared futures. At the same time our practice, while sharing elements of laboratory practice in the natural sciences, above all 'the time of the experiment', enjoys some freedoms unavailable to those spheres of research.

This has allowed us to follow a hunch that experiments and improvisations may prove more effective tools for thinking in a climate-changed world than attempts to perfect communications strategies or polish change agency models. We suggest that researchers and their creative partners could invest their ingenuity, freedom and distinctive skills in cultural mediations, rather than in simply amplifying a particular brand of 'approved thinking'. We recognised our notion of 'mediations' in Ciara Healy-Musson's account of the special nature of 'thin places', her description of her own practice as 'thin curating' and her pursuit of 'deeper engagements' between human and non-human (Healy-Musson, this volume). This (and the admission that she knew these experiments to be about friction – and to be professionally risky) resonated with our interest in improvisation and experimentation in a series of projects. Our initiatives have all gone under the banner of Culture and Climate Change, working in partnership with arts bodies, NGOs and charities, and also in RCUK-funded (Research Councils UK) projects, such as Interdependence Day and Stories of Change. Although we have throughout the life of these projects both contributed to and engaged with social science research into climate communication and engagement, this is an account of our 'learning by doing' rather than a portfolio of 'how to' guides. Furthermore, we are sceptical of the suggestion that the mass of people are wrapped in 'overwhelmedness' or 'apathy' (Moser, this volume). Similarly, we want to test Ford and Norgaard's (this volume) contention that there is a failure of popular mobilisation.

With consistent polling around the world that suggests across many years now that clear majorities believe climate change to be happening, and to be human caused, we are convinced that some key messages about climate change have been effectively shared, despite the difficult nature of this knowledge. Furthermore, the progress of the UNFCCC process, which has within a couple of decades established near unanimous political commitment to a programme of ratcheting actions that relate individual acts in the present to future outcomes for the global atmosphere, is a remarkable political achievement. This is particularly notable given that climate change emerges as an issue in an exceptionally lively period of economic, technological and societal transformations. In other words, as Krupnik noted in the context of the Alaskan village he studied, people have 'other things to worry about' (Krupnik, this volume). We think the cultural, and more importantly, political work ahead is messier but also more interesting than simply making sure everyone 'gets it'. Indeed, one way of understanding our work is to see it as a series of attempts to open out the political and ethical space in and around climate change knowledge rather than mobilising a particular kind of response to it. Quite apart from the latter approach being unlikely to work, it seems perverse to suggest that everything is about to change except the world view of a category of (mostly western) academics and policy analysts. Our projects start from the assumption that climate change changes us.

14.2 Interdependence Day: An Unruly Mix

The Interdependence Day project, 2005–2010, amounted to a programme of experimental events, publications and other interventions that could both test different framings of sustainability thinking and innovate in the forms of engagement between academics, publics and creative and policy partners. The activities were designed to probe the potency of the concept of interdependence at a time when the density of relations between the ecological, the social and the political was becoming so evident. Interdependence Day was politically explicit but frank about its experimental and uncertain status. The ideas were shaped by conclusions of much earlier social research that had confirmed that publics had a good nose for authenticity when it came to government encouragement for everyone to 'do their bit' in response to global environmental challenges (Smith *et al.* 1999, 2000). It was informed by work in human geography that addressed the ethical and political implications of 'thinking space relationally' and thus 'geographies of responsibility' (Massey 2004) and also by radical traditions of participation, interactivity and co-production in architectural design teaching (Tyszczuk 2007).

We sought to test means of navigating present and near-future environmental challenges 'in public and with publics'. Among other things, we were motivated to explore tones and approaches to publications and events that avoided both the monotonous 'too little too late' intonations of the environmental NGOs and also the hubris of other, and in our view naive, responses that stressed the availability of sustainable solutions and that emphasised the honing of 'correct' communications design. Our goal was to find ways of describing and responding to our state of global interdependence that respects but isn't confounded by its complexity. The Interdependence Day project started from the assumption that 'it is impossible to reach a viewing point from which we can fully account for myriad ecological and economic inter-relations: we are simply too enmeshed' (Tyszczuk *et al.* 2012:4). The events and publications all sought to contribute to a collage of careful but purposeful responses to this complex state of interdependency (Smith *et al.* 2007; Tyszczuk and Smith 2009; Tyszczuk *et al.* 2012; Smith 2012).

The Interdependence Day project acknowledged the complexity and seriousness of contemporary political problems and the way they have served to leave many people feeling disempowered. We therefore sought to try out new kinds of public event that would be both interactive and participatory. The three sold-out events tested a range of experiments in participatory exchanges around global themes. The first two were held at the Royal Geographical Society and the third at Queen Elizabeth Hall on London's South Bank. These all included 'unconference' elements, collaborative writing groups, workshops, installations and exhibitions, as well as more standard (short) talks formats. The interactive workshops included the creation of a new *mappa mundi* in the Map Room of the Royal Geographical Society as a 'living' and 'provisional' exhibition. Participants stitched their stories into a linen world map laid out on a table; their conversations were recorded and later transcribed and fed into publications (Tyszczuk 2012). We also devised 'Doctor's Surgeries' where small groups experienced and contributed to guided conversations about global themes 'in the company of experts' (academic researchers from a range of sustainability-related disciplines). The events revealed a strong and otherwise largely unmet appetite amongst attentive publics to talk through themes such as climate change, economic globalisation and biodiversity loss in the company of others.

The three Interdependence reports, co-written and co-published with the new economics foundation (NEF), were another element of this strategy of testing new framings (Simms *et al.* 2007; Simms *et al.* 2007, 2009). The qualitative and quantitative research that had explored community and household perspectives on sustainability (Smith *et al.* 1999, 2000) had left us convinced that the 'sustainability' policy idiom had little purchase on the public imagination. Indeed, it tended to encourage cynical responses about government and business failure to lead. Hence, we sought to find easy ways to communicate some of the perverse outcomes of an economic system that failed to place a value upon natural resources and ecosystem services. With a focus on honing concise news-friendly phrases and images, we translated complex arguments about perverse trade, low values on material and natural resources and future planetary-scale jeopardies into very contained narratives. The first report led BBC radio bulletins and also the ITN evening news, with giant gingerbread biscuits swapping places on a global map graphic. Our writing on off-shored carbon emissions in another of these reports, *Chinadependence* (Simms *et al.* 2009), was the first time the concept had appeared in wide circulation. A further popular and policy-facing publication was the edited book *Do Good Lives Have to Cost the Earth?* (Simms and Smith 2008). It included contributions from leading figures from all of the main UK political parties, as well as artists, writers, designers and others that gave their accounts of how strong environmental actions could deliver improvements in quality of life. Our editorial line and introduction and conclusion drove home arguments rooted in our academic research: that acting to mitigate climate change offered the best opportunity for generations to create a vision of better cities, work and everyday life. The themes of the book, following the design of the project as a whole, located action in visionary and strategic approaches to policy and politics, but framed these around the construction of mainstream cross-party consensus.

The extensive national broadcast news and print coverage of the reports, and the appearance of some of our concepts in political speeches, was only possible on account of our investment in relationship building with NEF, their skilled phrase making, and their work with media networks and designers. The Interdependence Day project demonstrated the centrality of generous, patient partnerships, full of give and take. It also demonstrated that a great deal can be achieved on very modest budgets indeed. We learnt from this work that the main currencies you need to invest in are ideas and collaborations. The main public-facing achievements of the project also required willingness to purposefully step away from 'conference mode', to stop worrying about academic reaction to the published work, and to behave like you want busy people, including government ministers and officials, journalists, family and friends, to engage with your ideas.

Our use of the term 'Atlas' to describe the main book publication of the project (Tyszczuk *et al.* 2012) – with all its implied completeness and dominion – was intentionally playful. Similarly, the title allowed us to nod towards Atlas, the fated hero, doomed to carry the weight of the world or hold up the heavens, depending on your point of view. The book allowed a glimpse of the ideas, art interventions, expert witness stories, and scientific responses to global environmental change of the project – what we characterised as an 'unruly mix'. It was a collection of responses for an unprecedented present and an unpredictable future. Many of the contributions recognised that small, niche-based gestures and practices could be understood as deft responses to uncertain conditions, or as seed-beds for testing alternatives to an unsustainable status quo. *Atlas* thus highlighted the value of 'tracings and probings of worlds which are currently in the making ... a guide to journeys that open new pathways; connections that may become networks; practices that could become effective institutions and niche experiments which might nourish purposeful change' (Tyszczuk *et al.* 2012:7).

Further insights from Interdependence Day – about the lack of continuity or short termism of most climate change–related projects – led to an ambitious project based around tracking how understandings of environmental change evolve over time. The Creative Climate project comprised a time-series online and broadcast diary project generated by the Open University and BBC World/ World Service. In addition to five TV documentaries and dedicated segments in nine radio programmes, a series of ten short films by young filmmakers were co-commissioned with BBC Comedy. The commissioned materials and the central device (diary keeping) were designed to also serve as higher level (upper school/ university) teaching and learning content and activities. These materials in particular reached big global audiences and also worked hard as teaching and learning materials. However, the participatory media elements were of very limited impact. Creative Climate taught us to contain expectations of 'the digital' as a realm of mass participation without appropriate institutional investments and commitments to social media. We recognised the need to anticipate and plan for institutional limitations in this area and to play to strengths. This led us to sharpen our resolve that our primary role as academics in much of this work, notwithstanding the news media and policy impacts of the Interdependence reports, or the direct value of the media seminars programme, was as incubators, experimenters and innovators rather than mass-communicators. These lessons directly informed the shape and purpose of our next collaboration on culture and climate change: Stories of Change.

14.3 Stories of Change: Prototyping Energy Transitions

The Stories of Change project allowed us to focus on the use of stories, narratives and storytelling in energy and climate change research. The project took decarbonisation of the energy system as its central theme. Our starting point was the simple fact that the ways in which humanity has lived with energy in the past has often changed – and will change again. The question is: what changes do we want and how do we tell these stories of change? Stories can help us rehearse for change. 'Stories do not just passively relate meaning – they create it, and they transform it. Ultimately they are like prototyping, a way of working out what to do next' (Smith and Tyszczuk 2018:103).

The project set out to support more dynamic public and policy conversations about energy by looking in a fresh way at its past, present and future. The project was shaped around the cross-party commitments to decarbonisation that sit at the heart of the UK Government's Climate Change Act of 2008 and was further energised by the UN Paris Agreement of 2015. Research has shown that many people feel disengaged, disempowered or actively hostile to changes to the United Kingdom's energy system required to meet the targets embedded in the Act. At the same time it is clear that there is wide acceptance that actions will be needed to reduce demand, decarbonise the energy supply system and prepare to cope with future environmental hazards. Stories of Change set out to experiment with novel ways to work through areas of concern and test shared ideas about energy system transformations.

By drawing on an unusually broad mix of history, literature, social and policy research and the arts, the project sought to encourage a more open approach to current and future energy changes and choices, and to explore elements of a collective vision. Above all, we have aimed to encourage a more imaginative and vigorous approach to future energy choices that takes much more account of the interests of people and places that are vulnerable to climate change now and in the future.

Stories of Change was organised around three research projects, or 'stories'. The first, Demanding Times, gathered together a novel mix of communities with interests around energy policy, mostly focused on London, often seen as the world's first 'global city'. It has generated new accounts of energy and politics past, present and future. The second, Future Works, was rooted in the English Midlands, unearthing fresh accounts of the long relationship between energy, industrial making and landscape, and exploring where it might go next. Everyday Lives examined the ways energy resources have continued to shape communities' lives in South Wales. Within the life of the project we saw young Londoners with little prior experience of policy, the media or environmental issues gain the confidence to interview leading policy figures and hold these experts to account for their role in shaping the future. Student and apprentice collaborators in the English Midlands worked with a wide range of businesses and institutions to devise industrial energy strategies for 'factories of the future'. A pop up storytelling studio in the South Wales Valleys helped reconnect people to their significant role in global-scale energy stories – whether coal mines or wind farms.

We have shared the very varied outputs publicly with performances and events, exhibitions, a web platform, and a free printed project book, *Energetic* (Smith and Tyszczuk 2018). All of the material has been presented publicly and was designed for easy sharing via Creative Commons licences. The project book, *Energetic*, gathers insights and images from across all of the work. Indeed, if the Stories of Change project were an exhibition, then this would be its catalogue. Following our scepticism about the 'openness' of publicly available academic texts, we followed the example of the *Atlas* Interdependence Day project book in bringing together short approachable pieces, and plenty of high-quality illustration and design. *Energetic* expresses the mix of creative writing, songs, photos and portraits, interviews, short films, performances, and museum and festival events that we co-produced in collaboration with our community, creative and research partners. A more comprehensive collection of material is held in the online Stories Platform. There it is possible to create new 'stories of change', by threading material together using the digital tools provided, browse individual items in the library or follow designed and edited pathways (stories) through the collection.

Just as with the Interdependence Day and Creative Climate projects, the design and ethos of Stories of Change was heavily influenced by the example of the Mass Observation movement's accounts of everyday life in mid-twentieth-century Britain (Hubble 2010). Their work combined a desire to give ordinary people a voice, radical innovations in social research, and bold new ideas about documentary media and the arts. They took an innovative approach to valuing and supporting lay social researchers and developed a groundbreaking blend of arts, social sciences and media applied to goals of social change. Mass Observation also made novel use of documentary tools to create a mould-breaking account of the life of people in the United Kingdom at work, at home and at play. One of the key members of the movement, Humphrey Jennings, had spent years developing a manuscript that amounted to an 'imaginative history of the Industrial Revolution' (Jennings 2012:xiii). This was published posthumously under the title *Pandaemonium* in 1985. Jennings had initially gathered various texts to support his regular Workers' Education Association lectures, and their collaged nature as a collection of what Jennings called 'images' offered further inspiration for the design of our book, web platform and its devices. Like *Pandaemonium*, both our *Energetic* book and web platform assume active readers and listeners who participate in sense-making and story-making rather than simply receive content. We designed many aspects of the project's work in such a way that people would not simply engage with the narratives generated but also see themselves as agents within them.

While energy systems change was a focus, our wider aim was to explore the degree to which playfulness, the imagination and the sharing of stories might play a profound role in preparing the way for the wider body of transformations that will be required if we are to respond to pressing environmental risks, from air pollution to climate change. All of this relates to the simple insight that one key feature of stories is that you can always change the ending. In other words, stories were understood to have agency. Our ambition was to extend storytelling beyond being understood as a form of communication into a mode of understanding and acting in the world.

Our experience suggests that creative and experimental methods rooted in the creation of, listening to and telling of stories can play a powerful role in energising engagement in policy issues that are not only important but also complex and at first glance uninviting. The approaches we have taken have drawn variously on fun, memory, emotion and connection to place, family, friends or work in order to expand the terrain of public conversations about energy systems change. It is not so much that stories in themselves drive transformations. Rather, we propose that stories have the capacity to invite many more constituencies to engage in imagining change and consequently have the confidence to participate in it. The key thing about encouraging people to tell and share energy stories isn't that there are transformative narratives waiting to be polished but rather that by being given the permission to participate in recalling the past or anticipating possible futures, participants feel they have both a stake and a potential role in positive transformations. Our goal has specifically not been to test and refine the 'right' transformative narrative and inoculate the population with it. On the contrary, we argue that, where a democratic system is faced with a complex or challenging topic such as energy transitions in spheres such as space heating, or personal mobility, the quality of public debate can be improved by anticipating and providing for people's need to hear their own ideas and concerns represented in public narratives (Smith *et al.* 2017).

14.4 Culture and Climate Change Scenarios: 'We Are All Climate Researchers'

The Scenarios project has been carrying some of the same principles but in relation to climate-changed futures. It is our most recent project in the Culture and Climate Change series and was launched in Paris at the UNFCCC COP 21 in

December 2015 with the ambition of bringing greater cultural depth to public conversations about future climate scenarios. Scenario thinking has long been a prominent strand in the work of the IPCC and the UNFCCC and draws on predictive scientific knowledge, based on computer models and simulations. Scenario and forecasting techniques have been widely applied in business and policy. Mahony *et al.* (this volume) note that the way society thinks about climate futures is 'informed and shaped by authoritative scientific projections of future environmental states, which oscillate between a disarming uncertainty about the near and far future, and a seductive offer of control over the global earth-system'. Our starting point is that, given the far-reaching influence of scenarios-based thinking in this field, it is vital to understand, engage with and open out this mode of thinking.

Scenarios are essentially stories of change and can thus be understood as collective acts of imagination about possible futures in human-natural hybrid systems. Moreover, their origins as a cultural form lie in the improvisations of *commedia dell'arte* street theatre in the sixteenth and seventeenth centuries. The term *scenario* here indicated the synopsis of a performance that responded to the complexities of the everyday. Scenarios presented to describe future climates tend by their nature to invite contestation. Bearing in mind that the root of scenarios is in improvisation and trial-and-error, rather than in the pursuit or definition of a complete 'solution' or answer, we have argued that these fundamental characteristics of scenarios should not simply be acknowledged: they need to be embraced.

The project involved the appointment of four artists (Teo Ormond-Skeaping, Lena Dobrowolska, Emma Critchley and Zoe Svendsen) who from July 2016 took part in an experimental model of 'networked residencies', which explicitly sought to both mirror and engage with the distributed but interconnected nature of climate research. The artists were challenged to explore and open up thinking on climate scenarios in the wake of the Paris Agreement. Across the year, their work on the residencies was detailed in monthly diary accounts and presented at public workshops and festivals (see the Scenarios project on the Culture and Climate Change website: http://www.cultureandclimatechange.co.uk).

The Scenarios project is another attempt to defy the widely held view of cultural responses to climate change that limit them to late-phase communications or engagement aids that come after the science and policy is done. The project started from the presumption that arts and humanities practices were not a response to, but rather an expression and component of, climate research. The experimental and co-productive elements of the Scenarios residency centred on the structuring of a sequence of hybrid and experimental encounters with different researchers and between different modes of climate change knowledge making and sharing. Over the year the artists engaged with a range of approaches to climate scenarios – including the models of research scientists, the designs of urban planners, and the forecasts of policy-makers. At the same time, working with moving image, photography, installation, theatre, and performance, they explored and extended the ways in which society might reimagine scenarios of climate change. The improvisational and reflexive intentions inherent in scenarios have served as a touchstone for the project. Our framing for the Scenarios residency was one of 'collective improvisations'. This referred to both the origins of scenario making in improvised street theatre and the 'collective experiments' of climate change. It drew on Bruno Latour's observation that laboratories had turned 'inside out' to become 'the worldwide lab' such that 'we are all engaged in a set of collective experiments' in the 'confusing atmosphere of a whole culture' (2003:30–31). This aligns with cautions regarding how the predictive knowledge of climate research tends to set the terms for running a worldwide sociocultural experiment, that is, 'bringing the worldwide emissions of greenhouse gases under directed management' (Hulme and Mahony 2010). With this context in mind, we proposed paraphrasing artist Joseph Beuys, that 'we are all climate researchers' (Tyszczuk and Smith 2018:59).

The Scenarios residency project gives an idea of the potential of a sustained collaboration between the natural and social sciences, arts and humanities in the public spaces of climate research. The varied projects are ongoing and iterative and hint at the multiple possible ways of responding to the complexities of climate change (Tyszczuk and Smith 2018). Ormond-Skeaping and Dobrowolska explored the scenario mode of their documentary photography and film practice in their project provisionally called 'Anthropocenes'. Their field-based research in Lao (PDR), Bangladesh, Uganda and the United Kingdom engaged with climate change adaptation in places where climate change is no longer a future scenario – and the impacts are intensifying. It explored the ways in which communities deemed most 'vulnerable' to climate change were also providing practical and intellectual leadership in demonstrating capacity to adapt to climate change. Their scenario making opens up a dialogue about a yet-to-be-determined future, asking important questions about political inequalities as well as new modes of governance and inhabitation in unsettled times. Who decides 'future scenarios' (when climate change is already here), who is involved, how and for whom are liveable futures worked out?

Visual and sound artist and diver Critchley's *Common Heritage* project engages with the 'frontiers' or thresholds of human reach, including the deep sea and deep space. The feature length film she is making asks why these spaces, and by implication the Earth, are treated like frontiers of conquest, rather than home. Critchley's scenarios are generated through collaborations with deep sea ecologists and climate researchers (Universities of Southampton, Plymouth, Cornell, Washington and Cambridge with the

British Antarctic Survey). Part of her research has been about acoustic pollution and its impacts on cetaceans/sound-oriented creatures. Sound here is not just an indicator of global environmental change but a powerful metaphor for climate change – something it is possible to be immersed in yet falls on different registers. *Common Heritage* not only considers the embodied and experiential aspects of change in the non-human natural world but also aims to show the inseparable relationships between that domain and the distinctively human world of international politics, resource exploitation and territorial ambitions.

Theatre maker Zoe Svendsen used the residency to develop *WE KNOW NOT WHAT WE MAY BE*, a performance installation at the Barbican (September 2018). Zoe was drawn to the economic and related social and cultural consequences of a climate-changed future. Her investigations have been rooted in a series of 'research in public' conversations with economics, politics, business and social science climate researchers who have been challenged to imagine what it might *feel* like to live in a society and economy designed in the best possible way to respond to climate change. The performance installation will involve audiences exploring these alternative economic futures, involving various economic measures (e.g. universal basic income, carbon tax), ideas about the future of food and land, the impact of robotics and AI, and the changing relationships to work. Participation in the event will lead to the creation of a collective vision of an alternative future, shared live and online.

Our ambition with the Scenarios project has been to support future imaginings that might better reveal a world where multiple, differentiated and uncertain futures are possible.

The collaborations around climate scenarios between the artists and their climate research community co-researchers (including ourselves as both convenors and participants) recognised the diversity and contested nature of climate change research, with its porous thresholds and 'indeterminate boundaries between science and its others' (Hulme and Mahony 2010). The 'collective improvisations' of the Scenarios residency explored ways of expanding the ethical, material and imaginative registers that living with uncertain climates might mobilise and to explore knowledge making in climate research in collaboration with others. Indeed, in the wake of reading Mahony *et al.* (this volume), we would argue that the initiative should be understood as experiments in co-production at the boundary of climate research and action. We argued that such collective scenarios could provide a 'rehearsal space' that might also result in more robust and considered responses in the near term to the prospect of surprising social transformations that are inevitably part of climate-changed futures (Tyszczuk and Smith 2018).

14.5 Conclusion

This volume has demonstrated the potential scope of cultural dimensions of climate change. We have sought to add some reflections on our body of work in this area. These projects had to be achieved 'in the gaps' between teaching and more 'traditional' academic publication and practice. Before terms like engagement, impact, and interdisciplinarity were considered respectable, indeed desirable, dimensions of academic life, most of the projects described here were often considered by others to be dilettante or displacement activities. Conditions are changing for the better, and there is a much more substantial community of practice and critique developing around cultural work on climate change. This leads us to want to share a few headline conclusions as to what we believe this work can do and how it can best be achieved.

We have characterised the complexities of climate change as an *unruly mix* of diverse knowledges, multiple framings, entanglements of human and non-human agencies, and unsettling responsibilities and vulnerabilities. These are seemingly incommensurable and yet, as Latour observes, 'there they are caught up in the same story' (1993:1). Our responses however can only ever be partial attempts at what Sheila Jasanoff describes as the 'reintegration between global scientific representations of and local social responses to the climate' (2010:235). At the same time, the unexpected nature of the process of engaging with climate change knowledges brings new skills, networks, and insights.

Like *prototyping*, our responses have been incremental and iterative, reflecting the processes of change as much as being involved with and within the change. And these changes are at once technological, social, political and cultural. Climate change understanding is itself evolving and changing. As Margaret Atwood opines, 'It's not climate change – it's everything change' (2015). Mike Hulme adds: 'Climate is therefore becoming everything, but also nothing' (2017:152). Prototyping shared climate futures in the current climate is therefore also about a commitment to careful risk-taking without guarantee and to learning through trial and error.

Climate research and how we make sense of this unsettling terrain can take on many forms – and should not be limited to the domain of natural and physical sciences or social sciences or even be confined to the academy: *we are all climate researchers*. Indeed, wider professional and public participation in climate research – doing research in public and with publics (Smith 2013b) – is not just a device for increasing engagement and commitment: it is reciprocal and changes the nature of the research. We propose that a diversity of perspectives, views and approaches is essential both to sense making and also to meaningful debate and stable decision-making, particularly within democratic systems.

Finally, we acknowledge the *time of our experiments and improvisations*. Some of the most lightly resourced projects described here have depended on investments of time and attention across many years. They were experiments that had to be left to run and sometimes required patient watching in order to see when the conditions could be right to take a particular theme or idea further. The most recent projects will likely continue finding their way through the world, sometimes with our help, sometimes not, for years to come. Experiments and improvisations in the sphere of culture and climate change require generosity of both time and spirit, time to get things right, redundant time, and time to let things unfold. This is not to suggest that the issue is not important or urgent. Rather, we want to suggest that insisting on one urgent fact, or one important figure, is in danger of making it more difficult for many people to attend and respond to this difficult new knowledge. Our experience leads us to argue that cultural responses to climate change will be all the more energetic, and ultimately effective, by building over time the many stories, of different voices, into waves of polyphony. Polyphonies are structures which support improvisation, yet still manage to bring many people together around a theme. The results can be moving, powerful and timely.

Acknowledgements

The Interdependence Day project, in partnership with the New Economics Foundation (NEF), was initiated by a network grant from the ESRC and NERC (Award Number RES-496–25-4015). It was further supported by The Open University's Open Space Research Centre, the Ashden Trust, and the Frederick Soddy Trust. The Creative Climate project was supported by the Open University as a means of piloting its Broadcast Strategy, and with Open University/BBC co-commissions. The Stories of Change project was supported by the Arts and Humanities Research Council (AHRC) (Award Number AH/L008173/1). The Culture and Climate Change series of projects have been funded by the Jerwood Charitable Foundation, The Ashden Trust, The University of Sheffield, The Grantham Centre for Sustainable Futures at the University of Sheffield and The Open University's Open Space Research Centre. We also acknowledge our project collaborators and participants and the support of many people in the administration teams at our universities over many years.

References

Atwood, M. 2015. It's not climate change – it's everything change. *Matter*. https://medium.com/matter/it-s-not-climate-change-it-s-everything-change-8fd9aa671804

Butler, R., Margolies, E., Smith, J., and Tyszczuk, R. (eds.) 2011. *Culture and Climate Change: Recordings*. Cambridge: Shed.

Castree N. 2016. Broaden research on the human dimensions of climate change. *Nature Climate Change*, 6(8), 731.

Culture and Climate Change: Scenarios www.cultureandclimatechange.co.uk/projects/#

Heymann, M., Gramelsberger G., and Mahony M. (eds.) 2017. *Cultures of Prediction In Atmospheric and Climate Science: Epistemic and Cultural Shifts in Computer-Based Modelling and Simulation*. London: Routledge.

Hubble, N. 2010. *Mass Observation and Everyday Life: Culture, History, Theory*. Basingstoke: Palgrave Macmillan.

Hulme M. 2017. *Weathered. Cultures of Climate*. London: Sage Publications Ltd.

Hulme M. 2009. *Why We Disagree about Climate Change: Understanding Controversy, Inaction and Opportunity*. Cambridge: Cambridge University Press.

Hulme, M. and Mahony, M. 2010. Climate change: What do we know about the IPCC? *Progress in Physical Geography*, 34(5), 705–718.

Jasanoff, S. 2010. A new climate for society. *Theory, Culture & Society*, 27(2–3), 233–253.

Jennings, H. 2012. *Pandaemonium 1660–1786: The Coming of the Machine as Seen by Contemporary Observers*. London: Icon Books Ltd.

Kossak, F., Petrescu, D., Schneider, T., Tyszczuk, R., and Walker, S. (eds.) 2009. *Agency: Working with Uncertain Architectures*. London: Routledge.

Latour, B. 1993. *We Have Never Been Modern*. Cambridge, MA: Harvard University Press.

Latour, B. 2003. Atmosphere, atmosphere. In: May, S. (ed.) *The Weather Project*. London: Tate Publishing, pp. 29–41.

Marshall, G. 2015. *Don't Even Think about It: Why Our Brains Are Wired to Ignore Climate Change*. London: Bloomsbury.

Massey, D. 2004. Geographies of responsibility. *Geografiska Annaler: Series B, Human Geography*, 86(1), 5–18.

O'Brien, K. 2012. Global environmental change II: From adaptation to deliberate transformation. *Progress in Human Geography*, 36, 667–676.

O'Brien, K. 2016. Climate change and social transformations: Is it time for a quantum leap? *WIREs Climate Change*, 7, 618–626.

O'Brien, K. and Selboe, E. (eds.) 2015. *The Adaptive Challenge of Climate Change*. Cambridge: Cambridge University Press.

Simms A., Johnson V., and Smith J. 2007. *Chinadependence: The Second UK Interdependence Report*. London: New Economics Foundation and Open University.

Simms A., Johnson V., Smith J., and Mitchell S. 2009. *The Consumption Explosion: The Third UK Interdependence Report*. London: New Economics Foundation and Open University.

Simms A. and Smith, J. (eds.) 2008. *Do Good Lives Have to Cost the Earth?* London: Constable and Robinson.

Smith, J. (ed.) 2000. *The Daily Globe: Environmental Change, the Public and the Media*. London: Earthscan.

Smith, J. 2005. Dangerous news: Media decision making about climate change risk. *Risk Analysis*, 25, 1471–1482.

Smith, J. 2011. Why climate change is different: Six elements that are shaping the new cultural politics. In: Butler, R., Margolies, E., Smith, J., and Tyszczuk, R. (eds.) 2011. *Culture and Climate Change: Recordings*. Cambridge: Shed, pp. 17–22.

Smith, J. 2012. Road map: Other ways of thinking about auto-mobility. In: Tyszczuk Tyszczuk, R., Smith J., and Butcher, M. (eds.) *ATLAS: Geography, Architecture and Change in an Interdependent World*. London: Black Dog Publishing, pp. 118–123.

Smith, J. 2013a. Mediating tipping points. In: O'Riordan, T. and Lenton, T. (eds.) *Addressing Tipping Points for a Precarious Future*. Oxford: Oxford University Press.

Smith, J. 2013b. Public geography and geography in public. *The Geographical Journal*, 179(2), 188–192.

Smith, J. 2014. Communication and media: Commentary. In: Crow, D. and Boykoff, M., *Culture, Politics and Climate Change: How Information Shapes Our Common Future*. London: Routledge/Earthscan.

Smith, J. 2017. Demanding stories: Television coverage of sustainability, climate change and material demand. *Philosophical Transactions of the Royal Society A: Mathematical, Physical and Engineering Sciences*, 375, 20160375.

Smith, J., Blake, J., Grove-White, R., Kashefi, E., Madden, S., and Percy, S. 1999. Social learning and sustainable communities: an interim assessment of research into sustainable communities projects in the UK. *Local Environment*, 4(2), 195–207.

Smith, J., Blake, J., and Davies, A. 2000. Putting sustainability in place: Sustainable communities projects in Huntingdonshire. *Journal of Environmental Policy and Planning*, 2(3), 211–223.

Smith, J., Clark, N., and Yusoff, K. 2007. Interdependence. *Geography Compass*, 1(3), 340–359.

Smith, J., Hammond, K., and Revill, G. 2018. Climate Change on Television. *Transactions of the Institute of British Geographers*, 43, 601–614.

Smith J., Butler, R., Day, R. J., Goodbody, A. H., Llewellyn, D. H., Rohse, M., Smith B. T., Tyszczuk, R. A., Udall, J., and Whyte, N. M. 2017. Gathering around stories: Interdisciplinary experiments in support of energy transitions. *Energy Research and Social Science*, 31, 284–294.

Smith, J. and Tyszczuk, R. (eds.) 2018. *Energetic: Exploring the Past, Present and Future of Energy* Cambridge: Shed; https://issuu.com/energeticbook/docs/energeticv09sp

Smith, J. Tyszczuk, R., and Butler, R. (eds.) 2014. *Culture and Climate Change: Narratives*. Cambridge: Shed.

Tyszczuk, R. 2018. *Provisional Cities: Cautionary Tales for the Anthropocene*. Abingdon: Routledge.

Tyszczuk, R. 2014. Cautionary tales: The sky is falling! The world is ending! In: Smith, J., Tyszczuk, R., and Butler, R. (eds.) *Culture and Climate Change: Narratives*. Cambridge: Shed, pp. 45–57.

Tyszczuk, R. 2012. Mappa mundi. In: Tyszczuk R., Smith J., Clark, N. and Butcher, M. (eds.) *Atlas: Geography, Architecture and Change in an Interdependent World*. London: Black Dog Publishing, pp. 10–15.

Tyszczuk, R. 2011. On constructing for the unforeseen. In: Butler, R., Margolies, E., Smith, J., and Tyszczuk, R. (eds.) *Culture and Climate Change: Recordings*. Cambridge: Shed, pp. 23–27.

Tyszczuk, R. (ed.) 2007. *Architecture and Interdependence: Mappings and Explorations by Studio Six*. Cambridge: Shed.

Tyszczuk, R. and Smith, J. 2018. Culture and climate change scenarios: The role and potential of the arts and humanities in responding to the '1.5 degrees target'. *Current Opinion on Environmental Sustainability* 31, 56–64.

Tyszczuk, R., Smith J., Clark, N., and Butcher, M. (eds.) 2012. *Atlas: Geography, Architecture and Change in an Interdependent World*. London: Black Dog Publishing.

Index

adaptation(s)
  cultural, 4, 300
  human, 212
  multi-temporal, 117
  technological and physical, 190
adaptive
  capacity(ies), 118, 119, 129, 130
  challenge(s), 309
  change(s), 142, 244
  governance, 129
  responses, 118
  strategies and activities, 120, 130
agency, 12, 94, 100, 282, 319
  and empowerment, 285
  and identity, 280
  and motivation, 273
  and structure, 272
  individual and collective, 287
agriculture, 36, 55, 69, 73, 121, 125, 127, 129, 130, 291, 295, 296, 302
  community supported, 293, 296
  globalisation of, 301
alternative
  cultures, 9
  ontologies, 2
Altiplano model, 122
ancestral law, or laws of origin, 78
Anchorage, town of, 170, 184
Ando Tadao, 296
Anglo-Eurocentric episteme, 76
animism, 94, 98, 100, 101, 102, 107, 304
Annwn, 93
Anthropocene, 6, 37, 75, 76, 243
anthropocentricism, 225
anthropogenic climate change, 23, 46, 244, 245, 246, 254, 255
apocalypse, 38, 149
archaic avant-garde, 97, 100, 101, 105
archives, 13, 191
Arts
  creative, 94
  traditional, 295
  visual, 94
atmosphere and ocean global circulation models (AOGCMs), 26
attachment theory, 273

Benesse Corporation, 296
Benjamin, Walter, 38
bio-indicators
  climate, 73
  cultural, 72
biotemporal indicators, 72
Bourdieu, Pierre, 6, 270
Buddhism, 243–262

capital
  cultural, 6
capitalism
  consumer, 233, 304
  cultural values of, 237
  fossil fuel, 9
  modern, 8, 9
  neo-liberal, 305
Carte des lignes isothermes, 54
Chukotka Peninsula, 170
civic epistemology, 34
climate
  denialists, 223
  injustice, 70
  justice, 78, 87
  models, 5, 13, 26, 27, 30, 60
  scepticism, 223, 225
  variability, 5, 70, 73, 74, 196
  visualisations, 46, 59, 64
Climate Analogues online platform, 61
climate change
  adaptation, 12, 68, 119, 142, 143, 181, 234, 292, 321
  mitigation, 68, 142, 181, 234, 254, 292, 302, 303
climate variability, 69, 77
climatic
  determinism, 51

327

climatic (cont.)
  difference, 63
climigration, 183
Club of Rome, 37
Cold War, 36
collective improvisations, 321, 322
community(ies)
  affected, 182
  Alaskan, 177
  climate-affected, 183
  epistemic, 2
  indigenous, 236, 237
  land and ecological, 236
  local, 117
  migration of, 183
  peasant, 10
  plant, 122
  privileged western, 234
  rural, xi, 10, 178
  Satoyama, 297
  scientific, 22
  sustainable, 304
  traditional, 8
  vulnerable, 10, 183
congregational domain, 100
core of the real, 94
cosmovisions, 71, 73
Creative Climate project, 316, 324
cultural
  artefacts, 95
  categories, 221
  governability, 79, 84
  individualism, 232
  invasion, 286
  synthesis, 286
  worldviews, 143, 220, 223, 226, 232, 237
culturally constructed, 71
cultures of prediction, 21, 22, 23, 26, 34, 35, 39, 42, 43, 309

decarbonisation, 317
decision-making
  challenge of, 36
  collective or individual, 78
  evaluation of, 251
  everyday, 5
  government and political, 40
  weather forecasting and farming, 126
dematerialisation, 253, 256
developmental psychology, 271
dialectical relations, 286
dialogic action approach, 275
disaster
  compensation, 182
  mitigation, 182
  relief, 182
discourse(s)
  Buddhist, 261

  colonial, 58
  cultural, 269
  Enlightenment-based, 94
  ethical, 305
  global, 68, 74
  learnt, 47
  moral, 305
  public, 47, 223
  social, 272, 281
  village, 181
Doctor's Surgeries, 314
documentary history, 195
Dove, Heinrich, 24
Dream of Scipio, 48
dukkha, 248

Earth Day, 293
eco-criticism, 304
ecological restoration, 299
economic liberalisation, 10
ecosophy, 101
emergency assistance, 181
emotions, 220, 221, 226, 236
Encyclopédie of Diderot and d'Alembert, 52
endogenous development, 302
energy, 25, 27, 30, 131, 141, 144, 150, 256, 298, 299, 302, 317
  Alaska's Renewable Energy Program, 178
  and climate change research, 317
  convergence of, 110
  efficiency, 267
  non-renewable, 291
  renewable, 237
  saving technologies, 246
  system, 317
  systems change, 319
  transitions, 319
environmental
  aesthetics, 299, 300
  justice, 87
  knowing, 3, 93
  self-determination, 79
environmentalism(s), 38, 76, 234, 241
estate records, 202
everyday experiences, 71
exchange networks, 69, 123
experience(s)
  accumulated, 259
  conscious, 248
  cultural, 13
  embedded, 3
  embodied, 3, 108
  everyday, 69, 93, 195
  geographical and historical, 292
  lived, 220
  of weather events, 12
  personal, 97
  sensory, 7, 13

transformative learning, 267
extreme weather events, 11, 120, 190, 191, 194, 195, 202, 210, 211, 212, 266

feminist perspectives, 86
Ferrel, William, 24
frigid zones, 48
frontlines, 148, 168, 182
future imaginings, 322

Gambell, Town of, 170–185
Gandhi, Mahatma, 306
gender inequalities, 77, 78, 301
geoengineering, 142, 266
geographical imagination, 48
Ghosh, Amitav, 298, 306
global imaginary, 54, 55
green
 economy, 256
 spiritualism, 244
greenhouse gas scenarios, 61

Hadley, George, 24
Haiven, Max, 303
hegemonic, 10, 40, 52, 75
historical climatology, 12, 59
holistic milieu, 99, 100, 101, 112
homesteading, 230–234
human–environment relationships, 196, 309
Humboldt, Alexander von, 24, 49, 54, 55, 64

ice ages, 24, 59
improvisus, 312
Indigenous people(s), 69, 70, 234, 235
Indra's net, 243, 247, 252
Interdependence Day project, 313, 314, 315, 318, 324
Inujima, 296
IPAT equation, 246, 255
IPCC (Intergovernmental Panel on Climate Change), 2, 5, 27, 34, 37, 68
Ishikawa Prefecture, 294
isothermal lines, 54, 55, 56

Jodo Shinshu, 304

Kamëntsá Biyá people, 79–84
kamma-vipaka, 247, 248, 252, 253, 254, 256, 259
Kanazawa, 294
Karuk people, 234
Keeling curve, 64
Khasnabich, Alex, 303
Kivalina, Town of, 183
Klein, Naomi, 300
klimata, 49
knowledge(s)
 ancestral, 83, 84
 Ancient Greek, 50
 Andean, 129

climate, 2, 3, 7, 13, 47, 64, 310
climate change, 323
construction of, 65
co-production of, 275
cultural, 72, 81
decontextualisation of, 74
differentiated, 77, 79
economic, 29
embedded cultural, 211
empirical, 23
environmental, x, 169, 221
esoteric, 36
evidence-based, 250
exchange of, 83
expert, 75
geopolitics of, 70, 76
global, 76
global environmental, 64
governmental, 34
historical, 212
indigenous, 10, 69, 70, 73, 75, 306
inscribed, 82
intergenerational, 127
local, 7, 69, 77, 78, 127, 169
meteorological, 32
of climate change, 230
of risk, 231
operationalised, 286
places of, 79
predictive, 21, 28, 37
production of, 68, 70, 77, 78, 79, 81, 82
recovery of, 80
scientific, 7, 77, 251
scientifically constructed, 5
systems of, 71
technical, 260
territorialised, 70
traditional, 119, 236, 296, 301
Köppen climate classification, 60
Kothari, Ashish, 301

Laki eruption, 196, 197
liminal, 8, 93, 150
Limits to Growth, The, 37
Little Diomede Island, 184
local worldviews and knowledges, 68
longue durée, 46, 47, 48, 53, 60, 61, 63, 64, 65

Macrobius, 48, 64
mappa mundi, 314
Maruyama-gumi, 297
Max Planck Institute for Meteorology, 34
meaning-making, 35, 150, 155, 310
Met Office Hadley Centre, 33
metamorphosis, 146, 150, 154, 161
mingas, or collective processes, 84
mirroring, 152
mobile climates, 61

modernity, 225–231
Murphy's winter, 204

Naoshima, 296
narratives
  climate change, 8
  cultural, 14, 221, 226, 237, 238
  documentary, 211
  experiential, 212
  folkloric, 93
  historical, 303
  meta, 94, 97, 112, 144
  neoliberal, 232
  political, 233
  public, 319
  weather event, 191
Nasa worldview, 73
National Center for Atmospheric Research (NCAR), 34
national memory of the weather, 194, 211
National Weather Service (NWS), 33
natural processes, 68
nature, 65, 71, 79, 231, 236, 252, 262, 291, 300, 301, 304
  and society, 130
  dominion over, 226
  duality of nature/culture, 76
  philosophies of, 94
  practices around, 77
  return to, 304
  sacredness of, 300
  society interactions, 65
  vision of, 76
neoclassical economics, 253, 262
networked residencies, 320
New World, 50, 52
Newtok, Town of, 183
Nome, Town of, 170, 176, 184
non-human elements, 69
Northcott, Michael, 304

oikoumene, 50, 51
ontological
  and epistemological plurality, 87
  politics, 68
  relationship, 75
  security, 230
  understanding, 98
Otherworlds, 93, 102, 103, 107

parametrisation(s), 27
parish registers, 196, 207, 209
Pasto people, 73–84
peak oil, 230, 298
perturbations
  climatic and non-climatic, 118, 126
  current and future, 130
  from multiple stressors, 130

non-climatic, 119
Philosophy
  Aristotelian natural, 51
  conservative political, 225
  East-West, 271
  European postmodern, 103
  Greek natural, 50
  Reggio Emilia, 109
  traditional natural, 53
polymorphous, 101
polytheistic thinking, 99
post-oil and post-affluent societies, 11
pragmatism, 101, 111
Ptolemaic ideas, 50

quantitative approach, 52, 61

reflexivity, 152, 159
relational interactions, 94
relational modes of being, 99
relocation costs, 181
resilience, i, 12, 129, 131, 181, 236
  collective, 117
  of local agricultural systems, 298
  social-ecological, 272
  to climate change, 131
rural
  areas, 234
  communities, xi
  economies, 180, 301
  health and housing, 131
  lifestyles, 293
  populations, 124
  re-vitalisation, 304
  tourism, 297

Satoyama, 291–306
Savoonga, town of, 170–180
scenario(s)
  and forecasting techniques, 320
  making, 321
  planning, 151
  residency project, the, 319, 320, 321
  thinking, 320
scientific
  and public authority, 29
  and theological viewpoints, 112
  and utilitarian paradigm, 4
  approach, 112
  archives, 13
  climate change, 64
  communities, 22, 205
  constructions, 68
  cultures, 2, 22
  data, 3
  discipline, 23
  framing, 228
  information, 219, 228

instruments or models, 13
knowledge systems, 2
language, 235
observations, 39
perspectives, 262
practices, 22, 28
predictions, 39
problem-solving, 262
procedures, 29
projections, 21, 320
representations, 323
research, 259
responses to global environmental change, 316
revolution, 52, 53
scepticism, 245
theory and research, 262
visualisations, 46
worldviews, 93, 244
Seikatsu Club, 293
sense of place, 39, 93, 108
Shaktoolik, Town of, 183
Shinto, 294, 303
Shishmaref, Town of, 183, 184
Shiva, Vandana, 301
Shrivastava, Aseem, 301
snow buntings, 176
social reality, 220, 272
social structure(s), 12, 117, 269
social-ecological
   interaction, 3
   resilience, 272
   systems, 117
socially constructed, 9, 83, 230
Societas Meteorologica Palatina, 23
socioeconomic and environmental stressors, 11
socio-environmental change, 129
St. Lawrence Island, 169–182
Stories of Change project, the, 317–320
subjective
   and relational experience of spirituality, 100
   life spirituality, 100
   wellbeing (SWB), 253
subjectivity
   and bias, 212
   and irrationality, 33
   impoverishment of, 94
   multifaceted nature of, 101
   rational, 13
subsistence, 125, 188
sustainability, i, 126, 132, 139, 144, 146, 149, 181, 236, 246, 247, 249, 251, 253, 257, 261, 262, 266, 267, 272, 273, 276, 287, 298, 300, 302, 305, 306, 313, 314
sustainable development, 76
symbolic
   constructs, 95
   kinship, 123
   practices, 81
   repertoire, 24
   representations, 84
   responses, 11
   strategies, 73
   systems, 222
symbolisms, 72

Teshima, 296
theories of happiness, 244
Thin Place, 93–113
Thistlewood, John, 197
tipping points
   cultural, 286
   social, 268
Tír na nÓg, 93
Tokugawa, 293
Tracking Extremes Of Meteorological Phenomena Experienced In Space And Time (TEMPEST), 191, 212
transformation(s)
   cultural, 267, 276
   deliberate, 309
   environmental, 69
   historical, 51
   individual psychological, 271
   social, 266, 272, 322
transformative narratives, 319
transition(s)
   climate change, xiv
   cultural planning for, 306
   culture(s) of, 292, 298, 299
   energy, 141, 319
   great, the, 298
   historical, 48
   little, 298
   model of, 300
   narratives of, 306
   process of, 298
   towns, 298
   Transition Movement, the, 298
Troutman Lake, 181

uncertainty(ies), 21, 22, 27, 28, 29, 254, 320
United Nations Framework Convention on Climate Change, 5
United Nations University (UNU), 292
United States Weather Bureau, 32
unruly mix, 316, 323
urban
   areas, 130
   design and theory, 4
   dwellers, 124
   farming, 231
   homesteaders, x
   homesteading, 230

Vibrant Matter, 94
visualising climate, 11

vital materialist, 94
vulnerability, 11, 69, 70, 119, 125, 129, 130, 131, 169, 171, 236, 298

walrus hunting, 175
weather diaries, 196
William Ferrel, 25

wind turbines, 171, 178
Woodbridge, William Channing, 56, 58, 59, 60, 61, 64

Yamagishi, 293

zonal distribution, 51, 53
zonality, 53, 55